Environmental Impact Assessment

PRINCIPLES AND APPLICATIONS

Environmental Impact Assessment

PRINCIPLES AND APPLICATIONS

Paul A. Erickson

New England Research, Inc.
Worcester, Massachusetts

and

Clark University
Worcester, Massachusetts

ACADEMIC PRESS New York San Francisco London 1979

A Subsidiary of Harcourt Brace Jovanovich, Publishers

ACADEMIC PRESS, INC.
111 Fifth Avenue, New York, New York 10003

United Kingdom Edition published by
ACADEMIC PRESS, INC. (LONDON) LTD.
24/28 Oval Road, London NW1   7DX

Library of Congress Cataloging in Publication Data

Erickson, Paul A

    Environmental impact assessment.

    Includes the text of the National environmental
policy act of 1969.
    Includes bibliographies.
    1. Environmental impact analysis--United States.
2. Environmental policy--United States. 3. Environmental law--United States. I. United States. Laws,
statutes, etc. National environmental policy act of
1969. 1979.
TD194.6.E74   333.7   78-22522
ISBN 0-12-241550-7

PRINTED IN THE UNITED STATES OF AMERICA

79 80 81 82   9 8 7 6 5 4 3 2 1

To Claire (Myers) Erickson

Contents

II
Assessing Impacts: The Physical Environment

6
Relating the Assessment to the Project

7
Introduction to Assessment of Impacts: Physical Environment

8
Abiotic and Biotic Impacts

17

Criteria of Adequacy of Assessment: Social Environment

IV

Assessing Impacts: The Total Human Environment

18

The Total Human Environment

19

The Role of the Interdisciplinary Team

20

The Role of the Public

21

The Assessment Report

22

The Review Process

23

Criteria of Adequacy of Assessment: The Total Human Environment

V

Environmental Impact Assessment: Conclusion

24

Present Trends and Identifiable Needs

VI

Appendices

Appendix A

Federal Agencies and Federal State Agencies with Jurisdiction by Law or Special Expertise to Comment on Various Types of Environmental Impacts

Appendix B

Agency Channels of Communication on NEPA Matters

Appendix C

The National Environmental Policy Act of 1969

Preface

This volume contains information and suggestions that will be useful to those who undertake the design, conduct, and/or management of comprehensive environmental impact assessments mandated by the National Environmental Policy Act of 1969 (NEPA). The specific information and suggestions offered are based on actual experience. While it is hoped that my experience will enhance the practicality of this volume, I also recognize that the scope of impact assessment under NEPA is so comprehensive that no one individual's or group's experience and expertise are as yet adequate to the task of codifying final approaches and techniques. The fact is—there is no single way of conducting an impact assessment which can be guaranteed to meet the wide-ranging national environmental goals of NEPA.

Yet, while our 8-year experience with NEPA has not resulted in the invention or discovery of a single "best way" of designing, conducting, or managing an impact assessment, that experience has clearly identified some attitudinal, informational, and procedural aspects of the impact assessment process that typically result in either an inadequate assessment or an inadequate utilization of impact assessment during project development. This volume attempts to give guidelines (including attitudinal as well as technical guidelines) that can obviate the mistakes that have already been made and which are too often repeated.

It would, of course, be quite presumptuous for a biologist to give guidelines to a hydraulic engineer for the conduct of a strict hydraulic analysis of a river meander. It would be just as presumptuous for a sociologist to give guidelines to an ecologist for the conduct of a strict ecological analysis of a salt marsh. It should be emphasized, therefore, that environmental impact assessment under NEPA is not the private professional province of any particular science or discipline.

NEPA clearly calls for interdisciplinary assessment. No one who has participated in an interdisciplinary study, either in a university, a private enterprise, or a governmental agency, would presume to have answers for all the technical and managerial problems that arise during the conduct of such a study. When this fact about interdisciplinary studies in general is coupled to a special fact about NEPA—namely, that NEPA mandates the inclusion of interdisciplinary assessment and public involvement in actual governmental decision-making processes—there can be no doubt that any guidelines for doing an impact assessment under NEPA must often go well beyond individual technical and scientific domains. Environmental impact assessment, conducted under the real constraints of time and money, and undertaken to meet comprehensive needs of the nation, simply has little room for disciplinary dogmatists; but there is a tremendous need for those who can efficiently organize, interpret, and act upon extremely diverse data, information, and concerns. The contents of this book have, therefore, been determined not only by actual experience with the technical requirements of impact assessment, but also by experience with the managerial needs of a large number of professionals, who, though they may have expertise in one discipline or even in several, nevertheless have day-to-day responsibility for eliciting, collecting, coordinating, and utilizing the informational inputs of a wide variety of other professionals and the lay public.

Because of the need to integrate diverse information, as well as to perform specialized analyses during the process of assessment, it is my opinion that impact assessment required under NEPA would not necessarily be improved, either in terms of the quality or efficiency of the assessment, if operating agencies were given unlimited funds to hire as many technical and scientific specialists as they desired. Rather, this volume is predicated on the belief that the quality and efficiency of impact assessment under NEPA is ultimately determined by the comprehensive, timely, and concise integration of diverse and specialized information, and that such an integration will be greatly improved if the conceptual prerequisites of comprehensive environmental impact assessment are (a) identified clearly and (b) organized into functional hierarchical schemes that continually focus the attention and efforts of both generalists and specialists on the following questions:

1. What specific social and physical phenomena and processes could possibly be affected by the proposed project?
2. What are specific mechanisms by which the proposed project (including planning, construction, operation, and maintenance activities, materials, and/or products) may affect these phenomena and processes?
3. What quantitative and qualitative tools are available for predicting changes in these phenomena and processes due to project development?
4. What data and information are necessary in order to utilize these tools, and where and how may the data and information be obtained?
5. How are identified social and physical phenomena and processes interconnected, and how will predicted changes in any one affect the others?
6. What are the specific procedures and criteria by which predicted changes may be evaluated for their desirability?
7. What are the specific means by which undesirable changes can be obviated and/or mitigated, and by which desirable changes can be introduced and/or enhanced?

The information and suggested guidelines contained in the following pages will prove useful to those with impact assessment responsibility only insofar as they provide either answers to these questions or practical mechanisms and approaches for finding the answers.

It should be emphasized that this volume is not meant to duplicate in-depth disciplinary information, which is, after all, best developed in respective professional literature. The reader of this volume should therefore consult the literature cited in order to pursue individual issues that are of particular relevance to his or her needs. Key words, concepts, processes, and issues are set in italics throughout the text. Many of these key words can be used in conjunction with the indexes of cited literature to locate appropriate materials. General texts are cited primarily for the benefit of the beginning student; technical and scientific monographs are also cited for the benefit of the more advanced student or professional.

This volume is composed of six sections.

Part I presents a broad overview of environmental impact assessment and discusses some of the general technical, social, bureaucratic, legal, and managerial aspects of impact assessment under NEPA. This section is not meant to give an exhaustive analysis of NEPA as law, but rather, to highlight NEPA as action. This section will be of particular use to those who might just be undertaking responsibility for the design, conduct, and/or management of impact assessment.

Parts II and III present information and guidelines for the assessment of impacts on the physical and social environments. In each of these sections, emphasis is given to the general approach to assessment, to the parameters and processes involved, predictive schemes, sources of data and information, and criteria of adequacy of assessment.

Part IV presents information and guidelines for considering interactions among physical and social environmental impacts and for ensuring that a consideration of these interactions becomes an integral part of the overall assessment process.

Part V reviews the overall assessment process and identifies current trends and future needs and possibilities.

Part VI contains appendices that will help the reader better utilize textual materials.

I gratefully acknowledge the professional contributions that numerous individuals and organizations have made to my knowledge and understanding of impact assessment under NEPA. Special acknowledgment must be given to: George Camougis and Edward Robbins of New England Research, Inc., who have directly influenced my thinking with respect to the technical and managerial aspects of impact assessment; the Federal Highway Administration and the American Right of Way Association, Inc., which have provided me with the opportunity both to teach and to learn from a large number of professionals who have day-to-day assessment responsibility. I am also indebted to the American Right of Way Association, Inc. for permission to utilize some materials in this book that were previously developed for the Association by New England Research, Inc.

Finally, this volume represents my attempt to integrate and apply a large diversity of disciplinary concepts and concerns to the practical problems and the intellectual challenge of impact assessment. Whatever success has been achieved is directly owed to the personal and professional guidance and instruction that I have been privileged to receive from George Pride, John Reynolds, and Frederick Killian.

Environmental Impact Assessment

PRINCIPLES AND APPLICATIONS

I

Structure and Function of the Assessment Process

1

An Overview of Environmental Impact Assessment

Introduction

There is a parable told among the Eskimos of a hunter who wrestled a god. The hunter's wife had not conceived though she had performed all prescribed ablutions, and for lack of any hope that they would finally have a son, the hunter and his wife decided there was little reason to continue performing pious acts. One evening, Moonman sent his sled dog to howl outside their igloo. But so impious had the hunter become that he killed Moonman's dog. And when Moonman himself appeared and called out, "Who is it that has killed my dog?" the embittered hunter become even more insolent. He told Moonman to go away. When the god would not go away, the hunter wrestled him to the snow.

The battle lasted many hours. However, the hunter fought like a wild animal and was finally able to get the better of Moonman, who, after all, thought he was fighting a man. In the end, the hunter was about to kill Moonman when Moonman shouted in desperation, "What? Then there will be no more ebb and flow of the sea! And there will be no more baby seal!" At that, the hunter become human again. He let Moonman up and promised to do his rightful penitence.

To any student of anthropology this story is not new. It is told in different ways, with different names, in different cultures throughout the world. Yet it is always the same. In societies where people live with only

3

the thinnest of insulation between themselves and their physical environment, the moral is clear—the measure of humankind is neither power nor the lack of it, neither plenitude nor the lack of it, and certainly neither peace nor the lack of it. But the measure of humankind is acceptance of the ultimate and unalterable dependence upon environmental processes and cycles. Our contemporary rediscovery of this truth is causing us to reconsider how we perceive our environment and what we do with it. In the process, we are changing the ways by which complex bureaucracies make and implement decisions.

Some Environmental Perspectives

There is a large variety of perspectives that can be brought to bear on our physical environment. Some of these perspectives focus on abiotic (or nonliving) components of the environment, such as its geology, climate, and hydrology. Some focus on biotic (or living) components, including approximately $1\frac{1}{2}$ million different kinds of plants and animals. Still other perspectives focus on specific interactions between abiotic and biotic components of the environment, as in horticulture, agriculture, forestry, and husbandry. Ecology also deals with interactions between abiotic and biotic components of the environment, but ecology has a much more general scope.

While ecology as a formal discipline is about 100 years old, it has only recently received public attention. Part of the reason for our contemporary interest in ecology is that—precisely because it deals broadly with interactions between biotic and abiotic aspects of the environment—it serves to organize the diverse data and information[1] from other disciplines into a conceptual whole.

For example, within a square mile of forest there are innumerable biotic and abiotic elements that may be studied. Trees, forbs, grasses, shrubs, insects, mammals, reptiles, soil chemistry, subsurface and superficial hydrology, macro- and microclimate, and bedrock geology are only a few of these elements. The individual who knows a great deal about grasses is likely to know very little, if anything, about snails, and neither the grass expert nor the snail expert is likely to know anything about the rock strata underneath his feet. It is for the ecologist to "put it all together." But the ecologist is not an expert on each component of the environment. He is, rather, an expert on how individual components interact. His task is to present an overall picture of how the plants and animals and all the abiotic components fit together. Such a picture of a

---

[1] Throughout this volume, date and information respectively denote quantitative and qualitative knowledge.

square mile of forest (or of any extent of the terrestrial or aquatic environment) is called an *ecosystem.*

Ecosystems share some basic characteristics with other dynamic systems, including one with which we are all familiar—social systems. Ecosystems and social systems, for example, include (a) individual component parts; and (b) interactions between those parts. In an ecosystem, the component parts are the plants and animals and the abiotic factors of their environment. In a social system (such as a metropolis), the component parts are people, their units of social organization, their machines, and raw materials.

Just as there are those functions and processes that people and machines carry on in a city and that give a city its characteristics, so there are similar functions and processes that plants and animals carry on in ecosystems. These include:

1. Transformation of energy from one kind into another
2. Modulation or regulation of energy flow within the system
3. Environmental adaptation
4. Environmental modification (Fig. 1.1)

In ecosystems, green plants (primary producers) transform radiant sunlight energy and simple nutrients into the chemical energy of plant tissue through the process of *photosynthesis.* This process depends upon chlorophyll. Animals, because they lack chlorophyll, transform the chemical energy of plant tissues or of other animals into the chemical energy of their own tissues (e.g., herbivores, carnivores). And still other biota (decomposers, scavengers) transform energy from dead plants and animals into the energy of their living systems. In the process, they return nutrients back to the environment where they are available, once again, to the green plants. The sequential consumption of plants by animals and of prey by predator and of dead material by living material is called a *food chain.* Food chains are mechanisms whereby energy is distributed throughout an ecosystem. The actual flow of this energy is modulated or regulated by the numbers, the kinds, and the ages of those biota that participate in a specific food chain.

In a city, machines transform the potential energies of fossil fuels and radioactive materials into kinetic or usable forms of energy, including electrical, heat, and mechanical energies. These transformed energies are likewise distributed throughout the city by other machines which regulate their flow through conduits and pipes, as in a power grid. The flows of all transformed energies in a city, including foods and manufactured products, may also be modulated by other devices, including economic and organizational mechanisms.

In order for plants and animals to act as energy transformers and modulators in their abiotic environment, they must be adapted to that

GENERALIZED DYNAMIC SYSTEMS

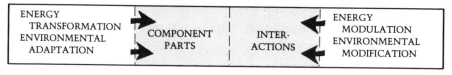

ECOSYSTEMS

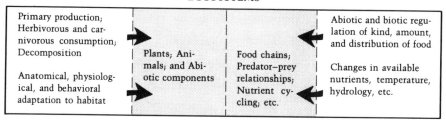

SOCIAL SYSTEMS

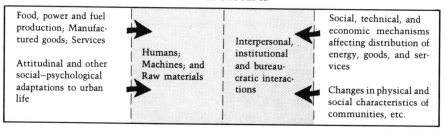

FIGURE 1.1. *Four processes that are common to ecosystems and social systems.*

environment. For example, a plant that floats in water is an energy transformer of the same kind (i.e., it carries on photosynthesis and is, therefore, a primary producer) as a plant rooted in the soil. While the ecological role of both plants (i.e., primary production) is the same, their structural characteristics are different and reflect the advantages and disadvantages of their respective aquatic and terrestrial habitats. Of course, people also adapt to a modern urban life. The machines upon which city dwellers depend, whether they are machines for transportation or for energy or food production and distribution, must also conform to urban realities.

Finally, because they are able to persist in the environments to which they are adapted, plants and animals modify those environments by their own activity. They can alter the nutrients and physical structure of the soils in which they grow and change the temperature in their immediate surroundings by casting shade. These and other environmental modifications may, in turn, make a particular environment more or less amenable to other biota. People and machines also continually modify the urban environments to which they are adapted. And it is also true that these modifications of both the physical and the social environments of metropolitan areas make cities more or less amenable both to other people and other biota.

The general processes of energy transformation and modulation and environmental adaptation and modification are not only useful tools for comparing ecosystems and social systems. They are also useful for understanding how the physical environment specifically influences social systems. Some of these effects are obvious. Others are more subtle.

POTENTIAL ENERGY SUPPLIES FOR HUMAN DEVELOPMENT

Biotic, abiotic, and ecological transformations of energy have provided humans with different types of energy throughout the development of human society. The sun, human and draft-animal muscle, wind and water, fossil remains, and the atom itself are the major energy sources. Some of these sources (e.g., sunlight, wind and water, and the atom) represent abiotic transformations of energies that have continued for billions of years of earth's history. Others (e.g., muscle power of living things, fossil fuels) have been made available through biological and ecological transformations that have continued over millions of years. Different stages of human development, including hunting and gathering stages, agricultural stages, and the contemporary industrial stage, may be characterized in a variety of ways by particular energy sources they each in turn have utilized (Table 1.1).

UNITS OF SOCIAL ORGANIZATION

The type of energy transformation and modulation that characterizes different development stages of human society depends upon specific technologies and, therefore, upon general sociological development. One way to measure such development in human society is to note the changing units of social organization.

In small food-gathering societies, for example, the nuclear family (i.e., husband, wife, and children) is an important regulator of energy flow from the environment (in the form of foods and materials) and into and throughout the social system. In early agricultural societies, which had a greater differentiation of labor and a greater variety of available energy sources, important modulators of energy flow included the extended family (i.e., the nuclear family plus selected relatives) and other kinship groups. Finally, in contemporary industrial society, transformation of potential energies into usable energies and the modulation of their flow

TABLE 1.1
Some Stages in the Development of Human Society[a]

	Type of society		
	Early hunting–gathering	Early agricultural	Contemporary industrial
Major energy sources	• sun • humans	• sun • humans • wind • water • draft animals	• sun • fossil fuels • atomic energy • water
Typical unit of social organization	• nuclear family	• extended family • kinship groups	• family • institutions • bureaucracies
Common attitude toward environment	Humans are at the mercy of the environment	Humans can manipulate the environment	Humans can control the environment
Type of environmental knowledge	Precise differentiation of types of biota that are important to human life	Understanding of general biotic and abiotic interrelationships	Sophisticated understanding of macro- and microenvironmental processes

[a] Adapted with permission from American Right of Way Association, 1976.

throughout society increasingly depend upon complex institutions and bureaucracies.

ENVIRONMENTAL ATTITUDES

All biota possess different structural, physiological, and/or behavioral adaptations to their environment. Humans exhibit mental adaptations as well. Such adaptations may be manifest in attitudes that individuals have toward their environment and which help ensure their survival in that environment.

In preindustrial societies, for example, it has been observed that a common attitude is that human beings are, by and large, at the mercy of the physical environment. Opportunistic utilization of environmental energies and maximum utilization of available but scarce supplies are, therefore, also common attributes of human behavior in these societies and help to ensure their longevity despite periods of deprivation. However, with the advent of agriculture, there was a growing appreciation of the fact that the human being can manipulate the environment to human advantage. In contemporary society, this attitude may take an extreme form—that humans can exercise absolute control over their environment. Wasteful energy transformations (e.g., driving a car) and inefficient modulations of energy flow (e.g., distributing foods and other goods through middlemen) are typical behavioral consequences. In a world of infinite resources, such behavior may indeed be adaptive, as in that peculiar form of social adaptation known as "keeping up with the Joneses." But as we are discovering, the real world has finite resources. Attitudinal adaptation in the real world will require us to change this behavior and those attitudes that sustain it.

ENVIRONMENTAL KNOWLEDGE

Attitudes toward the environment influence human behavior with respect to environmental resources. This behavior, in turn, restricts or expands our knowledge and understanding of that environment. It either confirms previous understanding or (as is currently the case) raises new questions about the efficacy of our behavior. Insofar as new knowledge about our environment influences what we do in it (or with it, or to it), then knowledge is an important tool by which human beings modify their environment.

In preindustrial societies, the smallest children are trained to identify precisely those plants and animals that mean sustenance to them. In agricultural societies, the children come quickly to understand some of those interrelationships between the biotic and abiotic components of the physical environment that mean the difference between a lush or a sparse harvest. In industrial societies, children learn the mechanical and organizational mechanisms that help to control their environment and that they, in turn, must learn to control. It has taken man probably on the order of a million years to proceed from a knowledge of what animal it was that eluded him on the hunt to a knowledge of how to use the principles of heredity and nutrition to make a "beef machine" that willingly goes to slaughter. In this same period of time, the role of humans as modifiers of earth's environment has also changed—from irrelevance to preeminence.

Reevaluation of Environmental Perspectives

Abiotic and biotic perspectives of the physical environment are typically analytical and are useful for identifying the pieces that make up the world. Ecological perspectives tend to be integrative and are useful for seeing how the pieces work together. Both analytical and integrative perspectives contribute to our contemporary understanding of the physical environment.

However, too narrow a concern with individual abiotic or biotic components of the physical environment conditions us to think about that environment in terms only of its products, such as "amount" of water, "quality" of water, "force" of water, "number" of birds, and "size" of deer. The ecological perspective, on the other hand, would have us concentrate upon those processes by which living and nonliving things interact and yield environmental products.

We use high-quality water as an environmental product when we brush our teeth. We want its "color," its "taste," and its "smell" to be of a certain kind. However, few of us think of the quality of that water as having been provided by an environmental process of natural purification, involving the interactions of innumerable biota and physical and chemical factors. As we are finding out, our consumptive rate of environmental products can outstrip or otherwise alter the rate at which environmental processes provide them.

The narrow conceptualization of the environment as a collection of individual products is historically associated with technological development. A technological society, after all, is a society in which reliance upon and consumption of manufactured goods and the consumption of natural resources seem to be mutually reinforcing phenomena. Natural resources are typically thought of as available material and energetic products of the environment.

We "consume" a rubber tire along our highways. We easily understand the tire as a manufactured item. Yet this manufactured item is a social recombination of material and energetic products of the physical environment, including such things as petrochemicals and metals and energies employed in the various phases of its manufacture. In a world in which there is an apparently infinite environmental reserve of petrochemicals and metals and energy, there is typically more concern with social means of production (e.g., labor pool expansion, capital investment, distribution, marketing) than with underlying environmental means of production (e.g., ecological processes). Also, in such a world of apparently infinite environmental products, the consequences of the actual use of tires are most often examined only in terms of consequences on social processes (e.g., more cars mean more highways; more highways mean more construction jobs). Consequences to underlying ecological processes (e.g., significant impacts of industrial wastes on natural purification in waters) are typically ignored (Fig. 1.2.)

Yet, as we have begun to realize, the world does not have an infinite reserve of environmental products.

In 1975, the total world population reached roughly 4 billion people. This represents an increase of 1½ billion (60%) over the previous 25 years. In the same period, the population of the United States increased by about 70 million people, or about 50% over the 1950 census. While these rates of increase are roughly the same, the consequences to us and

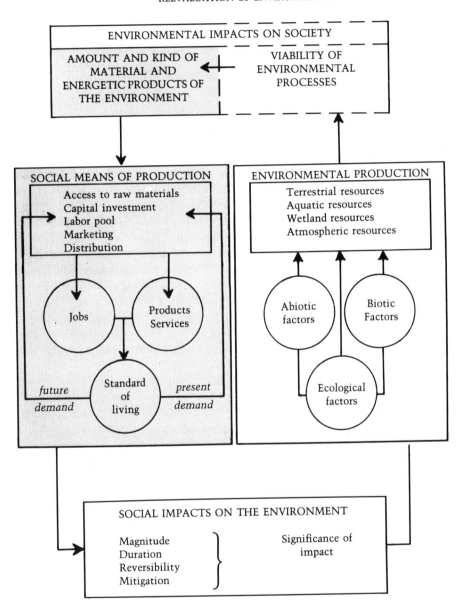

FIGURE 1.2. *Conventional (shaded) and more recent (unshaded) concern over the impact of human activities.*

to the world at large have been vastly different. For, coupled with our national population increase is the greatest increase in the standard of living that the world has ever known. This increase in standard of living is forcing us to reevaluate conventional environmental perspectives.

With respect to transportation alone, it has been reported that the United States uses three out of every five of the world's cars, and two out of every five trucks. In fact, our total amount of highway rights-of-way is 18,000 square miles, or about the total area of several of our smaller states. We also account for one-fourth of the world's total railroad freight, and one-half of the world's mileage for civil aviation. Total expenditures of energy in the United States for transportation amounted to almost 25% of our total energy expenditures in 1968. And as of 1963, our total energy expenditures accounted for 77% of the total world production of coal, 81% of the world's petroleum, 95% of the world's natural gas, and 80% of the world's hydroelectric and nuclear power. In short, the current high standard of living for roughly 5% of the world's population has been the result of an unprecedented availability of manufactured and processed goods and energies that, in turn, have been derived from an unprecedented harvesting of environmental products of matter and energy.

Of course, along with this harvesting of environmental products of matter and energy have been those measurable changes in air, water, and land that have been largely perceived as losses in environmental quality. These perceptions of environmental degradation have also coincided with perceptions of dwindling supplies of material and energetic resources of the environment. These dual perceptions have, in turn, generated not only general concern over how decisions affecting the environment are made, but also specific, far-reaching legislation governing many of the actual mechanisms of decision-making in our bureaucracies.

The National Environmental Policy Act of 1969 (NEPA), for example, clearly states that the responsibility of the Federal Government is to improve and coordinate Federal plans, functions, programs and resources to the end that the Nation may

1. Fulfill the responsibilities of each generation as trustee of the environment for succeeding generations
2. Assure for all Americans safe, healthful, productive, and esthetically and culturally pleasing surroundings
3. Attain the widest range of beneficial uses of the environment without degradation, risk to health or safety, or other undesirable and unintended consequences
4. Preserve important historic, cultural, and natural aspects of our national heritage, and maintain, wherever possible, an environment which supports diversity and variety of individual choice

5. Achieve balance between population and resource use which will permit high standards of living and a wide sharing of life's amenities
6. Enhance the quality of renewable resources and approach the maximum attainable recycling of depletable resources.

In order to achieve these objectives, Congress directed that each federal agency "shall include in every recommendation or report on proposals for legislation and other major Federal actions significantly affecting the quality of the human environment, a detailed statement. . . ." This statement, known as an Environmental Impact Statement, (EIS) must assess the environmental impacts of the proposed federal action.

However, NEPA is only a single piece of legislation that sets environmental objectives and guidelines. Other federal legislation (some of its recent, some of it older) affects actions that are environmentally significant with respect both to singular and general environmental issues. This legislation includes: the Federal Water Pollution Control Act Amendments of 1972, the Fish and Wildlife Coordination Act of 1958, the Migratory Bird Conservation Act of 1929, the Water Bank Act of 1970, the Land and Water Conservation Fund Act of 1965, the Endangered Species Act of 1973, the Wild Horse and Burro Protection Act of 1971, the Marine Mammal Protection Act of 1972, the Anadromous Fish Conservation Act of 1965, the Coastal Zone Management Act of 1972, the Wild and Scenic Rivers Act of 1968, the Wilderness Act of 1964, and the National Trails System Act of 1968.

In addition to this federal legislation, many states have passed specific legislation that similarly regulates certain kinds of activities that affect the environment. Some states have even adopted constitutional amendments that guarantee state obligations with respect to environmental resources and amenities. Finally, executive and juridical decisions and pronouncements have added to the legislated decrees calling for a reevaluation of the role of environmental values in decision-making processes that are critical to the planning and implementation of environmentally significant projects.

There are those who no doubt consider that many of the current legislative, executive, and juridical pronouncements establishing a place for environmental impact assessment in decision-making are overreactions to the emotional and unfounded forecasts of "doomsdayers"—or that "bugs-n-bunnies" have been elevated to the equal (if not the superior) of the "public good" as a factor in decision-making. However, NEPA, as the most comprehensive piece of environmental legislation to date, contradicts such an interpretation. As stated in the Act, "The purposes of this Act are: To declare a national policy which will encourage productive and enjoyable harmony between man and his environment; to promote efforts which will prevent or eliminate damage to the environment and biosphere and stimulate the health and welfare of man; to enrich the understanding of ecological systems and national resources important to the Nation. . . ."

We can reasonably doubt that the Congress of the United States was seriously concerned with "bugs-n-bunnies" concept of harmony between man and his environment, or with a "bugs-n-bunnies" concern for the health and welfare of man, or with a "bugs-n-bunnies" understanding of ecological systems and natural resources important to the nation. The National Environmental Policy Act is for people.

Total Environmental Perspective

As we have seen, the physical environment is composed of biotic, abiotic, and ecological components. Humans are part of earth's biota. We utilize, alter, and otherwise rearrange both biotic and abiotic components of the physical environment and, in so doing, we play a significant role in earth's ecology. We have also seen that the ecological processes of energy transformation and modulation, and environmental adaptation and modification influence similar social processes. Therefore, any comprehensive perspective of the earth's environment must take due account of humans, and of those elements of the human environment (e.g., human psychology, sociology) that cannot be adequately described in biotic, abiotic, or ecological categories.

As of January 1, 1970 it became (through NEPA) national policy to "use all practicable means and measures . . . to create and maintain conditions under which man and nature can exist in productive harmony, and fulfill the social, economic, and other requirements of present and future generations of Americans." This policy is clearly predicated upon concern for our *social environment* as well as our *physical environment*. It is a statement of policy about the *total environment*[2]—an environment that includes biotic, abiotic, and ecological components as well as the personal, interpersonal, organizational, and other institutional components of human life (Fig. 1.3).

The assessment of impacts of various human activities on the total environment is a complex task. Its complexity is compounded not only by the diversity of our social, physical, and biological disciplines, but also by several basic facts of our experience with our total environment.

---

[2] These categories are adapted from American Right of Way Association, 1976.

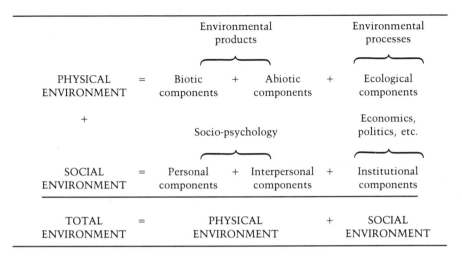

		Environmental products			Environmental processes
PHYSICAL ENVIRONMENT	=	Biotic components	+	Abiotic components	+ Ecological components
+					Economics, politics, etc.
		Socio-psychology			
SOCIAL ENVIRONMENT	=	Personal components	+	Interpersonal components	+ Institutional components
TOTAL ENVIRONMENT	=	PHYSICAL ENVIRONMENT		+	SOCIAL ENVIRONMENT

FIGURE 1.3. *Components of the total environment of humans.*

First, we simply do not know all the relationships and interactions between our social and physical environments. Second, we do not have precise understanding of how those relationships that we are aware of (e.g., groundwater supplies and urban expansion) may be disrupted. And third, the total environment includes subjective facets of human existence that cannot be circumscribed by the objective tools and devices of technical analysis.

NEPA deals with these facts directly in its charge to all agencies of the federal government to: (a) "utilize a systematic, interdisciplinary approach which will insure the integrated use of the natural and social sciences and the environmental design arts in planning and in decision-making"; and (b)"identify and develop methods and procedures . . . which will insure that presently unquantified environmental amenities and values may be given appropriate consideration in decision-making along with economic and technical considerations."

Another mechanism by which NEPA attempts to deal with our incomplete knowledge and the diversity of our social-selves is to require public access to decision-making itself. Once again, however, NEPA is not the only source of the public's right to participate in those decisions that have impact on the environment.

In general, any process of decision-making comprises several phases, including: (a) the collection and collation of appropriate data and information; (b) the evaluation of that data and information with respect to decision-objectives; and (c) the decision-making itself. Public participation in decisions affecting the total environment can occur in each of these three phases. Such participation is fostered by executive, legislative, and juridical pronouncements that define specific mechanisms for (a) the public's *access to information;* (b) the public's *contribution of information;* and (c) the public's *challenge of decisions.* In addition to many legal mechanisms that allow for public participation, there are important informal mechanisms by which the public can influence the wide variety of decisions that must be made during project development by governmental agencies.

Some federal legislation specifically defines mechanisms for public involvement in environmental decision-making. But similar mechanisms have also been defined in legislation that is not typically described as "environmental legislation." For example, the Freedom of Information Act of 1966 guarantees the access of any citizen to factual or investigatory information available on federal projects. Opinions bearing on policy matters are exempted. The NEPA also makes factual data available to the public, as well as policy information that is pertinent to the proposed project. More than this, NEPA provides a specific mechanism by which the informed public may act on this information. This may be by either or both oral and written comments and questions during public hearings. Such hearings must be held during what is called the Environmental Impact Statement Review Process. The Environmental Impact Statement must be available for public review prior to the public hearing.

Executive Order 11514 is another guarantee of public access to information. This order obligates federal agencies to maximize public information on environmentally significant actions. Also, Section 309 of the Clean Air Act requires that the administrator of the Environmental Protection Agency report on all proposed federal legislation, regulations or actions affecting air or water quality, pesticides, solid waste disposal, radiation, and noise, and that this report be available for public review.

In addition to these guarantees of public access to information on federally sponsored projects, there are legal mechanisms for public contribution and/or challenge to the decision-making process. The Administrative Procedure Act, for example, allows public participation in specified types of hearings. While the scope of public participation under this Act might be quite narrow, NEPA and pertinent guidelines by the Council of Environmental Quality (also created in the Executive Department by NEPA) expand public participation greatly.

Federal courts have moved to expand criteria used for establishing the public's standing in environmental suits. One example is the inclusion of conservation, aesthetic, and recreational "injuries" in the category of "injury in fact." This inclusion provides greater access to the courts by a larger portion of the public affected by projects. State courts have similarly expanded the right of the public to bring suit at the state level. Some states have also adopted constitutional amendments that give the public "environmental rights" and, therefore, access to the state courts whenever such rights might have been abrogated by projects that affect the environment. State utilization of existing public nuisance laws, as well as new legislation similar to NEPA, has also significantly increased the public's legal access to the decision-making process.

Public agencies are making increasing use of informal public hearings. During these informal hearings the public is encouraged to submit oral and written comments, questions, and suggestions and to participate in the decision-making process. Governmental agencies, as well as the private sector, often recognize that public participation in such informal hearings can facilitate the actual process of impact assessment. The public, for example, may be better able to identify regionally important environmental issues that should be considered. The public may also contribute data and information that are otherwise unavailable. And finally, the public may be able to suggest feasible project alternatives that could minimize or avoid adverse environmental impacts.

Overview

Contemporary concern for the environment has been described as an "environmental revolution." Many of the elements of this revolution, which were proclaimed over the years by conservationists and only recently catapulted into public acclaim, have been translated into executive, legislative, and judicial action. In the process, it is clear that what may have begun as a revolution against thoughtless exploitation of natural resources of our physical environment has evolved into an environmental ethos that calls for the evaluation of the effects of human activities on the total environment of man. Thus, if contemporary environmentalism is, indeed, revolutionary, the revolution does not lie merely in our recognition that humans are but one kind of living thing on this planet. It also lies in our recognition of a much more complex fact—that as a part of earth's ecology we both affect and are affected by the ecology; that human social relationships have as much an environmental reality as does the habitat of black ducks; that the nutritional needs of a white-tail deer is as much an environmental reality as a job for a person; and that these are not isolated realities of the total environment, but realities that can interact with one another, influence one another, and even determine one another. Of course, the Eskimo knows this. And people throughout the world who live in more direct contact than we with the adversities of the physical environment also know it. And it was not so long ago, before the advent of industrial society, that all people everywhere knew it.

The interdependence of social and physical components of the total environment is, therefore, not a new concept. But applying that concept

18

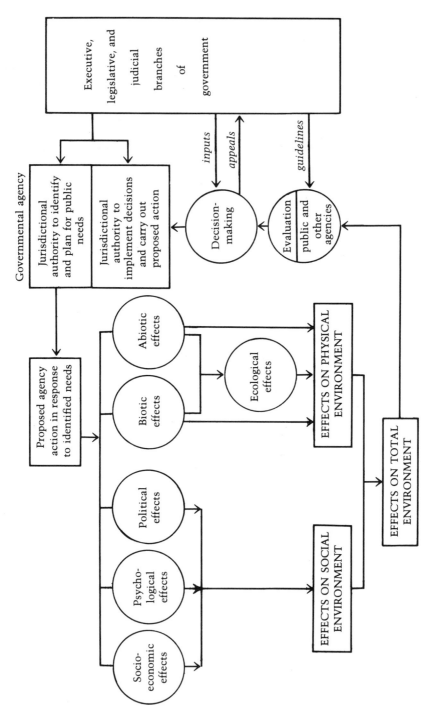

FIGURE 1.4. *Overview—putting the total human environment into decision-making.*

to the machinations of advanced industrial societies *is* a unique undertaking. A central issue in this undertaking is the inclusion of a total environmental perspective into the actual mechanisms of decisionmaking that are employed by complex, bureaucratic organizations.

As depicted in Fig. 1.4, and as discussed in this chapter, such an inclusion involves inputs from specialists as well as from the lay public, and a dynamic interaction between executive, judicial, and legislative branches of government. It is the quality of this involvement and these dynamic interactions that will determine the future quality of human life.

Reference

American Right of Way Association (ARWA)
   1976   *Student workbook, right of way and the environment.* Los Angeles: American Right of Way Association.

2

The Bureaucratic Basis of Environmental Impact Assessment

Introduction

Different federal, state, and local laws (and different interpretations of them) require different types of environmental assessment. Some require an assessment of particular components of the physical environment (e.g., thermal impacts on fisheries); others require assessment of particular components of the social environment (e.g., impacts on recreational areas). However, there is only one purpose behind any assessment process. It is to help decision-makers plan, design, manage, and regulate projects and programs. The ultimate products of any environmental impact assessment process are, therefore, decisions. Certainly those assessments which are made by federal agencies under NEPA, but which are not actually factored into real decisions that affect real projects and programs, constitute a misappropriation of public funds. And this is true regardless of any scientific sophistication or worth, or public-relations value of the assessment.

Individually, we make decisions every day of our lives. Sometimes we do it with concern for hardly more than immediate consequences, as when we choose between two items on a grocery shelf on the basis of cost and the amount of cash we actually have in hand. Sometimes we do it with concern for long-range consequences, as when we select a city or a region in which to live. Sometimes we decide only after long deliberation.

With respect to some decisions, we are hardly aware that we have deliberated at all. But, whether our individual decisions are made quickly or slowly, and whether they are made for long or short-term purposes, decision-making itself is a complex process that involves the integration of different kinds of information. The singular act of "making a decision" is the integration of specific information in a manner that gives us confidence that a particular kind of future will be realized if we act in a particular way. Government agencies and bureaucracies that have responsibility under NEPA must ensure that the components of their decision-making apparatus will be sensitive to and respond to data and information about the total human environment (Chapter 1).

Generalized Decision-Making Processes

As indicated in Fig. 2.1, a decision-making process may be analyzed in terms of its (a) experiential base; (b) prediction system; (c) value system; and (d) selection system[1]. The inclusion of an environmental ethos in contemporary decision-making processes affects each of these components.

The *experiential base* of a decision-making process is the sum of all information, data, and knowledge achieved up to the present time. For each of us, it consists of our experience with people and with the physical environment. For a governmental agency, it consists of its legal authorization and all previous actions and experiences in carrying out the mandates of its authorization.

At any moment the data and information of our experiential base help us to define the possible actions that we may undertake, as well as the goals and objectives that we may realize. Which of these possible actions, and which of these possible objectives we will, in fact, incorporate into the real world is the question that initiates decision-making.

The real world, in opposition to the dictum of a certain nursery rhyme, is not full of any number of things. At any one time, the real world is characterized by specifics, not by abstract possibilities. Individual, governmental, and corporate decision-making that affects the real world determines what those specifics are. Good decisions are those that result in the realization of a selected future. Bad decisions are those that do not result in the realization of a selected future. At least some portion of the difference between a good and a bad decision is, therefore, the difference between a good and a bad *prediction system*. A good prediction system is a tool that allows us to relate accurately present actions with future consequences of our actions.

---

[1] These components are adapted from Bross, 1959.

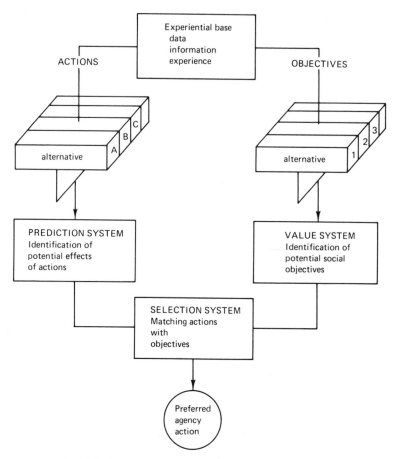

FIGURE 2.1. *Basic components of the decision-making process.*

Contemporary technobureaucratic societies may be described as pluralistic—that is, societies characterized by many different values and attitudes. In such societies, the *value system* promulgated by one institution (e.g., a particular religious institution) may often conflict with values promulgated by another institution (e.g., a particular corporation). Since values are important factors in our motivation to undertake specific actions, the identification of appropriate values is a complex process in pluralistic societies. Some decision-makers, of course, have to be concerned with the values of only a relatively small number of people (e.g., board of directors); others have larger constituencies (e.g., a congressional district). Decision-makers involved in environmental decision-making under NEPA typically deal with very diverse groups, organizations, and agencies, and therefore with a large number of diverse values.

The manner in which a plurality of values and predicted consequences of actions are integrated, and preferred actions and therefore alternative futures are selected constitutes the *selection system* of decision-making. This phase of decision-making may involve complex institutional and public interactions, as in a referendum or a public hearing. It may also be left to the judgment of an individual or a small, select group, as in the selection of technical specifications for certain construction materials and supplies.

NEPA Inputs into Federal Decision-Making

It is probably a good rule of thumb that whenever there is a disagreement about what a law says, it is because of what the law does not specifically say. The scope of NEPA's relevance to bureaucratic decision-making is certainly more easily approached in this manner. For example, NEPA does *not* specifically say that:

1. The historical experiential base of any federal agency shall be given absolute and exclusionary priority in planning current projects and programs.
2. Any particular prediction system shall be given absolute and exclusionary priority.
3. Any single social value, or any single criterion for selecting among conflicting social values, shall be given absolute and exclusionary priority in planning or decision-making.

What NEPA *does* specifically say with respect to these components of decision-making is that federal agencies, in fulfilling their legal authorization:

1. Will utilize interdisciplinary approaches
2. Will ensure an integration of natural and social sciences and environmental design arts
3. Will ensure that unquantifiable environmental amenities and values are given appropriate consideration along with economic and technical consideration.

In consideration of both what the Act does not say and what it does say, it is reasonable for an American citizen to expect the following with regard to current decision-making and planning by federal agencies whose actions may significantly affect the environment. Regardless of an agency's previous experiential base for decision-making, that base is supposed to be supplemented with interdisciplinary concerns; that regardless of the prediction system and technics formerly employed for planning and decision-making purposes, those systems and technics are supposed to be supplemented with methodologies and procedures that allow for the consideration of unquantifiable environmental amenities; and that, regardless of originally authorized objectives and regardless of value selection criteria historically employed by an agency, those values and criteria are supposed to be supplemented with the broad environmental concerns and goals of the nation as set forth in NEPA.

This is no less than a revolution in agency decision-making—a revolution that has resulted in very real frustrations and problems, but which was nevertheless undertaken because "previous practice was inadequate—agencies' planning and decision-making procedures had ignored or misjudged significant environmental impacts and would continue to do so [CEQ, 1975, p. 628]."[2]

Structural–Functional Thrust of NEPA

A *structural–functional* change in a dynamic process such as governmental decision-making may be viewed as one by which (a) a new structure is inserted into an existing decision-making system; (b) the operations of the altered system meet public needs efficiently; and (c) the altered system maintains itself for as long as pertinent public needs persist. Of course, changes other than structural–functional ones are possible, both in social and in physical processes. For example, one may institute simply a *structural change* in a clock by introducing a new gear. But unless the new gear meshes with those already present and in such a way that the time-telling capacity of the clock remains intact, the structural change will be counterproductive (i.e., dysfunctional) and will likely be "undone." Or, a pork-barrel job may be invented and filled, say, in a turnpike authority, and have little if any effect on the actual operation of the authority. Such structural irrelevancies are, regrettably, not easily "undone." They tend to persist until dysfunctions in the system (e.g., economic) are traced to them. Similarly, a merely *functional change* in any system, such as a new and added responsibility for an existing agency, will not necessarily generate desired results unless adequate structures (e.g., funding, staffing) are also provided. Unlike these examples, however, a *structural–functional change* is a change in which the proposed new structure and the self-maintenance potential of an effective new system are matched. The NEPA may be viewed as an attempt to accomplish such a structural-functional change in governmental decision-making.

In keeping with a social-science usage, we may consider that *structure* is operationally equivalent to a collection of people and powers that

---

[2] Reference citations of the Council on Environmental Quality are abbreviated to CEQ throughout this volume.

governs specific behavior. In terms of this definition, then, NEPA has inserted several new structures into governmental decision-making processes: (a) an environmental policy structure (i.e., Section 101 of NEPA, and the Council on Environmental Quality); (b) an action-enforcing structure (i.e., the Environmental Impact Statement and associated documents); and (c) a liaison structure (i.e., the EIS review procedure). Each of these structures is realized in specific arrangements of people who seek to accomplish national environmental goals by influencing the decision-making behavior of those who are entrusted with public funds. In particular, such behavior includes the processes by which people in agencies predict the consequences of agency actions, the process by which people in agencies select social values as appropriate values for agency action, and the processes by which people in agencies coordinate selected societal values and agency actions so as to define preferred agency actions. Whether such decision-making behavior is, in fact, regulated by the structural creations of NEPA, and whether consequent decision-making is adequate and efficient (i.e., functional) are central questions to any evaluation of NEPA (Fig. 2.2)

Some Results

The environmental policy structure of NEPA is manifest in the Council on Environmental Quality (CEQ) which attempts to set policy and establish guidelines with respect to the interpretation and administration of NEPA. The CEQ has no statutory authority to issue guidelines, but was given authority to do so by Executive Order 11514 (Druley, 1976, p. 1). Also, CEQ has no legislative authority to enforce compliance with its guidelines.[3] The courts have, therefore, acted to force compliance in both procedural and substantive aspects of the EIS process (Chapter 3).

The council's initial guidelines were issued on May 11, 1970 and were revised on April 23, 1971. Revised guidelines were again issued in 1972 and, finally, on August 1, 1973 (CEQ, 1973a, b). The final set of CEQ guidelines[4] . . . "establish the basic structure for the environmental impact process. Because Federal programs vary so substantially, [these] guidelines recognize that agencies must develop specific procedures tailoring the EIS process to their particular activities and they are authorized to do so [CEQ, 1976b; p. 5]."

As of April 1976, about 70 federal agencies and departments had developed EIS procedures based on the CEQ guidelines (Table 2.1). Depending upon an agency's role under NEPA (i.e., whether it is primarily

---

[3] See Chapter 4 for further discussion of this point.

[4] CEQ regulations, under development in 1978, will supersede these 1973 guidelines.

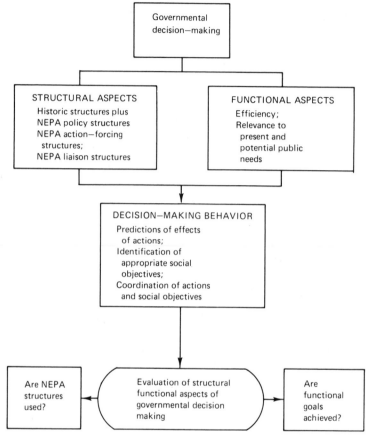

FIGURE 2.2. *NEPA as a structural–functional change in governmental decision-making.*

involved in reviewing EISs prepared by other agencies, or is typically a "lead" agency in the actual writing of an EIS), these guidelines give emphasis to different aspects of the EIS process.

Some agency guidelines have focused on locating specific assessment activities (including the identification of potential impacts, the evaluation of significance of impacts, and the mitigation of impacts) at appropriate phases of project development, such as planning, location, design, right-of-way acquisition, construction, operation, and maintenance phases (Fig. 2.3).

Some agency guidelines identify very specific types of information that should be included in EISs written by other agencies for various kinds of projects. For example, U.S. EPA (Region 10) guidelines identify minimum information to be included in EISs for highways, dredging,

TABLE 2-1.
U.S. Agencies Having NEPA Procedures as of March 1, 1976.

Agriculture
  Departmental
  Agricultural Stabilization and Conservation Service
  Animal and Plant Health Inspection Service
  Farmers Home Administration
  Forest Service
  Rural Electrification Administration
  Soil Conservation Service

Appalachian Regional Commission
[Atomic Energy Commission]
  Nuclear Regulatory Commission (regulatory)
  Energy Research and Development Administration (nonregulatory)
Canal Zone Government
Central Intelligence Agency
Civil Aeronautics Board

Department of Commerce
  National Oceanic and Atmospheric Administration, Coastal Zone Management
  Economic Development Administration
Department of Defense
  Army Corps of Engineers
Delaware River Basin Commission
Environmental Protection Agency
Export-Import Bank: no regulations
Federal Communications Commission
Federal Energy Administration
Federal Power Commission
Federal Trade Commission
General Services Administration
  Departmental
  Property Management and Disposal Service
  Public Buildings Service
  Telecommunications Service
Health, Education, and Welfare
  Departmental
  Food and Drug Administration

Department of Housing and Urban Development
  Community Development Block Grants
  Other programs

Interior
  Departmental
  Bonneville Power Administration
  Bureau of Indian Affairs
  Bureau of Land Management
  Bureau of Mines
  Bureau of Outdoor Recreation
  Bureau of Reclamation
  Fish and Wildlife Service
  U.S. Geological Survey
  National Park Service
  Interstate Commerce Commission
Department of Justice
  Law Enforcement Assistance Administration
Department of Labor
National Aeronautics and Space Administration
National Capital Planning Commission
National Science Foundation
U.S. Postal Service
Small Business Administration
State
  Departmental
  Agency for International Development
  International Boundary and Water Commission
Tennessee Valley Authority
Transportation
  Departmental
  Federal Aviation Administration Airport Development Act
  Federal Highway Administration
  U.S. Coast Guard
  Urban Mass Transportation Administration
  National Highway Traffic Safety Administration
Saint Lawrence Seaway Development Corporation
Treasury
  Departmental
  Internal Revenue Service
Veterans Administration
Water Resources Council

ᵃ Adapted from CEQ, 1976a; pp. 128–131.

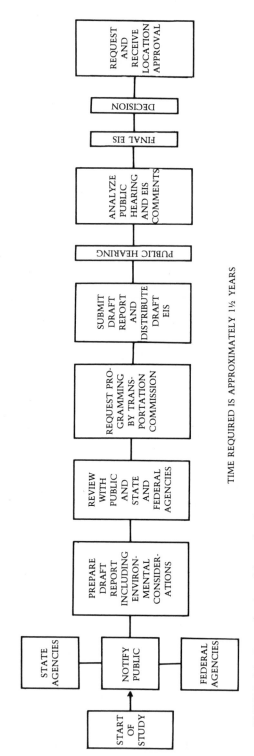

Programming and Corridor Location Study

TO REAFFIRM THE FEASIBILITY OF A PROJECT AND TO SELECT AND RECOMMEND AND PROGRAM THE MOST PRACTICAL ECONOMICALLY SOUND AND JUSTIFIABLE CORRIDOR BETWEEN IDENTIFIED TERMINI.

START OF STUDY

STATE AGENCIES

NOTIFY PUBLIC

FEDERAL AGENCIES

PREPARE DRAFT REPORT INCLUDING ENVIRON- MENTAL CONSIDER- ATIONS

REVIEW WITH PUBLIC AND STATE AND FEDERAL AGENCIES

REQUEST PRO- GRAMMING BY TRANS- PORTATION COMMISSION

SUBMIT DRAFT REPORT AND DISTRIBUTE DRAFT EIS

PUBLIC HEARING

ANALYZE PUBLIC HEARING AND EIS COMMENTS

FINAL EIS

DECISION

REQUEST AND RECEIVE LOCATION APPROVAL

TIME REQUIRED IS APPROXIMATELY 1½ YEARS

FIGURE 2.3. *Example of guidelines that locate assessment activities at appropriate phases of project development. (Adapted from Pennsylvania Department of Transportation Action Plans for the Commonwealth of Pennsylvania, September, 1975.)*

spoil disposal, land management, airports, water resource development, nuclear-power plants, and other projects (EPA, Region 10, 1973). Such guidelines not only establish review procedures within the regional EPA office, but also affect, of course, the content of EISs prepared in that region by the Federal Highway Administration, U.S. Corps of Engineers, U.S. Forest Service, U.S. Bureau of Land Management, U.S. Fish and Wildlife Service, and the Federal Power Commission. Examples of proposed and revised guidelines for representative federal agencies are referenced at the end of this chapter, and are also discussed in Chapter 4.

After reviewing its 6 years' experience with the environmental impact assessment process, the CEQ has recently summarized its major findings and recommendations (CEQ, 1976b, pp. 2–4) as follows:

1. Environmental assessment and EISs have substantially improved governmental decisions over the past 6 years . . . but the impact statement process can and should be more useful in agency planning and decision-making.
2. Agencies with major EIS responsibilities should support high-level, well-staffed offices charged with implementing NEPA and the EIS process efficiently.
3. In the early years of NEPA, delays in the EIS process resulted from a backlog of pre-NEPA proposals subject to impact statement requirements and from agency failure to conduct environmental analyses along with other necessary studies.
4. Agencies should continue to improve the efficiency of the EIS process so that it can more effectively anticipate environmental problems, resolve important environmental issues, correct a course of action in the light of better knowledge, and respond to changing public preferences with timely facts and analyses.
5. Guidance from CEQ is necessary to help agencies draw up EISs for broad federal programs or groups of projects.
6. Procedures that agencies use to notify other federal, state, and local agencies and the public of important EIS actions need improvement.
7. To improve EIS review, agencies, in consultation with CEQ, should clearly define their expertise and jurisdiction for purposes of commenting on EISs.
8. More interagency agreements are desirable to anticipate analytical needs and avoid unnecessary duplication in complying with NEPA.
9. The quality and content of EISs need continuing improvement.
10. Several special issues arising from the EIS process require attention and remedy, including federal permit actions, federal agencies operating abroad, water resource programs, and federal grants to states.

11. The adequacy of public participation in the EIS process needs thorough study and evaluation.

Environmental-Impact-Statement Process versus Environmental-Impact-Statement Document

It is important to emphasize that NEPA created a dynamic process. This process is consistently referred to by the CEQ as the *EIS process.* This process is not equivalent to the *EIS document,* which is only a single item included in the process and, as a static document, cannot achieve the basic objective of NEPA—namely, to improve governmental decisions by expanding agency and public knowledge about the total environmental consequences of actions and by coordinating the actions of one agency with the actions of other agencies. That NEPA is, in fact, a process within the federal bureaucracy, and not simply another document, is evidenced by the previously stated recommendations of the CEQ pertaining to one or another aspect of the use and function of the EIS document.

In an effort to ensure that the content and format of the EIS document will not undermine its functional utility in the EIS process, the following memorandum was distributed by the CEQ on February 10, 1976:

Executive Office of the President
Council on Environmental Quality
722 Jackson Place, N.W.
Washington, D.C. 20006

MEMORANDUM FOR HEADS OF AGENCIES

SUBJECT: Environmental Impact Statements

During the past year the Council on Environmental Quality has conducted a review of federal agency implementation of the environmental impact statement (EIS) requirement of the National Environmental Policy Act (NEPA). That review has indicated that federal agencies have increasingly used the EIS process successfully as a means to improve decisions affecting the environment. Nevertheless, situations continue to arise in which the impact statement process has been more an appendage to or justification for decisions already made than an aid to decision-making. Frequently these failures have been caused or aggravated by the inordinate and unnecessary length of EISs. Such documents at best obscure the intent of NEPA and can be extremely harmful to the environmental impact statement process. It is the purpose of this memorandum, therefore, to reemphasize to all agencies the Council's position on the appropriate focus, use, and length of environmental impact statements in the federal planning and decisionmaking process.

An unnecessarily large portion of many EISs has been devoted to descriptions of the proposed action and the existing environment. Frequently, EISs follow lengthy, detailed outlines in order to assure that at least some treatment, however brief, is given to every subject conceivably relevant to the proposal. In following this approach, agencies make little or no attempt to rank and then analyze in depth the most significant environmental impacts.

There are several reasons why EISs have taken this course: Some EIS authors believe that the EIS itself should be a comprehensive, highly technical, scientific document; the voluminous material received by an agency from an applicant or consultant may prove too time-consuming to edit; or an agency's lawyers may recommend coverage of every possible contingency, particularly if the agency should be sued. The adequacy of an EIS is then measured by its length.

These reasons, however, ignore the precept that the EIS is not an end in itself but is primarily intended to aid decisionmaking. The statement does not achieve this purpose when it has such prodigious bulk that, while it may serve some academic purpose, no one at the decisionmaking level in any agency will ever read it. Since its purpose is to clarify, not obscure, issues and to forecast and analyze significant impacts of a proposal and its reasonable alternatives, efforts must be made early in the EIS process to weed out unnecessary information. Then, by focusing effort and attention on meaningful analyses, the legal adequacy of an EIS will also be supported and enhanced.

It is the Council's position, therefore, that descriptions of the existing environment and the proposed action should be included in an EIS only to the extent that they are necessary for a decisionmaker to understand the proposal, its reasonable alternatives, and their significant impacts. The EIS should explain how the scope of the statement and its level of detail have been carefully delineated in accordance with the significant environmental issues and problems foreseen by the agency. Data and analyses in the EIS should consequently be commensurate with the importance of the impact as determined by the agency's environmental analysis. Less important material should be summarized, consolidated, or simply referenced.

These strictures are set forth in Section 1500.8 (a) (1) of the CEQ Guidelines on preparation of impact statements which states that descriptive material in an EIS should be:

> adequate to permit an assessment of potential environmental impacts by commenting agencies and the public. Highly technical and specialized analyses and data should be avoided in the body of the draft environmental impact statement. Such materials should be attached as appendices or footnoted with adequate bibliographic references. The statement should also succinctly describe the environment of the area affected . . .

Section 1500.8 (b) states that in developing the EIS

> agencies should make every effort to convey the required information succinctly . . . giving attention to the substance of the information conveyed rather than to the particular form, or length, or detail of the statement.

This section states further that each of the five points required by NEPA in an EIS need not always occupy a distinct section of the statement if it is otherwise adequately covered in discussing the impact of the proposed action and its alternatives—which items should normally be the focus of the statement.

In reemphasizing the policy behind these sections of CEQ's Guidelines it should be noted that the need for manageable and useful statements does not and should not imply a need or opportunity to reduce the quality or specificity of environmental research or study required for an informed decision. Environmental conclusions expressed in an impact statement must still be logically supported by references to standard texts, scientific literature, appendices, special studies, or textual material within the statement. Specific baseline inventories and environmental research will often be needed initially to determine if there are environmental problems that should be analyzed in an impact statement. While these studies should be made available to the public and, in the case of a legal challenge, to the courts, they should be referenced, rather than simply reproduced, in the EIS itself.

Although the value of the environmental impact statement process to federal agency decisionmaking has been demonstrated in the past, improvements in its application are necessary. Specific efforts to use the impact statement as a management tool, and to focus the statement on analyses of impacts of a proposal and its reasonable alternatives will require the attention and understanding of agency leaders at various levels. The Council will be glad to assist these efforts in any way that it can.

Russel W. Peterson
Chairman

By any standard, this memorandum is a strong affirmation of CEQ's intent that the structural and functional aspects of NEPA remain balanced. While this particular memorandum was directed to heads of agencies, its basic message was previously conveyed to the President, to the Congress, and to the public.

Properly conceived and written, the EIS is an extremely useful management tool. But too many statements have been deadly, voluminous, and obscure and lacked the necessary analysis and synthesis. They have often been inordinately long, with too much space devoted to unnecessary description rather than to analysis of impacts and alternatives [CEQ, 1975, p. 632].

Overview of NEPA

As discussed in the previous section, the EIS process under NEPA is a collection of guidelines (the environmental policy structure), a collection of documents (the action-enforcing structure), and a collection of review procedures (the liaison structure). And, according to policy statements and guidelines promulgated by the CEQ and, subsequently, by other federal agencies, these structures are supposed to function in such a way that governmental decisionmaking is (a) efficient; (b) relevant to present and potential public needs; and (c) responsive to the national environmental goals of NEPA. Ideally, then, the EIS process is a bureaucratic gear that, as depicted in Fig. 2.4, modulates the direction and progress of project development with the direction and progress of environmental assessment. One (i.e., project development) cannot proceed without the other (i.e., environmental assessment), and each affects the other. The more efficient the assessment and the coupling between environmental impact assessment and the structural–functional requirements of the EIS process, the more efficient project development becomes.

Other Legislative Inputs

While environmental assessment is coupled to project development by NEPA, various types of environmental assessments (not necessarily assessments of total environmental impacts) are either specifically required by or may be influenced by a number of other federal laws. These laws may be grouped according to their relevance to various objectives, including (a) protection of habitat; (b) protection of plant and animal species; (c) protection of recreational areas; (d) protection of historic and cultural resources; (e) land use control, (f) protection of water quality; (g) protection of air quality; (h) noise regulation, (i) pesticide regulation, and (j) solid waste disposal regulation.[5]

It is important for the efficient design and conduct of an environmental impact assessment under NEPA that all legislation pertinent to a specific project or program be identified, and that all environmental requirements (identified by competent legal counsel) be included in the overall assessment design. Such legislation may include (but not be limited to) the categories I will now discuss.

PROTECTION OF HABITAT

The Fish and Wildlife Coordination Act (1958) requires that any federal agency proposing a project which encroaches on a water resource must develop plans for the protection and enhancement of the wildlife associated with that resource. The Water Bank Act (1970) established a system by which the Secretary of Interior can take easements on wetlands to preserve and improve them as habitat for wildlife. The National Wildlife Refuge System Administration Act (1966) established a central control for the administration of the lands acquired as habitat.

---

[5] These categories and the following discussion are adapted from American Right of Way Association, (1976).

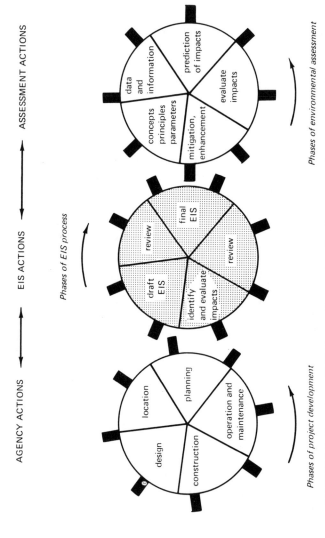

AGENCY ACTIONS ⟷ EIS ACTIONS ⟷ ASSESSMENT ACTIONS

Phases of project development

design
location
planning
operation and maintenance
construction

Phases of EIS process

draft EIS
review
final EIS
review
identify and evaluate impacts

Phases of environmental assessment

data and information
concepts principles parameters
prediction of impacts
evaluate impacts
mitigation, enhancement

FIGURE 2.4. *Overview of NEPA: coupling project development with environmental assessment.*

PROTECTION OF PLANT AND ANIMAL SPECIES

The Endangered Species Act (1973) requires that all federal agencies must take active steps to conserve endangered and threatened species. The Act includes flora as well as fauna, prohibits the harm, harassment, trade, or capture of endangered species, and provides for the protection of threatened species.

Congress has also passed legislation which protects individual species. Examples of this legislation include the Bald Eagle Protection Act (1940), the Golden Eagle Protection Act (1962), the Wild Horse and Burro Protection Act (1971), the Marine Mammal Protection Act (1972), and the Anadromous Fish Conservation Act (1965).

PROTECTION OF RECREATIONAL AREAS

The Wilderness Act (1964) recognizes the value of preserving certain areas that should be left "unimpaired for future use and enjoyment as wilderness." The Wild and Scenic Rivers Act (1968) authorizes the Departments of the Interior and Agriculture to study certain rivers and to acquire land along them for protection from development. The National Trails System Act (1968) provides for the development and protection of hiking trails. Congress has established a system of national scenic trails and national recreational trails. The Land and Water Conservation Fund Act (1965) provides funds to be matched by the state to plan, acquire, and develop land and water areas for recreational use.

PROTECTION OF HISTORIC AND CULTURAL RESOURCES

The Historic Sites Act (1935) enacted a national policy of preservation for historic properties. The National Historic Preservation Act (1966) expanded the national inventory known as the National Register of Historic Places, created a grant program to help preserve nonfederal properties, and created the Advisory Council on Historic Preservation. The 1935 Act also established the National Registry of Natural Landmarks, thus providing a procedure for the identification and registration of natural sites of outstanding significance.

LAND USE CONTROL

The Federal Aid Highway Act (1968) requires the preservation of parklands and refuge areas in highway planning. The Clean Air Act (1970) has placed constraints on industrial development. Section 208 of the Federal Water Pollution Control Act Amendments of 1972 provides for areawide waste treatment management plans. The Coastal Zone Management Act (1972) provides federal funds to assist states to develop land use plans in the coastal area.

PROTECTION OF WATER QUALITY

The Federal Water Pollution Control Act Amendments of 1972 replaced the Federal Water Pollution Control Act of 1965 which was the first significant federal response to industrial pollution of the nation's waters. The Act has three basic features: a grant program for municipal treatment facilities; receiving water quality standards for inter- and intrastate waters, and effluent discharge standards for all point sources including sewage treatment facilities. In passing the Act, Congress stated it wished to "restore and maintain the chemical, physical and biological integrity of the nation's waters." Congress set as national goals

1. The achievement of a water quality level which "provides for the protection and propagation of fish, shellfish, and wildlife," and "for recreation in and on the waters" by July 1, 1983
2. The elimination of discharge of pollutants into U.S. waters by 1985.

Other legislation in the water quality standard area includes the 1899 Rivers and Harbors Act in which the U.S. Army Corps of Engineers retains control of all dredging and filling activities. The Marine Protection Research and Sanctuaries Act (1972) regulates dumping in the ocean, provides funds for research to end all ocean dumping, and creates a system by which marine areas as far out as the edge of the continental shelf and the Great Lakes could be designated as sanctuaries and protected for incompatible activities. Finally, the Safe Drinking Water Act (1974) was passed as a reaction to problems in local drinking-water supplies which received national attention. The objectives of this Act are to increase control of the quality of drinking water and to control underground injections of pollutants.

PROTECTION OF AIR QUALITY

The Air Pollution Control Act authorized air-pollution control research and assistance to state and local governments. The Clean Air Act (1963) authorizes funds for research in a number of areas. It also authorized the first federal efforts at air-pollution control through the establishment of air-pollution abatement conferences. The Air Quality Act (1967) intensified the effort. By this Act, the federal government was to designate air quality regions and establish air quality criteria for each region. The states were to establish the standards and plans for implementation. Administrative failure led to ineffectiveness of this approach and, subsequently, to the Clean Air Act Amendments of 1970.

NOISE REGULATION

The Occupational Safety and Health Act (1970) is concerned only with the work place. It adopts the standards of the Walsh–Healy Act in which acceptable noise levels are based on duration of exposure. The responsibility of EPA in this area began with the establishment of the Office of Noise Abatement and Control under the Clean Air Act of 1970. EPA has responsibility for the full-day exposure to noise. Under the 1970 law, the EPA published a report on the effects of noise on humans, wildlife, and property. The Noise Control Act (1972) states that Americans must be free from noise that jeopardizes their health and welfare. Under this law, EPA's first action was to publish a series of documents establishing the relationship between types and quantities of noise and public health and welfare. Prior to setting of standards, EPA is identifying the major sources of noise.

PESTICIDE REGULATION

In 1972, Congress passed the Federal Environmental Pesticide Control Act which amended the existing Federal Insecticide, Fungicide and Rodenticide Act (FIFRA), and which controls pesticides through a variety of means. There is a large body of unregulated chemicals that are not subject to review under this legislation. Consequently, Congress enacted the Toxic Substances Act (TSCA) in 1976. The Office of Toxic Substances (OTS) of the U.S. Environmental Protection Agency (EPA) administers both TSCA and FIFRA programs.

SOLID WASTE DISPOSAL REGULATION

The Solid Waste Disposal and Resource Recovery acts (as amended in 1970 and 1973) were to encourage state action in this area. The EPA is working on guidelines for incineration, operation of sanitary landfills, storage and collection, source separation, beverage containers, resource recovery facilities, and increased use of recycled materials.

These kinds of legislation are federal legislation that may require some environmental assessment of a particular project and, therefore, close coordination between those having assessment responsibility and those having regulatory authority. Such assessments would be necessary regardless of the status of a particular project with respect to NEPA.

As of January 1973, there were several hundred laws relating just to conservation and development of fish and wildlife resources, environmental quality, and oceanography (Committee on Merchant Marine and Fisheries, 1973). Any single governmental department and its agencies may have specific responsibility under any number of these (and other) environmental laws (Table 2.2). Such agencies are important for determining what has to be done in the assessment of impacts for a particular project. They are also valuable sources of data and information.

State and Local Contributions

In addition to federal legislation, there are state and local governmental legislation and administrative actions that can be integral to the efficient environmental assessment of a proposed object or program under NEPA.

By 1976, 26 states had environmental assessment requirements.

TABLE 2.2.
Environmental Agencies and Their Responsibilities within the U.S. Department of the Interior[a]

Subdivision	Environmental functions
Office of Water Research and Technology	• Funding of research and development programs for water resources, including the conversion of saline water.
Fish and Wildlife Service	• Perpetuation of sportfish and wildlife resources • Fish hatcheries • Wildlife refuges • Research on environmental quality necessary to sustain wildlife • Endangered species program
National Park Service	• Administers system of parks, monuments, historic sites, and recreational areas
Geological Survey	• Topographical and geological mapping of all U.S. lands • Identification of water and mineral resources on federal lands • Identification of source, quantity, quality, distribution, movement, and availability of surface and ground waters • Hydrologic studies
Bureau of Land Management	• Management of 450 million acres of public domain (20% of the nation's total land base)
Bureau of Outdoor Recreation	• Planning and coordinating of state and federal programs • Administers Land and Water Conservation Fund
Bureau of Reclamation	• Irrigation projects for arid lands • Hydroelectric power

[a] Adapted from American Right of Way Association (1976).

Fourteen have enacted "little NEPAs" having general application; 5 have assessment requirements of limited application; and 6 have procedures adopted by administrative action and which are roughly equivalent to NEPA (CEQ, 1976a, p. 135).

The states have acted to implement the federal requirements for planning and regulation of air and water pollution control. For example, the states have developed "action plans" for achieving and maintaining ambient air and water quality. These plans are subject to the review of the EPA which has set minimum compliance standards. Monitoring of air and water quality are also state functions. Some states have taken over the National Pollution Discharge Elimination System permit program. Regional planning agencies have taken the lead role in the development of waste water management plans under Section 208 of P.L. 92-500 (Federal Water Pollution Control Act Amendments of 1972).

Noise regulation at the federal level is aimed at setting standards for new products. While the federal government has preempted this right, the state and local governments retain the right to establish and enforce restrictions on noise sources through licensing, and restricting local use, operation and/or movement of the product.

Zoning at the local level has been a basic form of land use control. Some state laws call for the identification and protection of areas of "critical concern" (Massachusetts' coastal and inland wetlands ban; Wisconsin shoreland and flood plain protection law; California's San Francisco Bay Conservation and Development Committee). Other states have adopted laws requiring state review of large-scale developments (Vermont's environmental control law, Maine's site location law, numerous power plant siting laws). Hawaii has developed a total statewide land use planning approach.

Conclusion

Regardless of the many problems and inefficiencies that have occurred as a result of translating environmental legislation into governmental bureaucracy, and regardless of the many improvements yet to be made, environmental impact assessment will in all likelihood continue to be the major mechanism for achieving national, state, and local environmental objectives. Environmental assessment does not appear to be a "fad left over from Earth Day which will blow away as soon as the kids go looking for a job." Nor is it likely to be tolerated as an academic exercise, or a "window dressing" for decisions already made. To the contrary, environmental impact assessment, and in particular—environmetal impact assesment under NEPA—is increasingly becoming a basic tool for decision-making throughout project and program development at the federal, state, and even local levels of government. And, as is true with any tool, it should be evaluated by what is actually achieved by its use.

There is, of course, only one thing that should result from the bureaucratic use of environmental impact assessment. Decisions! Decisions that affect projects and programs. We persist in the notion that good assessments help make good decisions (i.e., in terms of quality, and in terms of the efficiency of making them), and that good decisions make good projects and programs (i.e., in terms of meeting the needs of the total human environment). If, in a particular project, this is not the case, the bureaucratic basis for environmental impact assessment, however extensive it may be, is a functional irrelevancy at best.

References

American Right of Way Association
    1976   *Student workbook on right of way and the environment.* Los Angeles, California: American Right of Way Association
Bross, I. D. J.
    1959   *Design for Decision.* Macmillan, New York.
Committee on Merchant Marine and Fisheries
    1973   *A compilation of federal laws relating to conservation and development of our nation's fish and wildlife resources, environmental quality, and oceanography.* Washington, D C. Council on Environmental Quality (CEQ)
Council on Environmental Quality
    1973a.   Preparation of environmental impact statements, proposed guidelines. *Federal Register, 38,* (84, Part II). Washington, D. C.: U. S. Government Printing Office.
    1973b.   Preparation of environmental impact statements, guidelines. *Federal Register,* 38 (147, Part II). Washington, D. C.: U.S. Government Printing Office.
    1975   *Environmental quality: The sixth annual report of the Council on Environmental Quality.* Washington, D. C.: U.S. Government Printing Office.
    1976a.   *Environmental quality: The seventh annual report of the Council on Environmental Quality.* Washington, D. C.: U. S. Government Printing Office.
    1976b   *Environmental Impact Statements: An analysis of six years' experience by seventy federal agencies.* Washington, D. C.: U. S. Government Printing Office.
Druley, R. M.
    1976   *Federal agency NEPA procedures.* Monograph No. 23, *Environmental Reporter, 7,* Washington, D. C.: The Bureau of National Affairs.
U. S. Environmental Protection Agency (EPA), Region 10.
    1973   *Environmental Impact Statement guidelines.* Seattle, Washington: U. S. Environmental Protection Agency, Region 10.

3

Court Contributions to the Assessment Process

Introduction

There are three precepts which are useful for understanding the real-world implementation of NEPA and other environmental laws. They are that

1. The law is whatever the courts say it is;
2. What the courts say is a command;
3. Bureaucracies thrive by institutionalizing commands.

To undertake an environmental assessment of a project or program without due consideration of how courts have interpreted NEPA and what they have previously commanded is to invite unnecessary project delay. Project delay can come about directly and indirectly: directly, when a particular assessment (or lack of it) becomes the object of litigation; and indirectly, when agencies, groups, and organizations, which have review or regulatory input into the assessment process, modify their inputs in accordance with their interpretations of court cases involving other and different projects. The power of the court, after all, does not derive solely from the commands a court makes in a single case, but also from human judgments that the court will, if called upon, act in a predictable way. Accordingly, those who have assessment responsibility (whether in design, management, data collection and interpretation, and/or reporting tasks) should realize that any assessment can be suddenly and even drastically altered by court decisions even if, in their judgment, it is unlikely that their particular project will end up in court.

Of course, people who do have actual assessment responsibility are not typically lawyers and must, therefore, rely upon guidance from competent counsel in order to determine legal trends and the significance of these trends to the assessment of a particular project. Yet, legal trends and probabilities in contemporary environmental affairs are dynamically linked with technical environmental and social issues as well as with technical legal issues.

The nonlegally trained individual must therefore be sensitive and responsive to the law as that law is enunciated by the courts. But also, he or she should not be so overwhelmed by the complexity of legal argument and findings that he or she loses sight of the basic objectives of impact assessment. Certainly, NEPA was not written in order to keep the legal mind occupied. And just as certainly, court pronouncements concerning NEPA have not been handed down only to be ignored by technical and scientific specialists who become so engrossed in their own specialty that they confuse private research with national goals.

General Review of Litigation, 1970–1975

It has been estimated that NEPA has been cited in the courts about 10 times more often than other environmental laws (Young, 1975, p. 51). It has also been suggested (Druley, 1976, p. 3) that the role of the courts in the development of NEPA has perhaps been overemphasized, especially in comparison to the importance of individual agency regulations (Chapter 4). In order to make a realistic appraisal of court contributions to NEPA, it is, therefore, important to have a general perspective of the number and type of legal actions that have been completed or are pending.

For example, in the period 1970–1975, 6946 draft EISs were prepared and filed by the U.S. government agencies and departments. Of the total EISs filed about 4% were alleged in court to be inadequate. In the same period of time, there were 363 legal cases alleging that federal agencies had failed to prepare required EISs (CEQ, 1976a). About a quarter of all litigation under NEPA has involved the Department of Transportation. Other agencies involved in litigation include the Department of Agriculture, the Corps of Engineers, the Department of Housing and Urban Development, and the Department of the Interior (Table 3.1).

Of the total cases completed by June 30, 1975, about 18% resulted in temporary injunctions. About 1% of the total resulted in permanent injunctions (CEQ, 1976c, p. 123). Regardless of the final disposition of

TABLE 3.1
EISs and Litigations of Federal Agencies, 1970–1975[a]

U.S. federal agency or department	Total number draft EISs prepared (1970–1975)	Total EISs as percentage of all federal EISs (%)	Total number of cases[b]	Cases alleging inadequacy of EIS	Cases alleging need for EIS to be prepared
Agriculture	799	12	61	15	46
Corps of Engineers	1465	21	63	46	17
Environmental Protection Agency	92	1	24	10	14
Housing and Urban Development	173	2	94	27	67
Interior	485	7	83	52	31
Transportation	3049	44	157	77	80
All others	1883	27	172	64	108
Total	6946	100	654	291	363

[a] [From figures provided in CEQ, 1976a.]
[b] Includes both completed and pending cases.

both completed and pending cases, there can be little doubt that any litigation potentially reflects inefficiencies either in the design and conduct of an impact assessment, or in the bureaucratic determination of need for developing an EIS as an integral part of program or project development.

The Department of Transportation (DOT) has accounted for 24% (or 157 cases) of the total federal litigation under NEPA. In the same period of time, the DOT filed 44% of the total EISs filed by all federal agencies. Of DOT's EISs, about 3% have resulted in allegations of inadequacy. Similarly, the Department of Agriculture (DOA) has accounted for about 9% (or 61 cases) of the total federal litigation under NEPA. The DOA also filed 12% of the total EISs filed by all federal agencies. Of DOA's EISs, about 2% have resulted in allegations of inadequacy. From these figures, it would appear that both DOT and DOA have equally good records of writing adequate EISs *when they write EISs*, even though the DOT (for writing about four times more than has the DOA) has ended up in court more often.

As indicated in Fig. 3.1, the number of EISs alleged to be inadequate is generally a small percentage of the total number of EISs filed by each government agency, ranging from 2% for the DOA to 16% for the Department of Housing and Urban Development (HUD). However, agencies and departments have varied greatly with respect to the probability that

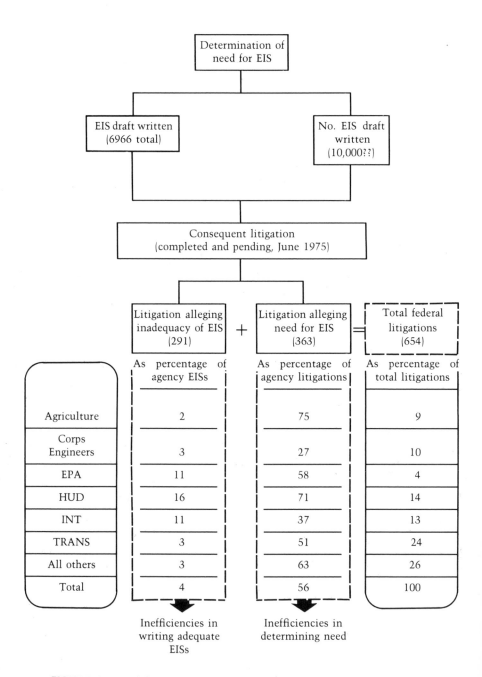

FIGURE 3.1 *Breakdown of Litigations (from figures provided in CEQ, 1976a).*

they will end up in court on the basis of alleged poor judgment as to the need for writing an EIS. For example, about 27% of the NEPA cases involving the Corps of Engineers (Corps) have alleged that the Corps should have written an EIS, whereas 75% of the NEPA cases involving the DOA have made the same allegation.

In light of (a) the large number of EISs that have been written; (b) the much larger number of federal actions that have been reviewed and determined not to require EISs; and (c) the relatively small number of litigations and injunctions that have ensued over the past 5 years, it is reasonable to suggest that the importance of court involvement in the implementation of NEPA is perhaps due less to the absolute number of judicial decisions that to the pervasiveness of those decisions with respect to the entire environmental impact assessment process. In fact, within only a few years of the enactment of NEPA, it became obvious that the courts were already actively defining judicial intent to review agency actions under NEPA from the time of earliest project development to the time of final agency decision and implementation, including the following five areas (Anderson, 1973, p. 289):

1. The agency's organizational mechanisms for deciding on courses of action subject to NEPA
2. The determination of need to write an EIS
3. The adequacy of environmental information developed in an EIS
4. The use of information developed in the EIS
5. The substantive correctness of final agency decisions with respect to Section 101 of NEPA.

Some Important Developments

Within a year and a half of the enactment of NEPA, there was a number of federal district court decisions that established citizens' right to sue for court enforcement of NEPA's environmental statement procedure (i.e., Section 102). In this formative period, suits brought by citizens involved the Federal Highway Administration, Department of Interior, U.S. Forest Service, Department of Agriculture, Corps of Engineers, the Farmers' Home Administration, and the Justice Department. Federal actions challenged in these suits involved construction projects (e.g., highways), regulatory permits (e.g., for the Alaskan pipeline and haul road), management plans (e.g., pertaining to National Forests), the use of pesticides (e.g., the use of Mirex against fire ants), water resource projects, and even grants for a prison facility (CEQ, 1971, p. 157).

While the first years after NEPA's enactment have been characterized as being primarily concerned with procedural compliance to

NEPA (Anderson, 1973, p. 292), a number of important cases in this period also established that the courts were likely to do more than merely examine procedural compliance. For example, in *Zabel* v. *Tabb* (1971) a federal district court questioned the authority of the Corps of Engineers to deny a permit to a developer of a trailer park (Coca Ciego Bay, Florida), finding that such a denial could only be issued in order to protect navigation. However, this decision was reversed on appeal when *the court found that NEPA gave federal agencies the authority and duty to promote environmental ends* (CEQ, 1971, p. 157).

A second indication that the courts would go beyond the narrow issue of procedural compliance and deal directly with the intent of Section 102 A and B, was the appellate court decision in the famous Calvert Cliffs case (*Calvert Cliffs' Coordinating Committee* v. *Atomic Energy Commission*, 1972). The decision in this case essentially meant that "agencies have to comply in earnest with Congress' intent to affect Federal agency decision-making," and that "final agency decision making would not in the court's view, be committed entirely to agency discretion [Anderson, 1973, pp. 256–259]."

By September 1973 the total number of NEPA lawsuits exceeded 400, and it was evident that litigation was focusing less on procedural questions and more on (a) the content of the EIS document; and (b) agency decisions that were made after the completion of the EIS document.

For example, in *Natural Resources Defense Council* v. *Grant* (1973) the court held that "an impact statement was inadequate because it omitted or inadequately described a number of important environmental effects and failed to disclose fully or discuss adequately alternatives to the project [CEQ, 1973b, p. 237]." In *Environmental Defense Fund* v. *Froehlike* (1972), the court declared that a "formal impact study supplies a convenient record for the courts to use in reviewing agency decisions on the merits to determine if they are in accord with the substantive policies of NEPA [Anderson, 1973, p. 253]."

Finally, in 1973 the Supreme Court accepted its first NEPA case (*SCRAP* v *United States*, 1973), and answered the basic question of who has the right to sue on environmental grounds—"The answer of the SCRAP decision is that where the public as a whole is affected, any member of the public may sue [CEQ, 1973b, p. 241]."

In light of even these few early cases, it is evident that the courts moved very quickly to establish judicial review of the implementation of NEPA. While the earliest cases focused on procedural compliance to NEPA—specifically, Section 102 (C)—the cases cited so far in this chapter were included in the first several annual reports of the Council on Environmental Quality in order to highlight judicial intention that (a) federal agencies utilize their authority under NEPA to promote environmental ends; (b) federal agencies utilize environmental considerations in their decision-making and not only in their EISs; (c) the adequacy of EISs would be reviewable by courts; and (d) the public would be given access to the courts on behalf of national environmental goals.

Of course, litigation under NEPA, even in the early years of its implementation, was not the only vehicle by which the federal agencies and the public were informed of their environmental obligations and rights. For example, Executive Order 11514 was issued on March 5, 1970 and amplified agency responsibility under NEPA with specific orders to heads of Federal agencies to:

1. Continually monitor, evaluate, and control agency activities so as to protect and enhance the quality of the environment
2. Develop procedures to ensure the fullest practicable provision of timely public information and understanding of Federal plans and programs . . . in order to obtain the views of interested parties
3. Ensure that information regarding existing or potential environmental problems and control methods developed as part of research, development, demonstration, test or evaluator activities is made available to federal agencies, states, counties, municipalities, institutions, and other appropriate entities
4. Review agencies' statutory authority, administrative regulations, policies, and procedures in order to identify any deficiencies or inconsistencies that prohibit or limit full compliance with the purposes and provisions of NEPA
5. Engage in exchange of data and research results and cooperate with agencies of other governments to foster the purposes of NEPA
6. Proceed with actions required by Section 102 of NEPA.

In addition to these directions to heads of agencies, the President also assigned 11 specific responsibilities to the Council on Environmental Quality. Among these responsibilities were (a) the issuance of "guidelines to Federal agencies for the preparation of detailed statements on proposals for legislation and other Federal actions affecting the environment, as required by Section 102 (2) (C) of NEPA"; and (b) the determination of "the need for new policies and programs for dealing with environmental problems not being adequately addressed."

The directive in Executive Order 11514 which orders CEQ to issue "guidelines to Federal agencies for the preparation of detailed statements" has, in turn, influenced court decisions under NEPA. Courts "generally have given some weight to the interpretation of NEPA by CEQ through its guidelines. . . . The established legal principle is that an administrative interpretation of a statute by the agency charged with its enforcement is entitled to great deference [Druley, 1976, p. 2]."

While it may appear to many that court involvement in the EIS process has been rashly precipitated by an emotional environmentalism of the late 1960s (which quickly came to be identified as a "doomsday syndrome"), other explanations are certainly possible. For example, NEPA might only have provided an environmental mechanism whereby courts have been able to continue an ongoing reform of judicial review of agency actions that goes well beyond a particular concern for the EIS process (Anderson, 1973, pp. 15–23). This suggestion is predicated by recent congressional concern with agency decion-making in various legislation (including the Federal Power Act and the Department of Transportation Act), and by court decisions involving both environmental and nonenvironmental litigations. It is succinctly summarized in the assertion that

> for several years the courts have been attempting to evolve better standards of review from existing statutory and case law. They have asked that agencies take a wider array of public interests into account in carrying out their statutory missions, that agencies consider alternatives to their proposals, that they articulate the grounds for their decisions, that they spell out procedures for principled decision making, and that they provide for public participation of various kinds. Congress' enactment of NEPA reinforced this trend by further defining the record that must be made in certain agency decision making . . . by requiring the study of alternatives, by detailing procedures for obtaining advance criticism from state and local governments and the public, by establishing routine, fair procedures applicable across the board to each agency, and by requiring more of a record in informal agency proceedings than [formerly required] [Anderson, 1973, p. 22].

From this perspective it would appear that the rapid comprehensive involvement of the courts in the first few years after NEPA reflected less a concern for a new environmental ethos than a general and growing concern for (a) the quality of all governmental decision-making in an increasingly complex governmental and social pluralism; and (b) the responsibility of the courts as one of the three branches of government.

It should be added that such a judicial concern about administrative decision-making may not only be independent of the phenomenon of environmentalism, but may also extend beyond the American legal system. For example, in a recent review of the challenges that face English law, it is rhetorically postulated that the contemporary challenge of complex societies (including but not limited to environmentalism) "can be left to administrative control *or an attempt can be made to use the legal system so as to ensure that the merits of decisions taken are subject to a measure of review* [italics mine; Scarman, 1974, p. 59]."

The Courts as Enforcers of NEPA

While it is likely that there can never be absolute agreement as to just why the courts moved so quickly in the area of NEPA implementation, it is clear that the courts have in fact done so, and that by so doing "they have enforced strict standards of procedural compliance, and in instances where Congress failed to specify how the Act should be implemented, they have imposed judge-made requirements which give it a wider scope. As a result, the courts are thought of as the principal enforcers of NEPA [Anderson, 1973, p. 16]."

Therefore, it is important for the person who has assessment responsibility to understand that certain characteristics of NEPA have been delineated not only by legislative intent and administrative action, but also by court demands. Among these characteristics are:[1]

1. NEPA is an environmental full disclosure law.

The Congress authorizes and directs that, to the fullest extent possible . . . all agencies of the Federal Government shall . . . include in every recommendation or report on proposals for legislation and other major Federal actions significantly affecting the quality of the human environment, a detailed statement by the responsible official.

—Section 102 (2) (C) of NEPA

The detailed statement required by Section 102 (2) (C) should, at a minimum, contain such information as will alert the President, the Council on Environmental Quality, the public and, indeed, the Congress, to all known possible environmental consequences of proposed agency action.

—*Environmental Defense Fund* v. *Corps of Engineers* (Gilham Dam)

The procedures established by these guidelines are designed to encourage public participation in the impact statement process at the earliest possible time. Agency procedures should make provision for facilitating the comment of public and private organizations and individuals by announcing the availability of draft environmental statements and by making copies available to organizations and individuals that request an opportunity to comment [CEQ, 1973a]

2. The EIS required by NEPA should be understandable to laymen and also sensitive to the technical and scientific needs of experts.

The impact statement . . . is not intended to be a scientific or technical document but should instead be based on and reflect a process of sound scientific analysis. The level and extent of detail and analysis in the statement should be commensurate with the importance of the environmental issues involved and with the information needs of both decision makers and the general public [CEQ, 1975, p. 633.

---

[1] See. V.T.N. Consolidated Inc. (Undated, 12–14) for additional documentation of these characteristics.

The EIS must be written in language that is understandable to non-technical minds and yet contain enough scientific reasoning to alert specialists to particular problems within the field of their expertise."

—*Environmental Defense Fund* v. *Corps of Engineers*
(Tennessee-Tombigbee Waterway)

3. The EIS is a decision-making tool.

"The EIS is not an end in itself but is primarily intended to aid decision-making [CEQ, 1976b].

"The EIS itself is intended to be, and often is, the tip of an iceberg, the visible evidence of an underlying planning and decision making process that is usually unnoticed by the public [CEQ, 1975, p. 628]."

[Sections 102 (2) (C) and (D)] seek to ensure that each agency decision maker has before him and takes into proper account all possible approaches to a particular project . . . which would alter the environmental impact and the cost-benefit balance. Only in that fashion is it likely that the most intelligent, optimally beneficial decision will ultimately be made.

—*Calvert Cliffs Coordinating Committee* v. *Atomic Energy Commission*

4. The EIS, as a decision-making tool, requires interdisciplinary inputs.

All agencies of the Federal Government shall . . . utilize a systematic, interdisciplinary approach which will insure the integrated use of the natural and social sciences and the environmental design arts in planning and in decision making which may have an impact on man's environment."

—Section 102 (2) (A) of NEPA

Agencies should attempt to have relevant disciplines represented on their own staffs; where this is not feasible they should make appropriate use of relevant Federal, State and local agencies or the professional services of universities and outside consultants. The interdisciplinary approach should not be limited to the preparation of the environmental impact statement, but should also be used in the early planning stages of the proposed action [CEQ, 1973a].

In the Gilham Dam case, the first set of memorandum decisions faulted a Corps statement for, among other reasons, giving insufficient attention to five specific environmental factors; the effects of flow fluctuation on downstream biological productivity and stability, the effects of the intrusion of other fish species, the effects of impoundment on existing food chains and downstream alluvial deposits, and the effects of occasional reservoir drawdown on shoreline vegetation [Anderson, 1973, p. 212].

[In the Gilman Dam Case] the court said that a systematic interdisciplinary approach must be used in evaluating the environmental impact to insure the integrated use of the natural and social sciences and the environmental design arts. Methods and procedures to assign values to presently unquantified amenities must be developed so that these may be considered along with economic and technical considerations [Young, 1975, p. 61].

Conclusion

There can be no doubt that NEPA litigation has been (and continues to be) an exciting and intellectually rewarding experience for those who are involved either in the practice or in the analysis of the American legal system. Yet, for those who are not legally trained and who, nonetheless, have assessment responsibility within the real constraints of time and money, the experience is most often one of confusion and frustration—confusion as to how esoteric, technical points of law can be realistically matched with the technical points of complex scientific, technical, and social disciplines; frustration with legal constraints and/or requirements that are often difficult for the legal layperson to understand, not to mention integrate into the already difficult task of assessment merely because a lawyer says it has to be done. The numerous litigations that have ensured NEPA implementation, the diverse jurisdictions and findings of American courts, and even the general lack of technical and scientific sophistication among lawyers, have probably exacerbated the situation.

Yet, it should be emphasized that, in spite of the numerous litigations and the diverse jurisdictions and findings, the four characteristics of NEPA discussed on pages 49–51 have been generally upheld in the nation's courts. The person who has assessment responsibility, but who is not legally trained, should consider the implication of these characteristics to his or her own assessment function. It is suggested that whatever that function may be, the courts of this country have said, in effect, that the person is participating in an interdisciplinary effort to provide decision-makers, the public, and government with comprehensive assessments of project impacts, and that these assessments will be available and understandable to both the lay public and to technical and scientific experts for the purpose of ensuring the consonance of final decisions with national environmental goals.

One might consider, then, the advice given by an environmental lawyer to federal agencies as being equally pertinent to the individual charged with assessment responsibility: "It is particularly important that [the individual] stop seeing the environmental statement as a means to prevent litigation . . . [the individual] should forget about litigation and prepare high quality, analytical environmental statements [Terris, 1976]."

References

Anderson, F. R.
  1973 *NEPA in the courts—a legal analysis of the National Environmental Policy Act.* Baltimore, Maryland: Johns Hopkins University Press.
Council on Environmental Quality (CEQ)
  1971 *Environmental quality, the second annual report of the Council on Environmental Quality.* Washington, D. C.:U. S. Government Printing Office.

  1973a *Guidelines, preparation of environmental Impact Statements. Federal Register,* Vol. 38 (147, Part II).

  1973b *Environmental quality, the fourth annual report of the Council on Environmental Quality.* Washington, D. C.: U. S. Government Printing Office.

  1975 *Environmental quality, the sixth annual Report of the Council on Environmental Quality.* Washington, D. C.: U. S. Government Printing Office.

  1976a *Environmental Impact Statements, an analysis of six years' experience by seventy federal agencies.* Washington, D. C.: U. S. Government Printing Office.

  1976b Memorandum for heads of agencies, re: Environmental Impact Statements. February 10, 1976. (Russell W. Peterson, Chairman).

  1976c *Environmental quality, the seventh annual report of the Council on Environmental Quality.* Washington, D. C.: U. S. Government Printing Office.
Druley, R. M.
  1976 *Federal agency NEPA procedures.* Monograph No. 23. Washington, D.C.: The Bureau of National Affairs.
Scarman, Sir Leslie
  1974 *English law—the new dimension.* London: Stevens and Sons.
Terris, B. J.
  1976 In *Environmental Reporter, Current Developments,* 7 (32).
V. T. N. Consolidated, Inc. Undated.
        Preparation of Environmental Impact/4 (f) Statements. Washington, D.C.: U. S. Department of Transportation, National Highway Institute.
Young, R. A. (Editor).
  1975 *Environmental law handbook* (3rd edition). Bethesda, Maryland: Government Institutes.

4

Agency Compliance

Introduction

Ideally, agency compliance with NEPA should be evaluated with respect to several basic questions:

1. How have agencies organized themselves for compliance (i.e., for conducting, utilizing, and/or reviewing impact assessments)?
2. How have agencies approached attitudinal and informational needs of their personnel (and personnel of other agencies) specifically charged with NEPA responsibility?
3. What have been the actual results of organizational, attitudinal, and informational adjustment in terms of meeting the comprehensive goals of NEPA?

Organization for compliance with NEPA includes the issuance of guidelines and/or procedures which identify and coordinate relevant, day-to-day activities of personnel, and the provision of money and manpower to ensure that the guidelines and/or procedures can, in fact, be carried out.

Attitudinal and informational needs of personnel charged with NEPA responsibility are directly influenced by training and educational courses and programs, by management techniques that reflect the importance (or nonimportance) given to NEPA by agency administrators, and by reorganization and/or creation of data and informational banks and other resource tools that facilitate the sharing, dissemination, and use of data, information, and environmental experience.

Overall results of organization and attitudinal and informational adjustment may be ascertained from governmental reviews, from other professional reviews (including academic and scholarly reviews), and from the effects of federal NEPA-related actions on state and local governments and agencies, and on the private sector.

While it can hardly be expected that the diverse observers of and participants in the assessment process will ever reach complete consensus in their evaluations of the quality and efficiency of agency compliance to NEPA, a number of issues with respect to their compliance can be generally agreed upon.

NEPA Guidelines and/or Procedures

According to the Council on Environmental Quality (CEQ, 1976b), "some agencies . . . have not established EIS procedures. However, most agencies have prepared or revised their procedures since the CEQ guidelines were revised in 1973 . . . [and] . . . most Federal agencies have established special high level offices to oversee and help implement agency responsibilities under NEPA [p. 5]."

As discussed in Chapters 2 and 3, the authority for CEQ to issue guidelines for the compliance of federal agencies to NEPA specifically emanates (up to May 1977) from Executive Order 11514, which, in CEQ's interpretation, orders the CEQ to "issue guidelines to Federal agencies for preparation of environmental impact statements and such other instructions to agencies and requests for reports and information as may be required to carry out the Council's responsibilities under the Act [i.e., NEPA] [CEQ, 1973]."

Because of the lack of express statutory authority to issue guidelines to other federal agencies there is some question as to the legal status of the 1973 guidelines and, therefore, as to the need for agencies to comply by establishing NEPA procedures within the framework of those guidelines. The question is: "Are the guidelines merely advisory or are they regulations? The language of the guidelines is mandatory and directive. However, the agencies have chosen to treat the guidelines as merely advisory. In fact, the courts also generally have considered the CEQ guidelines as merely advisory and have said that 'the CEQ has no authority to prescribe regulations governing compliance with NEPA' [Druley, 1976, pp. 1–2]."

On May 24, 1977, the President issued Executive Order 11991 which reiterates executive intent that the CEQ shall have such an authority. This order specifically directs the CEQ to:

> Issue regulations to Federal agencies for the implementation of the procedural provisions of the Act [i.e., NEPA]. Such regulations . . . will be designed to make the environmental impact statement process more useful to decision makers and the public, and to reduce paperwork and the accumulation of extraneous background data, in order to emphasize the need to focus on real environmental issues and alternatives. They will require impact statements to be concise, clear, and to the point, and supported by evidence that agencies have made the necessary environmental analyses. The Council shall include in its regulations procedures (1) for the early preparation of environmental impact statements, and (2) for the referral to the Council of conflicts between agencies concerning the implementation of the National Environmental Policy Act of 1969, as amended, and Section 309 of the Clean Air Act, as amended, for the Council's recommendation as to their prompt resolution.

The order also directs that federal agencies, "in carrying out their responsibilities under the Act and this Order, comply with the regulations issued by the Council except where such compliance would be inconsistent with statutory requirements."

Regardless of specific compliance with the CEQ guidelines of 1973, most federal departments and agencies published environmental impact procedures as of 1976 (CEQ, 1976a). Of those agencies (Table 4.1) that have published such procedures, slightly less than 50% did so within the period 1970–1973, and slightly less than 50% did so in the period 1974–1976.

The CEQ guidelines of 1973 specify basic considerations to be included in environmental impact statements, including:

1. A description of the proposed action, a statement of its purposes, and a description of the environment affected
2. The relationship of the proposed action to land-use plans, policies, and controls for the affected area
3. The probable impact of the proposed action on the environment
4. Alternatives to the proposed action, including, where relevant, those not within the existing authority of the responsible agency
5. Any probable adverse environmental effects which cannot be avoided
6. The relationship between local short-term uses of man's environment and the maintenance and enhancement of long-term productivity
7. Any irreversible and irretrievable commitments of resources that would be involved in the proposed action should it be implemented
8. An indication of what other interests and considerations of federal

TABLE 4.1
Agency NEPA Procedures as of March 1, 1976[a]

U.S. agency	Current procedures		Amendments	
	Date	Citation	Date	Citation
Agriculture				
Departmental	5/29/74	39 Fed. Reg. 18678		
Agricultural Stabilization and Conservation Service	5/29/74	39 Fed. Reg. 18678	12/20/74	39 Fed. Reg. 43993
Animal and Plant Health Inspection Service	1/29/74	39 Fed. Reg. 3696		
Farmers Home Administration	8/29/72	37 Fed. Reg. 17459		
Forest Service	5/3/73	38 Fed. Reg. 20919	10/30/74	39 Fed. Reg. 38244
Rural Electrification Administration	5/20/74	39 Fed. Reg. 23240		
Soil Conservation Service	6/3/74	7 C.F.R. Part 650		
		39 Fed. Reg. 19646		
Appalachian Regional Commission	6/7/71	36 Fed. Reg. 23676		
[Atomic Energy Commission]				
Nuclear Regulatory Commission (regulatory)	7/18/74	10 C.F.R. Part 51		
		39 Fed. Reg. 26279		
Energy Research and Development Administration (nonregulatory)	2/14/74	10 C.F.R. Part 11		
		39 Fed. Reg. 5620		
Canal Zone Government	10/20/72	37 Fed. Reg. 22669		
Central Intelligence Agency	1/28/74	39 Fed. Reg. 3579		
Civil Aeronautics Board	7/1/71	14 C.F.R. ¶399.110	8/25/75	40 Fed. Reg. 37182
		36 Fed. Reg. 12513	12/19/75	14 C.F.R. 312
				40 Fed. Reg. 59925
Department of Commerce	10/23/71	36 Fed. Reg. 21368		
National Oceanic and Atmospheric Administration,	8/21/74	36 Fed. Reg. 30153	2/4/75	40 Fed. Reg. 5175
Coastal Zone Management	11/13/74	39 Fed. Reg. 40122		
Economic Development Administration				
Department of Defense	4/26/74	32 C.F.R. Part 214		
		39 Fed. Reg. 14699		
Army Corps of Engineers	4/8/74	33 C.F.R. ¶209.410		
		39 Fed. Reg. 12737		

Agency	Date	Citation	Date	Citation
Delaware River Basin Commission	7/11/74	18 C.F.R. Part 401 39 Fed. Reg. 25473		
Environmental Protection Agency	4/14/75	40 Fed. Reg. 16813		
Export-Import Bank: no regulations				
Federal Communications Commission	12/19/74	39 Fed. Reg. 43884 41 Fed. Reg. 4722		
Federal Energy Administration	1/30/76	10 C.F.R. 208	11/18/75	40 Fed. Reg. 53391
Federal Power Commission	12/18/72	Commission Order No. 415-C 37 Fed. Reg. 28412	6/7/73	Commission Order No. 485 guidelines to applicants
Federal Trade Commission	11/19/71	16 C.F.R. 11.81 1.85 36 Fed. Reg. 22814		
General Services Administration				
Departmental	4/4/75	40 Fed. Reg. 15131		
Property Management and Disposal Service	12/30/71	PMD Order 1095.1A 36 Fed. Reg. 23704		
Public Buildings Service	3/2/73	PBS Order 1095.1B		
Telecommunications Service	6/30/71	TCS 1095.1		
Health, Education, and Welfare				
Departmental	10/17/73	HEW General Administration Manual, Chapters 30-10 through 30-16		
Food and Drug Administration	3/15/73	21 C.F.R. Parts 6, 601 38 Fed. Reg. 7001		
Consumer Product Safety Commission: no regulations				
Department of Housing and Urban Development				
Community Development Block Grants	7/16/75	40 Fed. Reg. 29992		
Other programs	7/18/73	38 Fed. Reg. 19182	11/4/75	39 Fed. Reg. 38922

Continued

TABLE 4.1
Continued

U.S. agency	Amendments Date	Amendments Citation	Current Procedures Date	Current Procedures Citation
Interior				
Departmental			9/27/71	36 Fed. Reg. 19343
Bonneville Power Administration			1/19/72	37 Fed. Reg. 815
Bureau of Indian Affairs			1/21/74	Bureau Transmittal Memo
Bureau of Land Management			5/4/72	Bureau Manual Release
Bureau of Mines			2/9/72	37 Fed. Reg. 2895
Bureau of Outdoor Recreation			3/24/72	37 Fed. Reg. 6501
Bureau of Reclamation			11/23/72	37 Fed. Reg. 24910
Fish and Wildlife Service			11/8/74	Bureau Transmittal Memo
U.S. Geological Survey			3/11/72	37 Fed. Reg. 5263
National Park Service			7/29/74	Bureau Manual Release
Interstate Commerce Commission			3/28/72	49 C.F.R. 11100.250
				37 Fed. Reg. 6318
Department of Justice			2/6/74	28 C.F.R. Part 19
Law Enforcement Assistance Administration				39 Fed. Reg. 4736
Department of Labor			3/15/74	29 C.F.R. Part 1999
				39 Fed. Reg. 9959
National Aeronautics and Space Administration			4/10/74	14 C.F.R. 11204.11
				39 Fed. Reg. 12999
National Capital Planning Commission			8/72	37 Fed. Reg. 16039
National Science Foundation			1/28/74	45 C.F.R. Part 640
				39 Fed. Reg. 3544
U.S. Postal Service			7/6/72	37 Fed. Reg. 13322
Small Business Administration			10/20/72	37 Fed. Reg. 22697

State				
Departmental	8/31/72	37 Fed. Reg. 19167		
Agency for International Development	8/1/75	Policy Determination 63 (supersedes 39 Fed. Reg. 22686-7 of 10/20/72)		
International Boundary and Water Commission	3/14/74	39 Fed. Reg. 9868		
Tennessee Valley Authority	2/14/74	39 Fed. Reg. 5671		
Transportation				
Departmental	9/30/74	39 Fed. Reg. 35232		
Federal Aviation Administration Airport Development	6/19/73	FAA Order 1050.1A		
Act	12/7/70	FAA Order 5050.2A; 36 Fed. Reg. 7724		
Federal Highway Administration	12/2/74	39 Fed. Reg. 41804 33 C.F.R. 771		
U.S. Coast Guard	10/22/75	40 Fed. Reg. 49383		
Urban Mass Transportation Administration	2/1/72	DOT Order 5610.1 37 Fed. Reg. 22692		
National Highway Traffic Safety Administration	11/10/75	40 Fed. Reg. 52395 49 C.F.R. 520		
Saint Lawrence Seaway Development Corporation	11/71	Procedure SLS 2-5610.1A	4/24/75	40 Fed. Reg. 18026
Treasury				
Departmental	4/26/74	39 Fed. Reg. 14796		
Internal Revenue Service	8/12/71	36 Fed. Reg. 15061		
Veterans Administration	6/17/74	39 Fed. Reg. 21016	8/25/75	40 Fed. Reg. 37126
Water Resources Council	2/10/71	36 Fed. Reg. 23711		

[a] CEQ, 1976a.

policy are thought to offset the adverse environmental effects of the proposed action.

Agency procedures for compliance with NEPA typically deal with these considerations, and with the following points which the CEQ makes in its 1973 guidelines.

The project description should be adequate "to permit an assessment of potential environmental impact by commenting agencies and the public," should "succinctly describe the environment of the area affected as it exists prior to a proposed action, including other Federal activities in the area affected by the proposed action," should include discussion of "the interrelationships and cumulative environmental impacts of the proposed action and other related Federal projects," and should "identify, as appropriate, population and growth assumptions used to justify the project or program or to determine secondary population and growth impacts resulting from the proposed actions."

In fulfillment of these requirements, CEQ guidelines also specify that all "sources of data used to identify, quantify or evaluate any and all environmental consequences be expressly noted," and that "the amount of detail provided in such descriptions should be commensurate with the extent and expected impact of the action, and with the amount of information required at a particular level of decision making (planning, feasibility, design, etc.)."

With respect to the relationship of the proposed action to land-use plans, policies, and controls for the affected area, the EIS should include "discussion of how the proposed action may conform or conflict with the objectives and specific terms of approved or proposed Federal, State and local land use plans, policies and controls," including those "developed in response to the Clean Air Act or the Federal Water Pollution Control Act Amendments of 1972." Where conflicts exist, the EIS should also discuss how the proposing agency "has reconciled its proposed action with the plan, policy or control, and the reasons why the agency has decided to proceed notwithstanding the absence of full reconciliation."

The discussion of the probable impact of the proposed action on the environment should include consideration of:

- Positive and negative effects on both the national and international environment,
- Secondary (or indirect), as well as primary (or direct) consequences for the environment.

In discussing alternatives to the proposed action, it is essential that the EIS include "a rigorous exploration and objective evaluation of the environmental impact of all reasonable alternative actions." Particular attention should be given to those alternatives "that might enhance environmental quality or avoid some or all of the adverse environmental effects." Alternatives that should be considered include:

- Taking no action (i.e. the no-build alternative)
- Postponing action pending further study
- Actions of significantly different nature (e.g., mass transit as an alternative to new highway construction)
- Different designs of the proposed action
- Measures to provide for compensation of fish and wildlife losses (including acquisition of land, water, etc., for such compensation).

Of special importance in the consideration of alternatives is CEQ's requirement that the assessment "be sufficiently detailed to reveal the agency's comparative evaluation of the environmental benefits, costs, and risks of the proposed action and each reasonable alternative."

Each EIS should include a brief section which summarizes in one place those effects that are adverse and unavoidable under the proposed action. This section should also include a concise statement of how other adverse effects will be mitigated.

The discussion of local short-term uses and the long-term productivity of the human environment should include a brief discussion of the extent to which the proposed action involves trade off between short-term environmental gains and long-term losses. There should also be a discussion in this section "of the extent to which the proposed action forecloses future options." The guidelines emphasize that "short-term and long-term do not refer to any fixed time periods, but should be viewed in terms of the environmentally significant consequences of the proposed action."

Irreversible and irretrievable commitments of resources should be identified from the list of probable adverse environmental effects that cannot be avoided. Each commitment should be discussed in terms of the "extent to which the action irreversibly curtails the range of potential uses of the environment." Commitments of resources include commitments of both natural and cultural resources, as well as the labor and materials devoted to the action.

Other interests and consideration of federal policies which are thought to offset adverse effects of the proposed action should be identified. Where cost–benefit analyses are prepared, they (or a summary) should be attached to the EIS, "and should clearly indicate the extent to which environmental costs have not been reflected in such costs. The EIS should "indicate the extent to which stated countervailing benefits could be realized by following reasonable alternatives to the proposed action that would avoid some or all of the adverse environmental effects."

According to Druley (1976)

> The most significant provision of the CEQ guidelines on the content of the EIS
> concerns the balancing of other interests against the environmental effects. . . .
> The effect . . . is to make the EIS an economic as well as an environmental
> document. Although these provisions do not require a quantified cost-benefit
> analysis, they apparently do require a benefit-to-cost analysis in qualitative terms.
> These provisions would make the EIS the decision-making document. Neither the
> agencies nor the courts have universally accepted the benefit-to-cost analysis as a
> part of the EIS, but perhaps of more significance is the extent of the acceptance by
> the agencies and the courts of the benefit-to-cost analysis [p. 15].

This overall interpretation of CEQ's guidelines is substantiated by CEQ's persistent stress on the basic role of the EIS in federal project development. "The environmental impact statement process needs improvements . . . the improvement most critically needed is to make the impact statement more useful to, and more a part of, agency planning and decision-making [CEQ, 1976a, pp. 122–123]." Also, "Agencies should review the ways in which analytical and technical requirements of the EIS process can be more thoroughly integrated into their administrative operations . . . agencies should consider ways to focus the EIS analysis on the impacts and alternatives that are most relevant to decisionmaking and the public [CEQ, 1976b, p. 25]."

Money and Manpower for the Environment

According to the CEQ [CEQ, 1976b, p. 43], it is not possible (both now and, possibly, in the future) to delineate the exact costs of federal compliance with NEPA. Several general and specific reasons have been presented:

1. The more integrated that NEPA becomes with agency operation, the more difficult it is to identify NEPA-related costs.
2. Some agencies have authority under legislation other than NEPA to analyze environmental issues. These costs have now become part of EIS related costs.
3. A variety of laws, including the Fish and Wildlife Coordination Act, the Federal Water Pollution Control Act, the Clean Air Act Amendments, and the National Historic Preservation Act, require analysis now intertwined with NEPA. Thus costs specific to NEPA are difficult to assess.
4. Agencies such as ERDA, the Forest Service, the Department of Transportation, and the Department of Health, Education and Welfare have so completely integrated the EIS process in adminis-

trative planning and decision-making that they can only approximate NEPA costs.

5. Various agencies determine EIS-related costs differently.

Because of these difficulties, Table 4.2 presents only estimated costs for NEPA related activities for various federal agencies.

As indicated in Table 4.2, estimated costs for preparing and reviewing EISs under NEPA are relatively minor percentages of annual agency budgets, ranging from 5.4% for the Federal Power Commission (FPC) to .0001% for the Department of Health, Education and Welfare (HEW). As an example of the minor cost of implementing NEPA, the CEQ has pointed out (CEQ, 1976b) that "the Nuclear Regulatory Commission estimates that it spent 14.9 million dollars on NEPA in 1975, which is about 2.2 percent of the cost of one nuclear powerplant [p. 45]."

While the actual costs of agency compliance with NEPA may never be determined, it is important to note the long-term trend of federal expenditures on behalf of the environment. As indicated in Fig. 4.1, expenditures for pollution control and abatement have undergone a rather constant increase from just over $1 billion in 1972 to just under $6 billion in 1977. Expenditures for environmental protection and enhancement, and for understanding, describing, and predicting environmental changes[1] have increased much less markedly. While there are many who would argue the necessity for much greater expenditures, it is reasonable to point out that different levels of expenditure from year to year for different environmental programs reflect priorities that are, after all, determined in large part by discrete pieces of legislation and political factors which influence both their passage and their enforcement. Regardless of any argument as to the particular use of environmentally targeted monies, however, it is obvious that there has been and continues to be a major federal commitment to environmental programs. The EIS process, regardless of the funds expended in its implementation, should rightly be seen as only one aspect of this federal commitment of public funds to the upgrading of the public environment.

Of course, money spent to achieve particular environmental ends is most often wasted money if those who utilize those monies have neither the interest nor the knowledge requisite for considering environmental values during project development. Many agencies have, therefore, sponsored training programs and courses that focus in on both attitudinal and behavioral aspects of environmental impact assessment. Agencies that can provide details on these programs and courses that are directly and indirectly related to impact assessment include: the Council on Environmental Quality, the U.S. Army Corps of Engineers, Department of

---

[1] The categories for environmental budgets are utilized by the Office of Management and Budget (CEQ, 1976a).

TABLE 4.2
Some Estimated Expenditures for NEPA-Related Activities [a,b]

Agency	Formal process for determining NEPA costs	Preparation FY 1974	Review and comment FY 1974	Total FY 1974	Total as percentage of operating budget FY 1974	Preparation FY 1975
COE	Yes	$21,933,832	$ 76,000	$22,009,832	1.2%	$27,057,447
USDA						
FS	No, estimate	27,000,000	225,000	27,225,000	2.7%	27,000,000
SCS	No-use estimates	3,500,000	360,200	3,860,200	1.0%	3,500,000
APHIS	No-use estimates	125,000	NA	125,000	NA	85,000
REA	No-use estimates	117,190	11,295	128,485	.73%	161,630
DOC	No-use estimates	430,000	3,174,000	3,604,000	.24%	526,680
DOD	No-use estimates	5,200,000	NA	NA	NA	4,200,000
DOI	No-use estimates	15,200,000	5,500,000	20,700,000	.1%	23,400,000
BLM	Yes	3,544,243	305,000	3,849,243	1.3%	NA
BOR	Yes	575,000	995,400	1,570,400	1.4%	600,000
BuRec	No-use estimates	3,796,000	390,000	4,186,000	.63%	4,050,000
FWS	No-use estimates	NA	1,849,000	NA	NA	NA
NPS	No-use estimates	1,900,900	545,000	2,445,900	.54%	2,013,600
USGS	No-use estimates	2,400,000	800,000	3,200,000	1.85%	5,000,000

DOT	No-use estimates	31,746,000	250,000	31,996,000	.18%	36,500,000
ERDA	No-use estimates	NA	NA	NA	NA	3,400,000
EPA	Yes	1,280,000	1,600,000	2,880,000	.048%	6,300,000
FEA	No-use estimates	NA	NA	NA	NA	330,000
FPC	Yes	1,481,419	70,000	1,551,419	5.4%	1,333,000
GSA	No-use estimates	1,600,000	10,000	1,610,000	.18%	2,235,000
HEW	No-use estimates	40,000	100,000	140,000	.0001%	80,000
HUD	No-use estimates	6,000,000	275,000	6,275,000	.07%	NA
NRC	Yes	NA	NA	NA	NA	14,900,000

[a] From CEQ, 1976b.

[b] Abbreviations are: COE, Corps of Engineers; DOC, Department of Commerce; DOD, Department of Defense; DOI, Department of the Interior; DOT, Department of Transportation; EPA, Environmental Protection Agency; ERDA, Energy Research and Development Administration; FPC, Federal Power Commission; FEA, Federal Energy Administration; GSA, General Services Administration; HEW, Department of Health, Education, and Welfare; NRC, Nuclear Regulatory Commission; HUD, Department of Housing and Urban Development; USDA, Department of Agriculture.

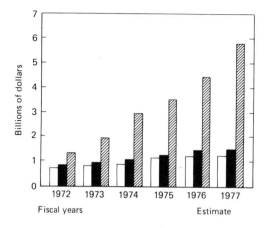

FIGURE 4.1. *Environmental outlays by category, 1972–1977* (CEQ, 1976a).

Agriculture, Department of Commerce, Department of Transportation, the Federal Power Commission, the U.S. Environmental Protection Agency, and the Training Branch of the U.S. Civil Service Commission.

The results of one such course (sponsored by the National Highway Institute of the Federal Highway Administration) highlights some basic issues involved in the effectiveness of these training programs and courses. This particular course, entitled "Ecological Impacts of Proposed Highway Improvements," was presented to 1169 highway professionals throughout the nation from December 1974 through April 1977. A final course evaluation (Federal Highway Administration, 1977) indicated that over 90% of those who had attended the course would recommend the course to their colleagues who also have impact assessment responsibility, and that well over 90% believed that such percentages reflect not only on the course itself, but also on the willingness of highway professionals to learn new skills in order to achieve the goals of NEPA. This willingness is also to be found throughout the federal bureaucracy.

However, an interesting comment was noted throughout the 40 individual presentations of the FHWA course—that higher-echelon decision-makers should also attend such courses. Such comments seem to reflect a frustration in those who are charged with day-to-day impact assessment tasks and who also feel that their sincere efforts to do a better "environmental job" are often thwarted by middle-management personnel.

According to the CEQ (CEQ, 1976b) "Particular attention should be directed to improving training programs both for those required to make and review the environmental analyses and for those required to apply the analyses to their plans and decisions. Agencies should invite State personnel to join Federal NEPA training programs to improve State participation in the EIS process to provide new perspectives on analytical and procedural problems [p. 11]."

Not all NEPA training courses, however, are sponsored by government agencies. For example, the American Right of Way Association has developed a comprehensive course on impact assessment (ARWA, 1976) focusing on the actual and potential contributions of right-of-way personnel to the assessment process and including biological, ecological, and social considerations. This course has been presented to several hundred right-of-way personnel in private industries throughout the nation and has been well received.

In addition to courses sponsored by governmental agencies and professional organizations, there are, of course, numerous academic courses and programs. Sometimes these are developed in connection with a government agency. For example, the University of Wisconsin–Extension (Milwaukee campus) sponsors a Professional Development Program within the Department of Engineering. Wisconsin Department of Transportation participates in the program when transportation topics are discussed (Mary O'Brien, Personal communication, 1977). With regard to the variety of courses and programs offered, it is reasonable to assume that, regardless of where one is, as long as one has access to a community college or to a college or university, it is likely that one can find one or more courses specifically geared to the environmental impact assessment process.

Some Concluding Comments

Whether agency compliance to NEPA today is a "boondoggle," or whether (as many state and federal agencies have declared) compliance has resulted in "substantially improved government decisions over the past six years [CEQ, 1976b, p. 2]," the fact is that the entire process can be improved and made even more useful in agency planning and decision-making. This is by no means a criticism of sincere federal and state attempts to implement NEPA. But NEPA is not just another piece of legislation. As discussed in earlier chapters, NEPA is no less than a revolution in government decision-making—and, as will be seen in following chapters, it presents major intellectual challenges. Individual failures and inadequacies should therefore be put in perspective of the overall effort.

There is no doubt that there are bad impact assessments and that agency compliance with NEPA is sometimes more in form than in substance. *But it is also true that there are excellent, comprehensive, and insightful impact assessments, and that agencies by and large have made sincere efforts to comply not only with the procedural require-*

ments of NEPA, but also with its overall goals. Agency staffs do include professionals in disciplinary areas that, prior to NEPA, were unheard of at the operating level. Money and manpower are in fact being expended for training as well as for on-the-job performance of environmental analyses. And, perhaps more importantly, the spirit of NEPA has become as institutionalized within agencies as have its substantive requirements. For example, in an army handbook for environmental impact analysis (Department of the Army, 1975) it is stated that

> Environmental impact assessment should be undertaken for reasons other than to simply conform to the law. According to the letter of the law, environmental impact must be assessed for activities with significant impact. However, the spirit of the law is founded on the premise that to utilize resources in an environmentally compatible way, it is necessary to know how activities will affect the environment and to consider these effects early enough so that changes in plans can be made if the potential impact warrants it [p. 9]."

With similar regard for the EIS process as a tool and not as an obstacle, a Federal Aviation Administration (1975) manual states that

> Environmental amenities and values shall be carefully considered and weighed in a timely manner in evaluating all proposed Federal actions in relation to airport planning and development, utilizing a systematic interdisciplinary approach. While environmental considerations are obviously not the only ones to be weighed, they are to be evaluated as fully and as fairly as non-environmental considerations. The FAA's objective is to avoid or minimize adverse environmental effects that might flow from any proposed Federal action [p. 3].

Of course, all this is for nothing if decisions are not in fact improved, and if the national environmental goals enunciated by NEPA remain only paper truths and do not become actual environmental truths. The first step toward these social objectives is, of course, to get major social institutions to agree that (a) decisions need improvement; and (b) NEPA is a basic tool for improving them. As evidenced by the quotations in this chapter as well as by the various guidelines and and regulations of federal agencies (see Table 4.1, pages 56–59), for citations of EIS procedures for each agency), that step has been taken.

Not only have government agencies largely admitted the need to improve decision-making and to utilize the EIS process toward that end, they have also apparently begun to view the task of impact assessment more and more as an in-house staff function, and less and less as a job for outside consultants (Chapter 5) who are to advise or to undertake special studies. Whether this is in fact true or is merely an artifact of my experience, it can be suggested that this is both a reasonable and a desirable direction to take. As I have discussed, there are many training courses available that deal with different aspects of impact assessment. If such courses are, in fact, worthwhile, agency personnel who attend them should be able to apply their content in actual assessments. Also, as the experience of agencies with project assessments increases, and as new staff, trained in relevant environmental disciplines, exert their influence, in-house staff should become increasingly self-sufficient in the various tasks of impact assessment. Finally, in-house personnel have both formal and informal lines of communication available to them that are typically not available to an "outside" environmental consultant. The judicious use of such communication lines (including those not within the defined chain-of-command) in order to achieve timely consideration of environmental impacts during project development can make the difference between a relevant and an irrelevant assessment. Outside consultants—if only by virtue of the fact that they are not physically present on a day-to-day basis where project decisions are being discussed and made— are less likely to be present at the right place and at the right time and with just that environmental consideration that can make the difference in a decision.

Of course, in agencies where decision-makers are sufficiently insulated from in-house personnel having impact responsibility, the probability of irrelevant assessment is also high. After all, the best support staff, the best funding situation, and the best available environmental training all count for precisely nothing if agency decision-makers continue to interpret a good environmental assessment primarily as one that does not interfere with the project. In short, institutional compliance with NEPA has been largely achieved; it is now a question of compliance by individuals.

References

American Right of Way Association
1976 *Right of way and the environment.* Los Angeles, California: American Right of Way, Inc.
Council on Environmental Quality (CEQ)
1973 *Preparation of Environmental Impact Statements, guidelines. Federal Register,* Vol. 38 (147, Part II). Washington, D. C.: U. S. Government Printing Office.
Council on Environmental Quality (CEQ)
1976a *Environmental quality, the seventh annual report of the Council on Environmental Quality.* Washington, D. C.: U. S. Government Printing Office.
1976b *Environmental Impact Statements: An analysis of six years' experience by seventy federal agencies.* Washington, D. C.: U. S. Government Printing Office.
Druley, R. M.
1976 *Federal agency NEPA Procedures.* Monograph No. 23, *Environmental Reporter,* 7.(10) Washington, D.C.: The Bureau of National Affairs.

Federal Aviation Administration (FAA)
  1975   *Instructions for processing airport development actions affecting the environ-ment.* Washington, D. C.: U. S. Government Printing Office. (Order No. 5050.2A)
Federal Highway Administration (FHWA)
  1977   *Final report for training course on ecological impacts of proposed highway im-provements.* Washington, D. C.: Federal Highway Administration, U. S. Depart-ment of Transportation.
Mary O'Brien
  1977   Staff Biologist, Wisconsin Department of Transportation, Division of Highways, Madison, Wisconsin. Personal communication.
U. S. Department of the Army
  1975   *Handbook for environmental impact analysis.* Baltimore, Maryland: U. S. Army AG Publications Center. (Pamphlet No. 2000-1)

5

Consultants and
the Assessment Process

Introduction

There is a parable that tells the story of a young missionary from Boston who took good advice, but who was nonetheless sorry for it.

The young missionary dreamed of the time that he would bring the word of God to the people of the South Pacific. But he promised himself that, unlike so many of his predecessors, he would interfere as little as possible with the traditional customs of the people he served.

Thus it happened that he was finally assigned as a minister to the inhabitants of several islands in the South Pacific and undertook to fulfill his dream. The people quickly came to love him, for not only did he not try to redo their culture, but he even showed a sincere respect of their ancient customs . . . and so they came also to love his God and worship Him.

So successful was the young missionary that his bishop could hardly believe the number of new parishioners that the missionary reported. In time, the bishop decided to go to the islands to examine his methods at first hand. He decided he would bring his wife along also and sent word as to when they would arrive.

At first, the young missionary was happy to hear of his bishop's intention. Yet, he soon became apprehensive. He would sit all day on a log looking out over the village, trying to understand the source of his concern. He could not quite put his finger on it.

Suddenly it came to him. Of course! The women of the village still wore their native costume—which was hardly little more than some straw tied about their waists. And they were as bare breasted as when he had first come. What would his bishop and his bishop's wife think!

Now the missionary became frantic and sought out the advice of a friend who had traveled throughout the world and whom the missionary respected for his vast experience. The friend laughed, and patted him on the shoulder.

"Why, there's no problem at all. Simply have the ladies make some blouses. Tell them it's a special occasion, because the bishop's wife, who is very flat chested, will be offended by their beauty. Tell them they can throw the blouses away when she's gone. It will be a great game!"

So the missionary distributed cotton cloth to the villagers and explained what he wanted them to do. They set about the job with much enthusiasm, giggling among themselves about the strange customs of the white people.

Finally, the big day came and the missionary met his visitors at the dock. He brought them up into the village in a little cart which passed through the villagers who were dutifully lined up on either side of the path. All the women of the village were dressed in their colorful blouses. The missionary sighed with relief.

Yet, so unaccustomed were the women to their blouses that the cloth actually made them more aware of their full breasts. And as they came more conscious of their breasts, they became more and more concerned about how offensive they must be to their holy, but flat chested, visitor. Thus, when the cart stopped only halfway through their lines, all the women reached down and pulled their grass skirts up over their heads in order to cover their breasts more completely.

Of course, one can imagine the missionary's dismay. He had never been advised to have them make panties as well!

According to CEQ guidelines (CEQ, 1973), "Agencies should attempt to have relevant disciplines represented on their own staffs; where this is not feasible they should make appropriate use of relevant Federal, State and local agencies or the professional services of universities and outside consultants." With respect to private companies, consultants have, of course, been consistently employed to give expert advice not otherwise available on-staff. In either case, whether in agencies or in companies involved in impact assessment under NEPA, the problem is essentially that faced by the young missionary—to secure advice which can be acted upon, but which is also relevant. Many times in impact assessment, the most precise and sophisticated scientific and technical advice can fall far short of what is required, and lead not only to personal embarassment and dismay, but also to needless project delay and environmental damage.

For example, in one typical impact assessment of a highway project in the northeast, the assessment team utilized (among others) a consultant on air quality and a consultant on biology. Both consultants conducted very professional studies in their respective areas, and their reports were ultimately included in the draft environmental impact statement for the highway project. During the public hearing on the draft assessment report, one lady from the community stood before the highway authority and presented her statement which was essentially the following:

> I realize that you are not going to be able to build this highway unless you have expert assurance that you will meet air-quality standards. I respect the expertise of your consultants in this area. I also know that you will not build this highway unless you have considered the ecology of the region. And I respect the national expertise of the biologists you have employed in this area. Now, would you please tell me what the experts on air quality and the experts on ecology have specifically identified as impacts on my vegetable garden?

Of course, there was complete silence among those who were responsible for the assessment. Finally they had to admit that they had not considered the effects of the highway on her garden. They had focused, instead, on "larger" issues.

One could not fault the expert advice of their consultants. But one could easily fault the relevance of that advice to the overall aim of impact assessment under NEPA, which, after all, is legislation that was enacted for the people of the United States, and not just for the experts primarily concerned with larger issues.

This is not to fault the consultants. Failure in this case (and probably in many others) lay within the relationship between expert consultants and those charged with project management responsibility. One cannot reasonably talk about one without talking about the other.

Some General Considerations

Consultants are often utilized, of course, not only for the special expertise and capability they may have, but also for a variety of other reasons, including (but not limited to):

1. The status a consultant may have, either locally, regionally, or nationally

2. In-house scheduling, manpower, and economic considerations that favor the use of outside help as opposed to in-house capability
3. The sense of objectivity which is usually associated with outside consultants and definitely not associated with in-house personnel and departments
4. The social–political goodwill engendered in local areas by utilizing local expertise.

While each of these reasons for utilizing consultants has its own historic validity within the realm of practical, applied social psychology and business management, they are not necessarily valid with respect to the final product of impact assessment—namely, the comprehensive consideration of total environmental effects in the overall decision-making process.

For example, there is no necessary direct correlation between the status of a consultant and the relevance or utility of his or her assessment input in a particular project; there is no guarantee that assessments and their utilization in decision-making will be improved by the use of outside experts or lessened by the use of in-house personnel; the supposed objectivity of outside consultants may in fact be imaginary, especially since the public is well aware of the economic interest of the supposedly objective consultant; and, finally, goodwill engendered by utilizing local expertise is hardly sufficient to guarantee good assessment of complex scientific, social, and technical issues.

For purposes of the following discussion we will make the assumption (which is, admittedly, naive with respect to the above considerations) that consultants in the assessment field are generally utilized for:

1. Data collection and/or generation, handling and evaluation
2. The identification of specific impacts
3. The identification of interrelationships among these and other impacts identified during the assessment process
4. Liaison services between agencies and the public that are necessary for the final integration of specialized studies with the broad goals of the National Environmental Policy Act
5. Performing these or other services within strict deadlines of time and budget.

It is vital that each of these consultant services be seen as services to an interdisciplinary team that is mandated by NEPA. In fact, once consultant activities become separate and independent from team activities, the kind of comprehensive assessment envisioned in NEPA, required by administrative guidelines, and underscored by court decisions is at serious risk.

Scope of Work

The specific place where assessment-project management and outside consultants first come together is, of course, the scope of work. This is a document which eventually becomes a legal document and that commits both of the contracting parties to certain actions to be conducted over some stated period of time and at a specific cost. The scope of work for consultants in environmental impact assessment has typically presented some serious problems.

For example, an engineer charged with broad impact responsibility often finds it an uncomfortable responsibility to be charged with setting a technical scope of work for a biological consultant. Even when proposed scopes of work are first solicited from prospective consultants and are examined by an in-house team prior to finalization, the problem remains—how to set a scope of work in areas in which the contracting agency or company or individual has little if any previous experience? Can historical criteria of performance in technical areas be applied in scientific areas? If the product of a scientific consultant is a report, should the report be, in fact, a scientific report? How does one specify scientific methodologies or technical methodologies when one is dealing with the "state of the art"? If experiments are included, what does one specify in a scope of work as follow-up action in case of inconclusive experiments? What is involved in a scientific experiment, anyway? How can one control time when the experiments have not even been conducted?

Some basic guidelines[1] for dealing with these problems are called for.

Whether the consultant is performing assessment work that is peculiar to the physical, to the social, or to the total human environment of the project area, the scope of work should not be finalized until the consultant employed for data collection has answered (and answered in detail) the following basic questions:

1. *What* is going to be measured?
2. *Where* is it going to be measured?
3. *When* is it going to be measured?

While these are standard reportorial questions, they are also basic to environmental assessment because they underly the recognition of any component of the environment (whether an ecosystem or a social system) as an integration of innumerable parameters that are dynamic over space and time. The selection of just what to measure, and where and when to measure it, is therefore a complex task and requires, at the minimum, a precise understanding of *why* something is to be measured at all. That is

---

[1] A number of these guidelines are based on U.S. Department of Transportation, 1975.

1. Why, in terms of *applicable standards* (e.g., air and water quality standards)
2. Why, in terms of specific *agency guidelines* (e.g., CEQ, EPA)
3. Why, in terms of *principles, concepts, and theories* (e.g., the concept of eutrophication, which gives particular emphasis to the long-term consequences of enhanced nutrient concentrations and the growth of aquatic plants), and
4. Why, in terms of *previous experience* with similar projects and resources (e.g., experience with case studies in which progressive eutrophication can be directly correlated with the decline in recreational opportunities).

Once appropriate standards, guidelines, theory, and experience have been identified, it is necessary to describe *how needed data and information are to be collected and/or generated.* In general, data and information for impact assessment may be generated in the field, in the laboratory, from existing literature, from government and private agencies and organizations, from ad hoc groups, and from individuals.

The specific best source of data and information should not necessarily be assumed to be the one that is most accessible to the consultant, but may, in fact, be most easily obtainable by on-staff personnel or other individuals. For example, rights-of-way personnel within a given region may have extensive negotiation experience with individual landowners and agencies within the region. Such individuals may be in the best position to identify specific sources of the very information required by an outside consultant. Such information may pertain to social or community solidarity and cohesiveness, to public attitudes and values, and to various components of the physical environment. Much of this information may be very expensive to generate by a consultant using surveys, questionnaires, etc.

A good rule of thumb is that once required types of data and information have been identified by the consultant, the team (and not the consultant alone) should identify the most efficient means of obtaining that information. This must always be done, however, in light of the methodological requirements of regulatory agencies responsible for promulgating environmental standards, as well as standard methods promulgated by recognized professional organizations. The scope of work should specify clearly the methodologies to be employed for information generation, collection, and analysis, and the responsibility of the consultant or others for obtaining and/or handling required data.

Provisions should also be made in the scope of work for obtaining any additional data and information that may be required as the input from the consultant is reviewed and refined by the full interdisciplinary team. For example, a particular consultant may be responsible for describing the existing sports fishery of a small impoundment. However, the interdisciplinary team may, in its deliberations, recognize that a declining number of angling trips to that impoundment (due perhaps to a recent increase in weeds which snarl lures and lines) may very well result not only in a decreased catch of sports species, but also a decreased harvest of nonsports species. Thus, one result of a continued low level of angling trips might be a long-term increase in the proportion of nonsports to sports species in the impoundment. The consultant, who is responsible for describing the game species, might therefore reasonably be requested to collect base-line data on nongame species as well. Although it may appear to some that a consultant would do this anyway, the fact is that this will typically not be done unless there is specific provision in the scope of work. Whereas specific data requirements of this kind cannot be foreseen and therefore specifically included in a scope of work, it is important that the scope of work requires (a) the consultant to participate in team deliberations of the overall assessment; and (b) provide such additional services (for additional fees) as may be reasonably assigned as a result of these deliberations.

In addition to the specific inputs required of the consultant, the following items would also be specified in the scope of work:

1. Budget, level of effort (i.e., man-hours) and timetable
2. Liaison required with agencies and other consultants
3. Appearances at public meetings and hearings
4. Types of submissions and reports.

While these are standard management tools in any enterprise, they have particular importance to the efficient and proper functioning of interdisciplinary efforts. *Cost overruns and delays by one consultant can modify, retard, and even stop meaningful interaction among all members of the interdisciplinary team and cannot be tolerated.* Liaison is an essential task, not only for the purpose of gathering information, *but also (under NEPA) for sharing information and coordinating efforts.* Appearances at public meetings and informational hearings are not incidental to the assessment process—*they are integral to it.* And, finally, the consultant report must be in a form that is directly usable to the layman as well as to the technical and scientific professional. *The interdisciplinary team must be able to count on receiving both progress reports and a final report which is sensitive both to laymen and to highly trained individuals.* This may very well require the consultant to supply a technical report and a summary report that is understandable to the layman.

Some Additional Considerations

Whatever else impact assessment under NEPA may be, it is the prediction of what will happen in the total human environment when certain actions are undertaken. Insofar as a consultant is employed for the specific purpose of making predictions, careful consideration should be given to his or her actual understanding of just what action will be undertaken. For example, a biologist may very well be capable of describing biological systems within a project area and also how those systems might react under particular conditions of stress. But what conditions will, in fact, be produced by specific project activities? Similarly, a sociologist may very well be capable of discussing community patterns and how those patterns can be modified by various factors. But just what factors will be introduced or altered by what specific project activities?

Any project may be subdivided into various component phases, the most general of which are the planning, location and design, acquisition, construction, maintenance, and operational phases. Each phase, in turn, includes specific activities (e.g., data collection, negotiation, relocation, dredging, excavation, mowing, and structural repair). Each phase also includes the use of specific items and materials in the project area (e.g., people, chemicals, vehicles). The consultant utilized for impact prediction should have some familiarity with specific actions that will in fact be undertaken, with how things actually will be done, and with regional variations of standard practices. After all, transmission lines are not installed in the same precise way as highways. Subdivision development requires different procedures and materials than pipeline installation. Dredging in Massachusetts is quite a different undertaking than dredging in Louisiana. If the consultant is not already knowledgeable about particular actions and materials and how they can differ from place to place, it is the responsibility of project management to ensure that the consultant does become aware of them.

Finally, the consultant should be held responsible for describing impacts in those terms that meet the criteria of adequacy defined by legislation, agency guidelines, the courts, and by the decision-making process itself. Such criteria include:

1. The *probability* of the impact occurring at all,
2. The *magnitude* of the impact, should it occur,
3. The *time frame* (i.e., long- or short-term) of the impact,
4. The *relevance of each impact to each alternative* of the proposed project,
5. Specific *acts of mitigation* (of negative impacts) *and of enhancement* (of positive impacts) that may be undertaken,
6. Documentation of findings and recommendations.

In some instances, the probability of an impact can be stated mathematically. For example, national statistics are available on highway accidents along improved and unimproved highways that involve the spillage of various chemicals, including potentially toxic materials. In many instances, however, it is impossible to give a statistical probability that something will happen. The consultant should, nonetheless, make the attempt to estimate the probability of an impact in qualitative terms (e.g., highly probable; possible, but not very probable). The importance of such judgments lies within the very purpose for identifying impacts in the first place—namely, to give decision-makers some meaningful guidelines for making their decisions. An estimate of probability, even if it is quite subjective (as long as it is not spurious), serves to highlight certain impacts for the decision-maker.

The significance of an impact, however, is not solely a function of its probability, but is related to the magnitude and duration of an impact as well. For example, it is highly probable that suspended particles in water (measured by the turbidity of water) will increase as a result of excavation in the vicinity of a stream or impoundment. Yet the consequences of increased turbidity to aquatic biota depend on (among other factors) how much suspended material gets into the water (i.e., the magnitude of the increase) and how long it stays in suspension (i.e., the time frame). Some impacts, such as toxic spills, have low statistical probabilities and low magnitudes, but may, nonetheless, have long-term effects on ecosystems. *There is no one consensual method whereby probability, magnitude, and duration can be consistently combined to define "highly significant" or "insignificant" impacts.* Rather, the consultant should work with the interdisciplinary team in making and refining these judgments.

With respect to project alternatives, the consultant should be contractually empowered to compare impacts on all project alternatives equally. Project alternatives include not only alternative alignments and locations within the same project corridor or area, but also other corridors and the no-build alternative (i.e., no project development at all) as well. It serves no purpose to concentrate the consultant's efforts on a single alignment and to have him or her evaluate impacts on other possible alignments in only a cursory fashion. This approach prejudices the findings of the entire assessment process.

Once the consultant (in close liaison with other members of the interdisciplinary team) has begun to identify probable impacts and to evaluate their significance in each alignment or possible project areas in terms of probability, magnitude, and time frame, consideration should be given to methods for mitigating undesirable impacts and for enhancing desirable impacts. Both mitigation and enhancement methods may involve changes in planning, design and engineering, and project management. *It is essential that suggestions for mitigation and enhancement be made as early as possible in overall project development.* This is because it is in the earliest phases of project development, before extensive planning and design have taken place, that changes may be more easily made. If consultants who are charged with the identification and evaluation of impacts do not have direct access (through the interdisciplinary team) to planners and decision-makers in these early phases, many impacts that might otherwise be avoided will, in fact, not be avoided and may have to be corrected later at greater costs, and probably only after serious project delay.

Finally, the consultant charged with responsibility for the identification, evaluation, and mitigation and/or enhancement of project impacts should document his or her findings and judgments. Documentation may include professional literature, experimental results, and previous experience with similar impact assessments and projects. Where possible, documentation based on state-of-the-art research should be minimized so as to avoid the danger of using data and conclusions that have, as yet, been untested by other professionals.[2]

Overview

As indicated in Fig. 5.1, the guidelines (pp. 75–80) that may reasonably be applied to the contributions of an outside consultant to an assessment project are applicable to the contributions of all members of the assessment team, including in-house personnel. While some consultants may be utilized for very specific and relatively narrow-scope purposes and others for general and broad-scope purposes, the assessment aprocess itself must govern every input into it. *The essential feature of assessment under NEPA is in its interdisciplinary and integrative nature.* Thus, one may characterize the assessment process as being "management intensive" because of the constant need to ensure that the flow of diverse information between diverse individuals will result in integrated judgments that can be utilized by decision-makers, and utilized when the decision-maker can best make use of them (see Chapter 6).

If outside consultants are utilized at all, they must be utilized (thus managed) as team members, all of whom (including outside consultants and in-house personnel) have ultimate responsibility to decision-makers and thus to the national goals of NEPA.

Of course, one may argue that outside consultants have been overused in impact assessment over the past several years. As discussed in Chapter 4, there is much to be said on behalf of maximizing the resources of existing agencies and in-house personnel, whether in an agency or a private company. The basis of this argument will be discussed more fully in the following individual chapters on the physical, social, and total human environment. But whether consultants have been used too often

---

[2] Each criterion discussed will be further examined and discussed in following chapters on impacts in the physical, social, and total human environment.

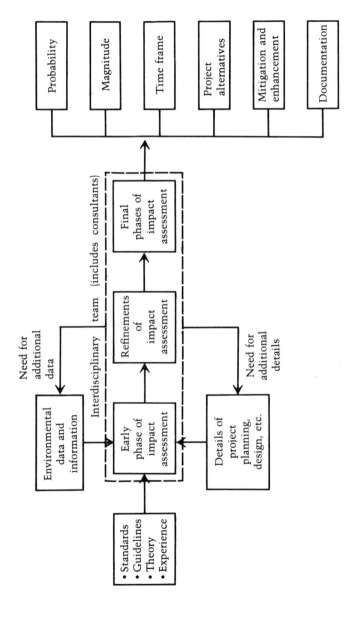

FIGURE 5.1. *Diagram of some basic steps in the impact assessment process. [Adapted from U.S. Department of Transportation, 1975.]*

or not often enough, another issue pertaining to the use of outside consultants is very important—specifically, just who are appropriate consultants?

University and college departments and individuals, as well as private consulting firms and research institutes, are, of course, extensively utilized by companies and agencies involved in project and program assessment under NEPA. The names of individuals, organizations, and companies are usually easily available in a variety of publications. Most notably, the U.S. Army Corps of Engineers has published a nine-volume *Directory of Environmental Life Scientists* (U.S. Corps of Engineers, 1974). Such compilations typically include recognized professionals who are qualified to perform specific consultant services in a variety of impact assessment projects.

One might consider, however, that an important group of potential consultants is often overlooked—possibly to the detriment of many impact assessments. This group included formal and informal organizations and individuals who have specific knowledge, not so much by virtue of professional training, but by virtue of long experience and deep-rooted interest in various components of the total human environment. For example, people who have been collecting wild flowers in an area for 20 years, and who know where they grow and under what conditions and who have invested their private time in increasing their knowledge about them, are in a far better position to provide useful botanical information and even to evaluate botanical impacts, than someone who may be a professional botanist but who is ignorant of local flora and conditions. Birdwatchers, fishermen, hunters, landowners, well diggers, local businessmen, personnel in local service agencies, realtors, and any other member of the public can often provide just that information and data, and just that local and regional insight, and just those skills that are required for the comprehensive evaluation of impacts on the total human environment. They may not qualify as expert witnesses in court. They may be restricted in terms of their availability and even in their interest. Yet, the requirements of NEPA exist not merely for the expert witness, not merely for those who make a full-time job out of the environment, and not merely for those whose interest in the environment transcends the necessities of everyday human life. Sometimes, a good assessment of project impacts requires the good advice, the good information, and the good sense of the nonprofessional.

Call such a nonprofessional a consultant in the legal sense or not— that person is potentially an advisor whose knowledge and skills are overlooked only at the risk of an assessment so overburdened with professional specialization and distance that it becomes irrelevant.

This is not to argue against the use of the outside professional consultant. It is only to argue that consulting under NEPA is ultimately a service to the people of the United States and that ordinary people often have extraordinary talents. Whether such people can be utilized as formal consultants in an impact assessment is a practical problem of bureaucratic flexibility and project management—it is not a conceptual problem inherent within the environmental ethos of NEPA.

Of course, another group of consultants are the governmental agencies themselves. Some of these agencies are specifically directed, either by legislation or administrative directive, to provide specific services which can be extremely important in a particular impact assessment.

For example, the U.S. Fish and Wildlife Service has created a new Biological Services Program which has (among others) the following missions: (a) to strengthen the Fish and Wildlife Service in its role as a primary source of information on national fish and wildlife resources, particularly in respect to environmental impact assessment; and (b) to gather, analyze, and present information that will aid decision-makers in the identification and resolution of problems associated with major land- and water-use changes.

The director of the U.S. Fish and Wildlife Service has specifically stated (U.S. Department of the Interior, 1975) that:

> Our overall objective is to make the Biological Services Program a major resource for timely, accurate information about the effects of resource development upon fish and wildlife resources and their supporting ecosystems. *This information is intended not only to assist planners in Federal, State and local governments and the private sector, but also to keep the American public full aware of the stakes involved in decisions affecting their environment.* [p. 2, emphasis added]

Similarly, other agencies including federal, state, and local regulatory agencies can usually be relied upon to supply advice and guidance as well as information critical to impact assessment.

Regrettably, agencies are typically viewed by those having responsibility (either in other agencies or in private companies) as adversaries. This attitude only serves to preclude important cooperative efforts early in the assessment process. As a general rule (and in keeping with the goals of NEPA), one should approach governmental agencies early in impact assessment with some assurance that they will typically do their best not only to provide useful information, but also to provide useful advice as to what kinds of impacts might be expected, the relative significance of these impacts, and how might one avoid the undesirable impacts and enhance the positive ones. When such advice from governmental agencies is available to the assessment team, it may be possible to reduce the team's reliance on private consultants and/or to better direct the efforts of outside consultants. Directories of agencies are available that describe their areas of interest, publications, and information services provided (including consulting services). These directories include separate directories for the federal government (National Referral Center, 1974), and for government, educational organizations, and societies dealing with the social sciences (National Referral Center, 1973), the biological sciences, physical sciences and engineering, general toxicology, and water.

References

Council on Environmental Quality (CEQ)
  1973 *Preparation of Environmental Impact Statements, guidelines. Federal Register, 38* (147, Part II). Washington, D.C.: U.S. Government Printing Office.
National Referral Center
  1973 *A directory of information resources in the United States: Social sciences (Rev. ed.).* Washington, D.C.: U.S. Government Printing Office.

  1974 *A directory of information resources in the United States: Federal government, with a supplement of government-sponsored information analysis centers* (Rev. ed.). Washington, D.C.: U.S. Government Printing Office.
U.S. Corps of Engineers
  1974 *Directory of environmental life scientists* (9 vols., by geographic region). Washington, D.C.: U.S. Government Printing Office.
U.S. Department of the Interior
  1976 *Fiscal year 1975, biological services program.* Washington, D.C.: Wildlife Service, U.S. Department of the Interior.
U.S. Department of Transportation
  1975 *Ecological impacts of proposed highway improvements.* Washington, D.C.: Federal Highway Administration, National Highway Institute.

II

Assessing Impacts: The Physical Environment

$$6$$

Relating the Assessment to the Project

Introduction

Impact assessments are predictions about systems—physical systems, social systems, and total environmental systems. Whether one is dealing with project impacts on fisheries, or on regional economics, or on the relationship between the availability of natural resources and future industrial growth and the standard of human life, one is dealing with systems—small and large systems, simple and complex systems. Systems are comprised of (a) individual component parts (e.g., fish, temperature, oxygen); and (b) interactions between those parts (e.g., job availability in industry that relies on proper management of renewable resources, such as timber). Impact assessment, as predictions about systems, are, therefore, predictions as to how projects will affect these component parts *and* their interactions. When one considers the seemingly infinite number of component parts to the human environment and the various ways in which they may interact, one must conclude that any impact assessment is essentially project-specific.

Of course, previous experience with similar types of projects in similar types of environments can and does focus the attention of the assessment team on certain impacts as being highly probable and important. Yet, no two projects, however similar, are precisely alike—just as no two project areas, however similar, are precisely alike. Thus, just because a specific impact occurred as a consequence of one particular project in one particular area is no guarantee that the same impact will occur or have the same significance as a consequence of another and similar project in another and similar area.

While this fact is easily recognized in the abstract, it is often overlooked in the actual conduct of impact assessments. People as well as ponds, and forests as well as neighborhoods become neatly categorized, and the categories are individually applied in assessment after assessment, even though analytical categories utilized in one project may be irrelevant or misleading in another project. This is not to criticize the reductionist approach (i.e., the breaking of complex and variable systems down into standard analytical categories) in any intellectual effort. After all, we learn about the world by simplifying it, reducing it down to apparent essentials, and inventing analytical categories that can be individually applied in situation after situation.

Yet, the validity of such an approach (as evidenced by the tremendous successes of analytical sciences) does not require that we discard a contrary approach—namely, that one may also learn about the world by focusing on the uniqueness of peoples, places, and things.

While impact assessment under NEPA is a relatively new human undertaking, it is obvious from the guidelines promulgated by government agencies (Chapter 4) that categories of important environmental factors, typologies of impacts, and environmental standards are often utilized in a checklist approach to impact assessment. Also, a great deal of attention is currently being given to conducting impact assessments of whole programs, rather than of each individual project that falls within a program (CEQ, 1976). This approach may well enhance the checklist approach to impact assessment.

Nevertheless, the person charged with impact assessment responsibility should realize that careful and precise adherence to checklists designed for specific purposes does not absolve one of any failure for having overlooked what is not yet in the guidelines and standards, but what may actually occur. Meeting the "letter of the law" in terms of currently promulgated standards, guidelines, and other environmental checklists is no guarantee for meeting the comprehensive national goals of NEPA. And the brief experience with impact assessment under NEPA is even less of a guarantee that our experience and knowledge are as yet sufficient to allow us to follow checklists (whatever their legal basis) blindly.

In a 1977 article, Eugene Odum expressed the view

that science should not only be reductionist in the sense of seeking to understand phenomena by detailed study of smaller and smaller components, but also synthetic and holistic in the sense of seeking to understand large components as functional wholes [and that] an important consequence of hierarchical organization is that as components . . . are combined to produce larger functional wholes, new properties emerge that were not present or not evident at the next level below.

Professor Odum's remarks highlight a basic problem in any impact assessment—namely, whether to proceed (within the strict constraints of budget and time) more in the direction of *analysis*, or more in the direction of *integration*. For example, one may easily spend large amounts of money and time collecting specific analytical data in an impoundment (e.g., pH, alkalinity, temperature, trace metals, inorganic and organic nutrients, aquatic flora and fauna). In fact, in order to follow guidelines established by specific legislation (e.g., the Safe Drinking Water Act of 1974, the Federal Water Pollutional Control Act Amendments of 1972) it may be necessary to focus on selected and highly specific water-quality criteria for purposes of impact assessment. Yet, under the comprehensive mandate of NEPA, money and time also have to be spent integrating analytical information—"pulling the pieces together" into judgments and recommendations about systems (physical, social, political, economic, etc.) that are useful to decision-makers.

It is, of course, a good deal easier to proceed in the direction of analysis. After all, as long as we can focus only on the analysis of individual components of either physical or social systems, we can utilize specific techniques, protocols, standard procedures, and analytic equipment and devices that have been specifically designed to generate precise and accurate data. In fact, given the infinite diversity of environmental components, we can spend our professional lifetimes generating such analytical data. Yet, when it comes to integrating individual environmental components (i.e., constructing the system that they describe) we often find that there is less certitude as to how to proceed, more disagreement, and more confusion.

Still, NEPA requires "the integrated use of the natural and social sciences and the environmental design arts in planning and in decision making which may have an impact on man's environment." Thus, regardless of any propensity we may have toward the analysis of individual environmental components, *sufficient time and money has to be spent on the integration of these components into systems which have direct and indirect influence on the total human condition.*

Basic Approach

A major task in any impact assessment under NEPA is to organize and coordinate specific analytical and integrative efforts with overall project development. These efforts may be summarized as:

1. Identifying all relevant regulatory standards and guidelines, concepts, principles, and previous experience which may be applicable to the proposed project and the general region in which the project will be undertaken
2. Generating and collecting technical data and information on the proposed project and all environmental data and information that will aid in (a) describing baseline environmental conditions prior to the project; and (b) identifying possible impacts on the physical and social environment
3. Identifying interactions between impacts on the social and physical environment
4. Evaluating the significance of all impacts on the physical, social, and total human environment; this requires consideration of the probability, magnitude, and duration of individual and interrelated impacts
5. Identifying and evaluating (in terms of effectiveness and other secondary impacts) specific mitigation and enhancement techniques, including design, engineering, and management alternatives
6. Making recommendations for specific actions to be taken during the proposed project development

It is important to stress that these efforts should be organized and coordinated in such a manner that *recommendations for specific actions can be made and factored into decision-making at each phase of project development,* including early planning, location, design, acquisition, relocation, construction, and operation and maintenance phases. Poor organization and poor coordination of these efforts will result in a useless assessment from the point of view of decision-making.

As indicated in Fig. 6.1, the first three efforts are basically analytical, requiring attention to individual parameters, processes, and interrelationships among environmental and project components. The last three are essentially integrative, requiring objective and subjective evaluations of complex systems, as well as judgments as to required action.

The relative periods of time required for each of these efforts cannot be stated in the abstract. Of course, with infinite amounts of time and money it would be possible to concentrate on each of these efforts equally. However, within the real-world constraints of both money and time, it is strongly recommended *that the major effort in impact assessment should be invested in integrative efforts.* If this is not done, then even the most sophisticated of analytical information will never accomplish what NEPA requires—to affect decision-making during project development. It is in the evaluation of the significance of

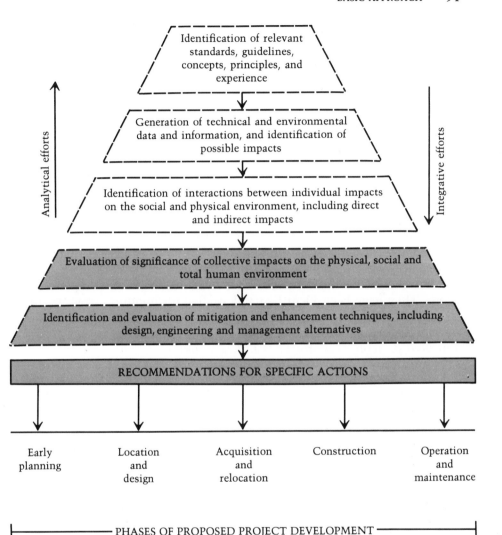

FIGURE 6.1. *Analytic (unshaded) and integrative (shaded) tasks required to affect decision-making during project development.*

impacts and, therefore, in the selection of impacts for the major attention of decision-makers that the assessment process demonstrates the usefulness of impact assessment as a decision-making tool. Contrarily, if the assessment process never gets beyond a picayune examination of individual environmental components, the decision-maker will be faced with an overload of highly specialized information, none of which he or she knows how to use.

Some Important Issues

Analytical and integrative efforts typically required in the assessment of physical, social, and total environmental impacts will be specifically discussed in following chapters. However, it is important to highlight here some important issues which are generally pertinent.

IDENTIFYING VARIABLES AND PROCESSES

In the process of lecturing on impact assessment to hundreds of professional engineers throughout the nation, I have often been told: "You can't start impact assessment until the design work has been done!" This is not really so. The first analytical effort in any impact assessment is to identify the basic variables and processes that, in one combination or another, describe the real project. Important social and physical variables and processes were not invented on Earth Day, 1969, nor will they be newly discovered by the engineer when he completes his work. They have been identified and studied for hundreds of years. Impact assessment begins with identifying which of these innumerable variables should be considered in a particular project. *This effort can be initiated as soon as a project is talked about.*

As already indicated, some of these variables are defined by environmental standards and agency guidelines. Others are easily identified through previous experience with similar projects in similar environmental surroundings. The inputs of competent legal counsel, and the inputs from team members with diverse experience are essential inputs early in the assessment process.

Yet, regardless of the legal expertise of counsel and regardless of the diversity of experience among team members, other inputs are also necessary. These inputs derive from disciplinary knowledge and understanding of basic processes that characterize the physical and social environments and their relationships to humankind. It is doubtful whether any assessment team will include individuals trained in each of the relevant disciplines. The team may, of course, choose to employ outside consultants (Chapter 5).

Assuming that consultants, however, are not utilized, and in lieu of team expertise in all possible relevant disciplines, project management should ensure that team members have the following basic attributes: They should have good communication skills; they should be methodical in their approach to issues; and they should enjoy learning.

If someone can communicate, is methodical, and also enjoys learning, that person is in the best possible position for tapping that environmental knowledge which has been collected for hundreds of years and which is easily available in lay as well as professional publications and in the heads of university and governmental professionals available in the community. This is not to say that such a team member is expected to become a "mini expert"—it is only to say that *team effort begins with the assumption that its members do not know everything, and that, as individuals, they must begin dealing with other professionals, with other agencies, and with other individuals in order to undertake their assessment responsibility.* The assessment team that satisfies itself that it has identified all required kinds of information and data without ever having recognized the depth of its collective ignorance is precisely that team with the greatest potential for failure.

Of prime importance in the earliest analytical phase of impact assessment is the *need for having one or more team members who have some technical knowledge and understanding of the various phases of the proposed project development.* An environmental assessment team composed of all kinds of expertise except the kind required for designing, locating, and constructing a project will likely produce an assessment that is largely irrelevant to the project.

GATHERING DATA AND IDENTIFYING POSSIBLE IMPACTS

As soon as variables and processes that are relevant to the project and the project area are identified, the collection and/or generation of data and information and the identification of *possible impacts* should begin. Some general guidelines for conducting these activities can be summarized as follows[1]:

First, *all data and information collected should be utilized for making actual predictions.* Data and information that are collected but that are never utilized are a waste of time and money. Any waste of time and money during analytical efforts will detract from important integrative efforts in later phases of assessment. It must be emphasized that a compilation of data is a means to an end (i.e., a prediction) and not an end in itself. Therefore, careful consideration should be given to *just what level of precision and accuracy is required in order to make predictions.* For example, arguments about precision to two or three decimal points are probably interesting intellectual exercises among technicians, but if the predictive tools that are utilized are sensitive only to integral values, the exercise is misdirected and totally irrelevant.

Second, it is generally much more expensive and time-consuming to generate new data in the field and/or laboratory. Therefore *new data should only be generated when the data are required for making predic-*

---

[1] Each of these will be discussed further in subsequent chapters on the physical and social environment.

tions and when they are not already available from other sources, including government agencies, educational institutions, previous environmental studies, organizations, and individuals. In most cases, it is best to make the assumption that most required data exist and, therefore, to devote some time to finding and collecting them.

Third, technical data include design, engineering, and other information that become available as the proposed project proceeds through its various phases. Continual liaison between the assessment team and project planners, engineers, negotiators, etc., ensures that data files used for assessment purposes will be continually updated and refined.

Fourth, the identification of possible impacts in the physical and social environment should proceed by the systematic analysis of individual project activities, including all activities associated with each phase of project development. It is important to identify as many possible impacts as early as possible in impact assessment in order to maximize the amount of time and budget available for refining and evaluating impacts, and for formulating timely recommendations pertaining to those particular impacts judged to be the most significant.

INTERACTIONS BETWEEN IMPACTS

It is generally best to assume that an impact on the physical environment can affect the social environment and vice versa. For example, turbidity raised during construction can have direct and long-lasting impacts on important sports fisheries in a local area. This may, in turn, affect recreational fishing in the area; and this may, subsequently, have economic impact on local small businesses that manufacture or distribute sporting equipment. Similarly, relocation of homeowners in a proposed right-of-way may require some replacement housing to be constructed in previously nondeveloped areas. Such new development may, in turn, affect wildlife population in these areas.

The identification of interactions among individual impacts should proceed within the following minimum guidelines:

1. Each individual environmental impact should be examined for its long- and short-term consequences to other components of the environment, and should include consideration of direct and indirect effects.
2. The interdisciplinary team (and not isolated individuals) should conduct these analyses in order to maximize the use of diverse experience and understanding of complex interactions.
3. The team should not disregard too easily the possibility of some interactions merely because they sound extreme. It is better to check the reasonableness of possible interactions with professionals, the literature, and actual experience with previous projects.

EVALUATION OF SIGNIFICANCE

The evaluation of the significance of identified impacts involves both *objective and subjective considerations*. It is, therefore, extremely important to recognize that the evaluations made by the interdisciplinary team may differ strikingly from the evaluations of agencies, individuals, and the general public, even when all groups are dealing with the same basic facts. The assessment team should therefore *consider alternative points of view* so that decision-makers will ultimately receive recommendations that have a broad base in both scientific and social fact.

It is not the function of the team to decide which contrary evaluation of impact significance is correct; but rather, to make its own judgment openly, and in light of all relevant information. Thus the team should refrain from the use of so-called "absolute criteria" implicit in such statements as: "The increase in jobs associated with this project has to take precedence over any negative impact." Such statements of judgment do not become statements of fact merely because of the vehemence with which they are said. Local agencies, organizations, and citizen groups may say the opposite with just as much vehemence. For example, in some areas, regions, and municipalities, there may be a policy of "no growth," including the kinds of growth that may occur as a result of "boomtime enterprise."

MITIGATION AND ENHANCEMENT

In the first years after the enactment of NEPA, many assessments (especially those for projects already underway when NEPA was passed by Congress) dealt solely with mitigation techniques, and then only in a cursory fashion. Mitigation was typically handled in an EIS by statements purporting that a particular negative impact "will be mitigated by standard good engineering practices." This approach is no longer adequate.

Any act of mitigation and/or enhancement, including planning and design, engineering and management actions, is an act that should be analyzed for all of its possible impacts on the physical, social, and total human environment.

For example, in a particular case it may be decided that borrow pits should be located and designed in such a way that they can be utilized as shallow impoundments and wetland areas after completion of the project. This decision may mitigate the loss of other wetland and wildlife habitat in the area due to project location and/or design. However, it is important to consider all consequences of such a decision. How will the new wetland habitat affect public health by serving as habitat for mosquitoes? Will the new wetland attract waterfowl in such numbers that their fecal droppings into both the new and existing water supplies will lead to problems of water quality and aesthetics in the area? What of the safety of children in the area? Also, if the new impoundment is specifically designed and managed so as to provide recreational fishing, how will access provided for anglers affect other components of the local social and physical environment?

These are only representative of the kinds of questions that should be considered before making specific recommendations to those decision-makers who must, after all, choose a course of action.

In addition to the identification and evaluation of all effects of mitigation and/or enhancement actions, the interdisciplinary team should consider the following:

1. Can the effectiveness of the proposed act of mitigation and/or enhancement be documented (e.g., in the professional literature, previous projects, or by undertaking a monitoring program in the current project)?
2. What will be the consequences if the action fails to achieve its objective?
3. What type of maintenance (if any) will be required?

The interdisciplinary team should not assume that mitigation and/or enhancement will require new technology and/or new approaches. Frequently, mitigation and enhancement of an environmental impact can be accomplished by the timely application of practices already known and tried. Of course, sometimes special actions may have to be taken. The interdisciplinary team should, therefore, consult first with agencies and organizations having broad experience with different national regions and types of projects in order to identify the need for undertaking extraordinary actions, and what such actions might be.

In my experience, it is evident that assessment teams (whether in governmental agencies or private companies) considering mitigation and/or enhancement of impacts in one project are often unaware of what has been or is being done by their own colleagues in different parts of the nation. Therefore, increased communication and liaison between agencies, organizations, and companies is essential for the proper selection and implementation of mitigation and/or enhancement techniques.

RECOMMENDATIONS

The ultimate utility of any assessment under NEPA is in the timely presentation of clear, concise recommendations to those decision-makers who are authorized to act on them.

Depending on agency or company policy, such recommendations may be presented in the form of informal or formal written reports (e.g., an in-house environmental report; an EIS), memoranda, or oral presentations. Regardless of the vehicle utilized, the importance of such recommendations is in the effects they have on decision-making.

It can hardly be expected that recommendations will be transformed into meaningful action if:

1. They are written in such specialized and technical language that decision-makers cannot understand them.
2. They are presented in such voluminous detail that a decision-maker could not possibly take the time to read them.
3. They are presented in such abstract, generalized form that the decision-maker cannot immediately understand how they can be actually implemented in the specific project for which he or she has responsibility, the manpower and timetable required, and the costs and benefits expected.

There is often a strong tendency among members of an interdisciplinary team who do not have the same understanding of overall project development as they do with individual aspects of physical and social phenomena to make recommendations based primarily on their own specialized knowledge, with little or no regard for the economic, political, and bureaucratic realities of the real world. *Although such realities, under NEPA, do not necessarily take precedence over environmental considerations, it is equally true that NEPA does not give precedence to environmental considerations over all others.* Thus, because impact assessment is meant to affect real decisions, any recommendations made as a result of impact assessment must include consideration of all real requirements for the implementation of the recommendation.

Management Issues

Environmental impact assessment for a particular project requires inputs from and cooperation between a number of different specialists. There is, therefore, the temptation to define the requirements of impact assessment in terms of the requirements (data, time, budget, etc.) of each individual specialty. Within the real constraints of time and money and manpower, such a perspective is likely to lead to serious and disruptive confrontations between various contending interests. Whenever this happens, the broad national goals of NEPA are typically laid aside.

A contrary approach to defining the assessment process is more in keeping with congressional intent—namely, that impact assessment (whatever else it requires) minimally requires the cooperation of individuals toward a common goal (i.e., informing decision-makers of the full consequences of their actions). Such an approach to defining the assessment process, which emphasizes the basic task of getting things done through people (and not specialties), highlights the assessment process as a management intensive process (Chapter 5).

The manager of any project plans, organizes, staffs, directs, and controls the activities of others. Thus, it is the manager of the impact assessment process who is ultimately responsible for (a) the efficiency; and (b) the relevance and adequacy of the assessment process with respect to the proposed project.

As discussed in this chapter, there are basic analytical and integrative efforts that must be undertaken in the assessment process. Each of these efforts may, in turn, be subdivided into specific tasks and subtasks. For example, specific tasks performed as part of the overall effort to collect environmental data may include the collection and compilation of data and information on such individual environmental components as:

- Surficial and subsurface hydrology
- Soil types
- Regional flora and fauna
- Macro- and microclimatology
- Special and unique habitats
- Demographic patterns and projections
- Recreational areas
- Water quality and quantity

The performance of each task requires manpower, time, and money. In making decisions affecting the allocation of manpower, time and money, the manager of an impact assessment must continually balance two considerations. First, what are the total available budget and manpower resources available to the total assessment effort? Second, what does each expenditure of money and effort specifically buy in terms of useful inputs into decision-making processes in the development of the proposed project?

It is easy to allow individual specialists to go off on a data-collection spree. It is difficult to organize and control data-collection tasks so as to ensure that, upon their completion, a sufficient budget remains to undertake the other analytical and integrative tasks that must be performed in the process of impact assessment.

While there is no one best way of managing an environmental impact assessment, certain attributes of the well-managed assessment may be identified and summarized as follows:

1. Individual tasks (within both analytical and integrative efforts) are identified and described early in the assessment process. The description of each task includes specifically what is to be done, how it is to be done, and by whom. Task description also specifies the product or deliverable

which will be in-hand upon the completion of the task (e.g., a report, a map, a graph, a file index). Finally, a task description should include the expected number of man-hours or days for task completion, and the time frame over which those man-hours are to be expended.

2. Before any specific task is actually undertaken, a summary is made and compiled of man-hours and time frames required for all analytical and integrative tasks. This summary will allow the interdisciplinary team to visualize the overall assessment project, and to make whatever revisions and adjustments are obviously necessary in order to ensure an efficient, comprehensive, and useful impact assessment. Once refined, the summary can be used to compile a schedule for the actual assessment process.

3. The schedule for the overall assessment process will also specify man-hours and time frames for in-house team conferences, internal and external liaison, and any public meetings and hearings.

Task summaries and schedules may be depicted in a number of ways. One example is provided in Fig. 6.2. It is important that some graphic representation of these tasks be made available to the total team so that each individual can easily identify (a) his or her individual responsibilities; and (b) how his or her responsibilities and those of other team members are interrelated and interdependent.

It is true, of course, that impact assessments deal with many unknowns; one may, therefore, question the importance of careful consideration of schedules early in the assessment project.

The answer is obvious to anyone who has had experience with any effort constrained by time and money and by the necessity of providing a product which is directly useful to someone else (i.e., decision-makers). *It is just because we cannot foresee everything that might happen that we have to schedule what we do know will happen—otherwise, we cannot reserve sufficient budget or time or manpower to deal adequately with problems as they arise.* The good manager of an assessment is not an idealist. The manager assumes that things will go wrong; that required data will not be as easily available as someone said they would be; that certain real impacts will be completely overlooked and will require careful consideration; and that differences in personalities, values, and styles among team members and others utilized for assessment purposes can cause serious problems and delays. These things can and will happen, regardless of whether there is a carefully planned schedule of carefully defined tasks. But without that schedule, they are more likely to happen—and to happen when there is a greater probability that time, funds, and other team resources have already been expended or committed.

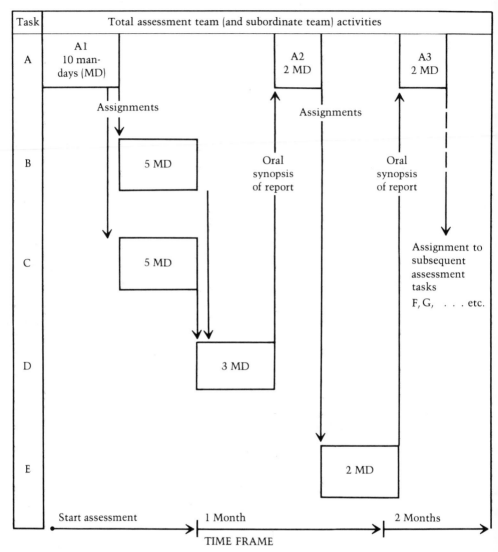

FIGURE 6.2. *Example of assessment task summary. Task A: Total team conference (five members):A1*–Identify appropriate standards, concepts, legislation, etc., and make assignments for initial data collection; *A2,A3*–Evaluate written reports and make subsequent assignments. *Task B: Collection of existing data on area hydrology, geology, soil types, and industrial, residential, and commercial development. Task C: Collection of existing data on area habitats and biological communities. Task D: Integration of information from B and C into a written report describing relationships between various environmental components. Task E: Refinement of report from D, including additional data collection and consideration of other issues identified in A2. Total man days: 29; Total time frame: 2.5 months; Products at end of time frame: (1) Written report on physical system in project area; (2) base-line data file.*

Some Special Issues

I have emphasized that impact assessment is a management intensive effort to organize and coordinate analytical and integrative efforts
that provide meaningful environmental inputs into the decision-making
apparatus of a specific project. In addition to those issues already discussed as being typically important issues in specific analytical, integrative, and management efforts, there are two special issues that are relevant to each of these efforts and which therefore require attention.

REGIONAL DIVERSITY

The social and physical diversity of the United States is so extreme
that it typically suffers the fate of other facts that are generally known but
not directly experienced by people in their day-to-day lives. Like chastity,
people talk about it and then mostly ignore it for all practical purposes.

But to ignore regional diversity in impact assessment is to run the
risk of expending a great deal of effort on an irrelevant exercise.

Projects are not abstracts, not even when they exist only on a draw-
ing board. They are real activities and real structures and real processes
that will have a specific site location.

As a general rule, the more site-specific are our environmental data
and information (social and physical), the more realistic our predictions
of impacts become. Also, the more site-specific our evaluations of the
significance of impacts are, the more useful they become to those
decision-makers who, after all, do not have the responsibility of the entire
nation, but only that for a piece of it—a piece that contains very real
people, politics, and local concerns and resources. Thus, the more local
our approach to assessment, the more efficient our management of the
total assessment process becomes.

To underscore the importance of site specificity in project assessment is certainly not to argue that broad regional, state, and national
interests should be ignored. It is, rather, to argue that a comprehensive
assessment includes local as well as national considerations, and state as
well as regional considerations.

Some EISs appear never to have considered any issue of less than
national or even worldwide scope (e.g., the balance of nature, the importance of agricultural productivity in a world of hunger, the qualitative
diversity of nature, the national and world need for energy). Others appear
never to have considered any issue which has not been voted on in a local
election in the project area over the past several years (e.g., job opportunities, zoning, recreational use of local surface water, school population
pressures). If an impact assessment is truly comprehensive, including due
consideration of national as well as local environmental concerns, it is

important to utilize inputs into the assessment process from both local and broad-scope sources and to utilize people with local experience as well as those with experience in different and diverse regions.

It is my impression (and one that cannot be objectively documented) that too many assessment teams (especially in the first few years of NEPA) ignored local expertise and local experience and relied almost entirely on outside expertise. Of course, outside experts (unless they have extensive experience with assessment in very diverse physical and social surroundings) can easily overlook important local environmental issues and thereby fail to provide decision-makers with meaningful recommendations.

ENVIRONMENTAL STANDARDS

It is necessary, of course, to utilize all pertinent standards and regulations for the purpose of identifying project impacts on air and water quality, noise, environmental concentrations of potentially toxic substances, etc. However, it is also necessary that the assessment team understands that environmental standards are dynamic components of environmental quality. In order to appreciate the dynamism of environmental standards and consequent effects on impact assessment, it is useful to differentiate between environmental parameters, criteria, standards, and guidelines and/or regulations.

An *environmental parameter* is any single characteristic of the total human environment that can be measured. Many of these characteristics (e.g., pH, turbidity, species diversity, property values) can be measured by basically objective methodologies. Other parameters (e.g., aesthetic value, attitudes) have to be measured by obviously subjective methodologies. A total listing of environmental parameters would, in effect, give us a total compilation of everything in our environment that can be identified, measured, and analyzed.

Environmental criteria are, essentially, selected parameters. The selection is based on our current understanding of just what are the most important parameters for understanding environmental phenomena, processes and/or uses. For example, pH (a measurement of acidity) is an important criterion of water quality because of various relationships between pH and the chemistry of water. Different ranges of this criterion determine the suitability of water for various uses, including human consumption, swimming, and wildlife management. A total listing of environmental criteria would, therefore, be a smaller list than that of environmental parameters and would reflect the current state of our understanding of qualitative and quantitative interactions, and our capacity to control or otherwise affect them.

Environmental standards are those environmental criteria that have

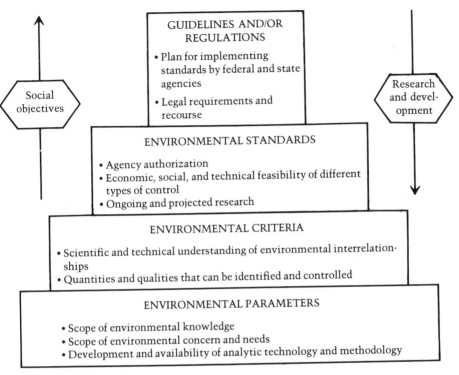

FIGURE 6.3. *The dynamic components of environment quality standards.*

been specified for control within certain limits which are believed to be requisite for the achievement of different social objectives. Standard setting is essentially the process of identifying which criteria shall be legally controlled by an authorized government agency and deciding what limitations are to be imposed for each criterion for the purpose of meeting desired social objectives.

Environmental guidelines and/or environmental regulations describe the overall plan for control, including enforcement procedures, methodologies and techniques and schedules for monitoring, and punitive and/or corrective actions to be taken whenever promulgated standards are exceeded.

As summarized in Fig. 6.3, these components of environmental quality are dynamic linkages between social objectives and scientific research and development. As social objectives and our scientific and technical understanding of our physical, social and total environment change, so will promulgated standards and regulations. In fact, one can reasonably expect individual standards to change within a matter of only a few years.

Because of the inherent dynamism of environmental standards, the assessment team should consider that *compliance with currently pro-*

mulgated standards is minimal compliance with the long-term objectives of NEPA. If this attitude is not a controlling attitude during impact assessment, project compliance with existing standards will only serve to enhance complacency within the assessment team. This, in turn, may well cause the team to overlook or purposely ignore impacts which, regardless of current standards, can actually occur and significantly affect the human environment.

References

Council on Environmental Quality (CEQ)
   1976   *Environmental Quality 1976: the Seventh Annual Report of the Council on Environmental Quality.* Washington, D.C.: U.S. Government Printing Office.
Odum, E. P.
   1977   The emergence of ecology as a new integrative discipline. *Science, 195*(4284): 1289–1293.

7

Introduction to Assessment of Impacts: Physical Environment

Introduction

Six basic tasks of impact assessment, including analytical and integrative tasks, were identified and discussed in Chapter 6. The following chapters will focus on the practical problem of completing these tasks—first, with respect to the physical environment; second, with respect to the social environment; and third, with respect to the total human environment.

It is worthwhile to emphasize the phrase "practical problem." This is not to demean the importance and relevance of general theories or even of grand philosophies, but rather, to focus our attention on the on-the-job difficulties of translating general environmental considerations into specific recommendations for action and decision-making as mandated by NEPA. For example, one may intuitively grasp the sense of the statement that "all things are connected to one another." But, in practical terms, we know that all things are not connected equally—that everything in the world does not change because one part of it does. Just as I can reasonably expect that my stomachache will not likely affect the blossoming of violets, so I can reasonably expect that some components of the environment can change or be altered with no measurable effect on some other components. In both instances I can accept the doctrine of holism—that my body is a whole, and that the environment (which includes my stomachache) is a whole. But practical experience tells me to discern the degree of relationships or connectedness between the individual parts of any whole.

Practical experience also tells me that I can easily focus on irrelevant issues and fail to recognize some important ones, whether they be issues pertaining to my individual health or issues pertaining to environmental quality. Certainly neither the well-being of our bodies or of our total environment is a direct function either of the logic or of the knowledge of any one individual. We have to assume that, individually, we do not perceive how all the various individual components of the environment are, in fact, interconnected. Therefore, we require as much guidance from others as we can get.

These two perspectives—namely, that environmental components are not interconnected equally, and that no one individual has sufficient knowledge to identify precisely all the interconnections—are not presented here as idle reflections; rather, it is suggested that they are necessary practical tools for undertaking any impact assessment. *The first perspective requires the analyst to be selective in what he or she analyzes and how he or she does it. The second requires the analyst to conduct the assessment in such a way as to maximize informational input from a large diversity of practicing professionals.*

In many assessments, it would appear that those with assessment responsibility have expended entire budgets of money, time, and manpower collecting bits and pieces of data and information on almost every conceivable aspect of the project area—with the result that no one knew, finally, how to put it all together in order to make realistic recommendations (or, if they did know how, they did not have the money or time left to do it!).

Assessments have also been conducted with the apparent conviction that those with impact responsibility had "total capability" (a favorite phrase!) and thus had little need or inclination to solicit opinions, information and suggestions from government agencies, academic scientists or other professionals, and the public as a whole. Assessments conducted in such a manner are not in keeping with the overall intent of NEPA and should not be tolerated.

While selectivity and openness of approach cannot, of course, guarantee the value of any impact assessment, the absence of these qualities is probably a good indication that the assessment is a pro-forma exercise. Those who have assessment responsibility should, therefore, consider just how these qualities can be actually utilized to conduct and improve impact assessment as a decision-making tool.

Selection of Environmental Components

Much emphasis was given in previous chapters to the fact that impact assessment under NEPA ultimately requires the assessment of impacts on systems. It was also noted that systems are ultimately defined by components *and* interactions between components. How does one identify those components (other than by legal counsel) that are most useful, therefore, for assessment purposes? A first impression might be that such a question is fatuous, yet that is by no means the case. For example, Weinberg's (1975) work on general systems thinking identifies a basic problem in the analysis of any system: "[Other authors] . . . rightly emphasize 'relationships' as an essential part of the system concept, but fail to give the slightest hint that the system itself is relative to the viewpoint of some observer. The idea of set [for example] is very common in mathematics, but contrary to the impression of precision it gives, it is one of the 'undefined primitives' in most theories. The mathematics of sets . . . tells us much about the properties of sets, *but tells us nothing about how observers might choose them* [p. 63 emphasis added]."

This is, of course, the same problem that has been identified by various authors, including the more poetic:

> *A waggon passed with scarlet sheels*
> *And a yellow body, shining new.*
> *"Splendid!" said I. "How fine it feels*
> *To be alive, when beauty peels*
> *The grimy husk from life". And you*

> *Said, "Splendid!" and I thought you'd seen*
> *That waggon blazing down the street;*
> *But I looked and saw that your gaze had been*
> *On a child that was kicking an obscene*
> *Brown ordure with his feet.*

> *Our souls are elephants, thought I,*
> *Remote behind a prisoning grill,*
> *With trunks thrust out to peer and pry*
> *And pounce upon reality;*
> *And each at his own sweet will*

> *Seizes the bun that he likes best*
> *and passes over all the rest.*
> "The Two Realities"
> —Aldous Huxley

That any two individuals who look at a project area will focus on different components of that environment, or even focus differently on the same component cannot be questioned by anyone who has actually conducted an impact assessment. In fact, the recognition of a qualitative infinity to how the environment (both physical and social) is organized has long been recognized.

For example, Schultz (1969, pp. 77–93) refers to *homomorphic models* of ecosystems as those models that reflect various levels of discrimination of system parts. One homomorphic model of a physical system may focus on individual atoms, molecules, cells, etc. Another homomorphic model of the same physical system may focus on various combinations of these components, such as individual organisms. Still another homomorphic model of the same physical system may focus on higher-order combinations, such as whole species or genera. *Each combination is a simplification of the real, total system. Each reflects the discrimination or the interest of the observer (Jacob, 1977). And each is only as useful as it serves to organize our information and increase our understanding of the whole system.*

A recognition of the diversity of possible homomorphic models of any given project area is a prerequisite of impact assessment. For example, if the project area includes a surface-water resource, just what compartments of that resource should be investigated? A respect of the diversity of possible homomorphic models underscores the necessity of discriminating from among numerous possibilities. Is a homomorphic model to focus on individual chemical species found in the water? If so, there are literally hundreds. Is another model to focus on larger categories (e.g., combinations of certain chemical species) as, for instance, nutrients and toxins? This might be necessary, depending on pertinent legislation and other considerations. But what of even higher-order homomorphic models, including a model built on food chains and food webs? Such a model is important for identifying information and data required for various biological and ecological analyses of project impacts.

The importance of the concept of a homomorphic model is not that any one such model will be sufficient for purposes of assessment, but that the various possible models reflect how existing data and information are organized, what general kinds of information will be required for purposes of analysis, and possible sources of that information. In the example of a surface-water resource, a recognition of homomorphic models of individual chemical species is a recognition of the need for information typically utilized by an aquatic chemist. However, homomorphic models of nutrients and toxins are typically not the province of chemists, who are interested in the overall chemistry of water, but of biochemists, toxicologists, and plant and animal physiologists. Similarly, a homomorphic model of that same resource, but one based on food webs, indicates the need to deal with the literature and in the terminology of aquatic ecology.

One way of looking at the practical usefulness of homomorphic models is to compare the informational needs of impact assessment with an individual's need when he or she goes to a vast library. What would be the probability of that individual's finding just what he or she needed by wandering randomly through the library stacks? On the other hand, what would be the probability of finding what is needed by utilizing a library file index? Homomorphic models are just that—they are the file index to existing information and understanding.

Of course, another important function of homomorphic models is that they allow the assessment team to obtain an overview of the total analytical effort. If it is obvious that the total budget of time and money is being expended on only one homomorphic model of a resource (say, the individual chemical species in surface water) it is obvious that no money or time will be left over for considering project impacts at the level of organism or community. Such an assessment could hardly be described as comprehensive under the mandate of NEPA. Thus, homomorphic models are not only useful for directing information gathering activities, but also for managing the overall assessment (Chapter 6) so as to conform to the requirements of NEPA.

Maximizing Inputs

Various homomorphic models for different types of environmental resources are readily available in the general literature and will be discussed in following chapters. Most (if not all) such models are easily understood and utilized by the layman in the respective disciplines. This is because homomorphic models are essentially lists of components; and because we typically have to deal with such lists in our day-to-day lives, whether they be lists of the components of our economic status (as defined by the IRS), of our physical health (as defined by our physician), or of marital harmony or disharmony (as defined by our spouse).

A more difficult type of model for the person who has assessment responsibility is what has been called an *isomorphic model* (Schultz, 1969, p. 82). An isomorphic model of an environmental resource is essentially a map of how various components (defined by different homomorphic models) are interrelated—that is, how a change in one component can instigate a change in another, the conditions that must be met for the change to occur, and the rate at which it occurs. In short, *an isomorphic model depicts the dynamics of the system.*

While it is true that our overall understanding of total environmental dynamics is essentially infantile, it is also true that we do know a very great deal about a number of things pertaining to these dynamics. In fact, it has perhaps been a disservice to practicing professionals in a large variety of disciplines to emphasize during recent years how little is known and how much we have to learn about the environment. To the layman, it must appear that scientists and engineers and other professionals operate solely on luck. If such an attitude is allowed to persist in the minds of those who undertake assessment responsibility, the result will be that such people will quickly make fools of themselves. Simply, if one is not a professional biologist, it is best for one to allow the biologist to bemoan biological ignorance with respect to isomorphic models involving biology. Similarly with the climatologist, the physiologist, the limnologist, and soil microbiologist, the hydrologist, and all the other specialists who collectively make up our current bank of technical and scientific knowledge and understanding of the dynamic processes which take place in the environment. It is generally a good rule of thumb that *the person with assessment responsibility should assume that isomorphic models of different resources in the project area do exist and that, regardless of his or her own conception of their value or usefulness, he or she should consult those individuals who are technically and scientifically qualified to discuss their relevance and/or irrelevance to the assessment of impacts in the proposed project area.*

This rule of thumb is not offered casually. It is offered out of respect of actual assessments and because of the excesses of uninformed opinion on the part of persons with impact assessment responsibility. For example, an isomorphic model of an aquatic resource may identify possible relationships among (a) degree of shading; (b) surface area of impoundment; (c) amount of aquatic photosynthesis; and (d) food chains depending on living aquatic vegetation. Such a model may indicate that shading due to bridge superstructure may reduce rates of photosynthesis of aquatic vegetation and, subsequently, the amount of vegetation that can be used as food by various aquatic organisms. The idea that shading water with a bridge can have serious impact on downstream fisheries may seem (and it has seemed) ridiculous to those who do not have the technical or scientific training required to evaluate its pertinence in a particular situation. Such an individual may therefore ignore the whole issue as "foolishness." Such a person may also find the entire project threatened with court action and himself explaining to technical people why he thought the phenomenon not worth considering.

Similarly, unless one has specific training, the potential effects of security lighting on the environment may appear again to be one of those imaginary crises invented by crazy environmentalists. Yet, it is well documented that certain isomorphic models can relate intensity and type of security lighting to vegetative health and, therefore, have direct relevance to particular types of projects (U.S. Department of Agriculture, 1973).

Note on Assessment Methodologies

Assessment methodologies are essentially rules or procedures that identify particular homomorphic and isomorphic models and use these models for predicting and evaluating impacts. Five basic categories of methodologies have been reviewed by Warner and Preston (1974). These include:

1. Ad hoc methodologies, which typically identify only broad areas of possible impact. Such areas include large biotic and abiotic environmental components (e.g., impacts on forests, wildlife, lakes)

2. Overlay methodologies, which result in overlays of individual maps of the social and physical characteristics of the project area (e.g., soil types, community patterns, geological features, area hydrology)

3. Checklist methodologies, which itemize environmental parameters that should be investigated for possible impacts

4. Matrix methodologies, which correlate cause–effect relationships between particular project activities and impacts

5. Network methodologies, which define a network of possible impacts that may be triggered by project activities and which require the analyst to trace out appropriate project actions and direct and indirect consequences.

According to Warner and Preston (1974) "in only a few cases are [these methodologies] full-blown methodologies developed specifically for impact statement preparation. More commonly, they are more limited ideas borrowed from other fields with potential application to NEPA environmental assessments [p. 1]." They also emphasize that "there is no single 'best' methodology for environmental impact assessment . . . only the user can determine which tools may best fit a specific task [p. 1]."

In addition to the 17 methodologies included within these 5 categories, several dozen other methodologies (some of which can also be included in one or more of these categories) have been described and utilized either for specific project assessment or for assessments of types of projects (Canter, 1977, pp. 302–310). And perhaps this is only a small representation of all the methodologies that have been actually designed, examined, and discarded, or are currently being employed. Authors of these assessment methodologies include government agencies (federal and state), corporations, research institutes, academic researchers, and other groups and organizations having regulatory, educational, monetary, and/or other interest in environmental quality and assessment. Thus the assessment team for a specific project is not only faced with a large number of possible homomorphic models of the environment and a large number of possible isomorphic models of the environment, but it is also faced with an already large and evidently increasing number of methodologies for selecting and utilizing these models!

The situation in which such a team can actually find itself is depicted in Fig. 7.1. In this figure, (from a UNESCO document on aquatic weeds) one biotic component (i.e., an aquatic weed) is depicted in relationship to other biotic and abiotic components of the environment. The total number of these biotic and abiotic components, and any combination of them, represents the possible homomorphic models of a particular physical situation. Lines drawn between each component represent isomorphic models (or relationships between components). Thus the task of the assessment team is to select the appropriate components and combinations of components (i.e., homomorphic models) and the appropriate relationships (i.e., isomorphic models) for the specific project activities it is examining for impact. Add to this problem one other—there are dozens of different rules currently available (i.e., assessment methodologies) for selecting appropriate components and relationships.

The extreme diversity and plurality of homomorphic and isomorphic models and of assessment methodologies currently available may force an assessment team into one or more of several directions, including:

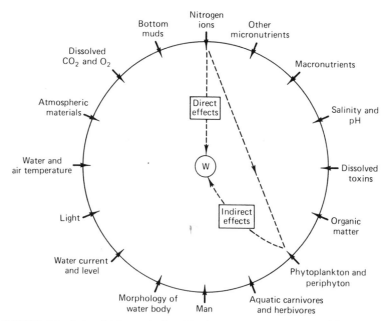

FIGURE 7.1. *Potential interrelationships between an aquatic weed (W) and environmental components.* [*Adapted from Mitchell, D.S. (Ed.) 1974.* Aquatic Vegetation and Its Use and Control. *Paris: UNESCO.*]

1. Ignoring the real complexity of their task, and doing as little as possible and/or doing only what lawyers specifically require
2. Using the real complexity of their task as an excuse for (a) collecting vast amounts of data; and (b) never reaching any conclusions or making any recommendations based on the data
3. Spending time and effort to simplify and organize the task so as to achieve meaningful analyses and recommendations

The first two are perfectly understandable reactions to what must appear as an impossible task. They are also perfectly nonprofessional and cannot be tolerated by any responsible agency or individual. The third approach is the only acceptable approach. But it requires going back to some basics.

Back to Some Basics

Any one or combination of individual components of the physical environment participates in and/or contributes to or otherwise influences dynamic processes. While these processes are identified and technically described within various disciplinary frameworks or paradigms, many may also be categorized and identified in nontechnical terms by their causation of and/or dependence on (a) the movement of materials and energies through the environment; and (b) the alteration of these materials and energies in the progress of their movement throughout the environment.

The movement and alteration of materials and energies through the physical environment are influenced by social factors as well as by biotic (living) and abiotic (nonliving) factors. People as well as fish introduce nutrients into water. People as well as clouds can change the amount of solar energy impinging an area of land. People as well as other predators can influence the rate at which browse is converted to deer flesh. And people as well as winds and rains can affect the rate of erosion in particular areas.

Whether they are social factors or physical factors that affect the movement and alteration of materials and energies in the physical environment, they exert their effects in at least five basic ways by altering:

1. The *introduction* of materials and energies into environmental compartments
2. The *transformation* of materials and energies within environmental compartments
3. The *translocation* of materials and energies from compartment to compartment

4. The *sequestration* or trapping of materials and energies within specific compartments for various periods of time
5. How materials and energies are *dissipated* or lost from one or more environmental compartments (and then introduced into others)

The alteration of material and energetic introductions into environmental compartments may be manifested by changes in (a) types or kind of materials and energies; (b) their concentration or amount; (c) the timing and duration of their introduction; and (d) the place of their introduction.

The alteration of how materials and energies are transformed within environmental compartments include alterations of (a) rates of transformation; (b) agents of transformation (i.e., biotic and abiotic transformers): and (d) the location of transformation processes within the environment.

Alterations of the translocation of environmental materials and energies are changes in how these materials and energies are moved from one place to another without undergoing physical, chemical, or biological modification. Such translocative alterations include changes in rates of movement, directions of movement, and periodicity of movement.

Sequestration of environmental materials and energies may involve numerous physical, chemical, and biological factors. Regardless of the specific mechanisms by which materials and energies are trapped or sequestered in environmental compartments, project-mediated changes in ongoing environmental sequestrations may be described in terms of changes in (a) rates; (b) duration; (c) location; and in (d) agents accomplishing the sequestration.

Finally, materials and energies are dissipated or lost from one or more environmental compartments by numerous mechanisms, including losses through evaporation, physical conveyance out of a compartment (as in the flushing of sediments out of an impoundment), and the migration of organisms out of a particular region. Project-mediated changes in dissipative processes may, therefore, be described in terms of changes in rate, timing, and agent.

The five categories of dynamic processes in the physical environment, as well as those factors identified as measures of change in these dynamic processes, are not offered as absolute, fully comprehensive categories that define the entire complexity of the physical environment. Rather, it is suggested that they can be useful to the assessment team for organizing and conducting the six tasks of impact assessment discussed in Chapter 6. It is vital to the accomplishment of these tasks that the assessment team proceed selectively and in such a manner as to maximize professional and other input into both the analytical and integrative phases of assessment. The categories of physical environmental dynamics and their descriptors are particularly useful to persons having assessment responsibility but limited (if any) professional experience in relevant sciences. *Their utility lies in the various ways that they can guide the overall efforts of assessment (i.e., promote selectivity) and facilitate (i.e., by maximizing meaningful input) both the collection and analysis of appropriate data:*

1. *It does not take an expert* on soils in a particular region *to ask* such questions as: What are the materials and energies that ordinarily get into soils, and how do they do it? How are these materials and energies transformed within soils? How are they (and their products) translocated from one layer or place to another? How can they be sequestered? And how can they be dissipated? *Nor does it take an expert in limnology to ask precisely the same questions about a lake* in a project area; *nor an expert in wildlife management to ask precisely the same questions about a wildlife community* in the region surrounding a proposed project area. Of course, to give answers to these questions—answers that will utilize different jargon and different concepts, according to the discipline—does require expertise. Thus, the assessment team can ask the questions, but unless the team includes a broad spectrum of in-depth environmental expertise it will have to seek out the answers from the scientific literature and community, organizations and agencies, and from knowledgeable individuals.

2. Once the assessment team has begun to collect answers to these questions, the team is in a position to identify important components and relationships (as well as factors that typically influence those relationships) in the project area. Attention can then be turned to identifying the various mechanisms by which project activities can influence these systems. Again, we begin with questions such as:

-   How can construction alter ongoing material and energetic introductions (i.e., the types of materials and energies being inputted, their concentrations)?
-   Will operational phases alter typical transformations (i.e., rates, agents, location)?
-   What design features might alter ongoing translocations in the area (i.e., rates, periodicity), and can such alterations affect environmental systems outside of the project area?
-   What maintenance and construction activities can alter ongoing sequestration (i.e., rates, agents, duration) in the soils and in the water in the project area?
-   How do project structures and other design features facilitate or

retard the dissipation (i.e., rates, agents, timing) of different material and energetic inputs into the present system?

3. In the progress of asking questions with respect to these basic categories of environmental dynamics, the assessment team should quickly come to learn the basic jargon used in individual disciplines to identify and discuss relevant processes. For example, transformations of materials and energies (and respective disciplinary jargon) include *primary productivity* (ecology), *oxidation and reduction* (chemistry), and *metabolism* (biology). Examples of translocation of materials include *infiltration* (hydrology, soil science), *precipitation* (climatology), *sedimentation* and *aggradation* (hydrodynamics), and *diurnal* and *seasonal migration* (wildlife biology). Once the team learns specific disciplinary applications of basic dynamic phenomena, it is in the position to increase the precision of its questions about environmental processes and to identify specific factors (social and physical) that play determinant roles in the specific environmental manifestations of these processes. These factors can then be examined for their direct and indirect relationships with proposed project activities and final design.

Conclusion

In each of the following chapters on the physical environment, impact assessment will be primarily discussed within the framework of a single scheme. This scheme (Fig. 7.2) is not to be viewed as a precise methodology for identifying impacts or evaluating them, but as a general procedure for organizing the efforts of the assessment team so that any one or combination of the existing methodologies may be effectively and efficiently utilized. This scheme is, therefore, not a substitute for these other methodologies; it complements them. Furthermore, this scheme (and the use of it in following chapters) is based on certain assumptions about the individuals who might use it, including that:

1. They do not have in-depth disciplinary knowledge in all of those areas that they will have to consider
2. They are working within the constraints of finite time and budget
3. They approach impact assessment as a positive contribution to decision-making, and not as a legal obstacle to project development
4. They are not afraid to ask what might appear to be stupid questions or discouraged when they get apparently stupid answers.

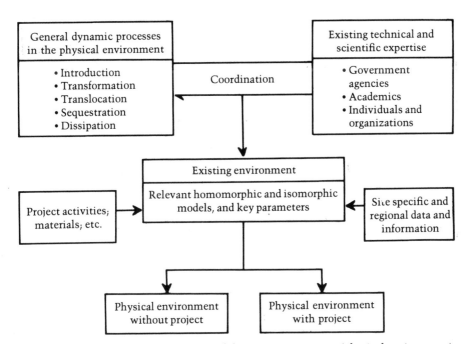

FIGURE 7.2. *Important components of the assessment process (physical environment).*

As depicted in Fig. 7.2, the initial step for the assessment team is to "discover" how general dynamic processes in the physical environment can be translated into specific technical and scientific jargon. The team accomplishes this by asking questions—questions directed to people (in agencies, universities, etc.) and to professional literature. Of course, the general dynamic processes described do not take place in the abstract. They take place in water (hydrosphere), in the air (atmosphere), in the earth's crust (lithosphere), and in living populations (biosphere). Thus the team must direct questions about general dynamic processes to those individuals and to that literature dealing specifically with these basic compartments of the physical environment and with their interactions.

As the team begins to learn how the limnologist, the chemist, the geologist, the biologist, the ecologist, and all the other specialists translate general dynamic processes into specific environmental processes, the team should also ask questions about the key factors in these processes. For example: What is it that affects the movement of water through soils? How does heat affect metabolism? Is it possible for sound to affect the movement of organisms? What factors influence chemical transformations in water?

As a result of first its general and then its more specific questions, the assessment team should move quickly to construct a general overview of the existing physical system of the project area and its environs. This overview will include the components (homomorphic models) and inter-relations (isomorphic models) identified by the respective specialists (and/or literature), as well as key factors that typically influence these components and interrelationships. It is generally desirable to double-check the technical and scientific reasonableness of such an overview before proceeding to the next step.

The next step is to begin to "flesh out" the general overview of the project area and its environs with site-specific and regional data and information. This step, in effect, should transform the theoretical over-view into a real physical environment. Key parameters, which were identified by homomorphic and isomorphic models, can now be mea-sured or otherwise valued in order to gain an understanding of just what the system is doing and how. Parameters that do not contribute to such an understanding and that are not otherwise required by specific en-vironmental regulations should not be measured.

Of course, as soon as the general overview begins to gain substance with facts, the assessment team can begin to evaluate both direct and indirect effects of project activities on key components and interrelation-ships, and to evaluate those impacts in light of NEPA requirements as discussed in previous chapters. As discussed in this chapter, there is a variety of methodologies that may be employed. However, the minimal objective in any assessment methodology should be to define clearly and precisely just what the physical environment would be like without the project and what it would be like with the project.

Before proceeding with the application of this approach to various aspects of the physical environment, it is necessary to reiterate two points.

First, there is no real dividing line between the physical and social environments in the real world. Yet, there are effective dividing lines (if only communicative) in our compartmentalized information on and understanding of the real world. In the following chapters, which discuss impacts on the physical environment, we will, therefore, withstand the urge to connect everything absolutely and immediately with everything else and concentrate largely on rather compartmentalized relationships. In later chapters on the social and the total human environments, we will focus more specifically on interrelationships between the physical and social environments.

Second, impact assessment (even when dealing with the physical environment alone) must often proceed on the basis of judgment—as opposed to quantitative certainty. Attention will, therefore, be given in the following chapters to the importance of qualitative as well as quan-titative considerations. Where possible and where relevant to the con-straints of time, budget, and other factors that must influence impact assessment, objective methodologies for predicting or evaluating impacts will be discussed or pointed out. But where such methodologies are primarily of academic interest, attention will be given instead to practical rules of thumb and to even less scientifically proper techniques.

References

Canter, L. W.
   1977 *Environmental impact assessment.* New York: McGraw-Hill.
Jacob, F.
   1977 Evolution and tinkering. *Science, 196* (4295): 1161–1166.
Schultz, A.
   1969 A study of an ecosystem: The Arctic tundra. In *The ecosystem concept in natural resource management,* edited by G. M. Van Dyne. New York: Academic Press
U.S. Department of Agriculture
   1973 *Security lighting and its impact on the landscape.* Beltsville, Maryland: Agricultural Research Service, Plant Genetics and Germplasm Institute, Agricultural Environmental Quality Institute. (Publication CA-NE-7)
Warner, M. L., and E. H. Preston
   1974 *A review of environmental impact assessment methodologies.* Washington, D.C.: U.S. Environmental Protection Agency. (Publication No. EPA-600/5-74-002)
Weinberg, G. M.
   1975 *An introduction to general systems thinking.* A Wiley–Interscience Publication, New York: John Wiley & Sons.

8

Abiotic and Biotic Impacts

Introduction

Abiotic impacts[1] of project development are those environmental consequences of project construction, operation, and maintenance that may be measured, evaluated, or otherwise defined in terms of physical and chemical parameters. Such effects as increasing turbidity in water, altering groundwater flow, changing cool-air flow patterns in a local area, and altering chemical species in surface soils, are examples of potential abiotic effects of various project activities. Abiotic impacts may have biological consequences (e.g., turbid particles may settle out to coat fish gills and cause suffocation); they may also be influenced or even caused by biological agents and processes (e.g., the turbid particles may be algae or other microorganisms). But the abiotic impact itself, regardless of any biological causes and/or consequences it may have, is strictly describable in terms of physical and chemical parameters.

Conversely, biotic impacts are those environmental consequences of project development that may be measured, evaluated, or otherwise defined in terms of the analytical and conceptual categories of the profes-

---

[1] The word "impact" is not used in the same quasi-legal sense of "significant negative effect of project activity" as it is in various governmental agencies. "Impact" is used synonymously with "effect" and/or "consequence."

sional biologist. Such effects of project activity as disrupting animal reproduction, increasing the number of organisms, and changing the amount (or productivity) of vegetation in the project area are examples of potential biotic effects of various project activities. Biotic impacts of project development may have abiotic consequences (e.g., the growth of vegetation in a previously nonvegetated area can cause drastic changes in relative humidity close to the ground); they may also be influenced or even caused by preceding abiotic agents and processes (e.g., construction changes ambient noise characteristics in the project area, and these changes can, in turn, affect reproduction of some animals). But the biotic impact itself, regardless of any abiotic causes and/or consequences it may have, can be described in terms of biological categories.

The importance to be given to the distinction between abiotic and biotic impacts is not related to any absolute division in the real world between living and nonliving components of the environment. *The distinction is only important insofar as it facilitates (a) the gathering of relevant assessment data and information; (b) the assessment team's efforts to contact the "right" people with the "right" questions; and (c) the effort to define as precisely as possible cause-and-effect relationships.*

For example, one is not likely to find usable data and information on the sedimentation of turbid particles in a biology text; nor is one likely to find usable data and information on the propensity for gill lesions in certain fish species in a text on soil mechanics or hydrodynamics. It is necessary, in this particular case, to deal with the sedimentation of turbid particles as an abiotic impact and, by going to the literature and/or to professionals who deal with this abiotic phenomenon, identify (*a*) the general dynamic processes involved (Chapter 7); and (*b*) those parameters and factors that professionals (i.e., soil scientist, hydraulic engineer) use to discuss the phenomenon. Once this has been done, and once appropriate measurements and calculations have been made with respect to expected turbidity and sedimentation patterns in the project area, then (and only then) would one go to the fisheries biologist. For it is only then that one can give the biologist what he or she needs to know (e.g., size of turbid particles, abrasiveness, time of release, estimates of total loading of stream, location of downstream fallout) in order to make a judgment of the probable effect of turbidity on fisheries.

While differentiating between abiotic and biotic impacts and phenomena is stressed as an important first step in information gathering, it is often a difficult one. In fact, there are two basic problems in organizing the physical world into abiotic and biotic categories. First, the real-world coupling between the abiotic and the biotic is extensive, and numerous disciplines and much professional literature reflect this real-world coupling. Second, the terms *abiotic* and *biotic* are extremely general descriptors and do not themselves identify, or relate one-to-one to the many specialties and subspecialties that may have to be considered in the progress of impact assessment.

Abiotic and Biotic Coupling

One of the more comprehensive examples of abiotic and biotic coupling is the model of *biogeochemical cycling* that includes the total dynamic relationships between the lithosphere (surface layers of earth), the atmosphere (air), the hydrosphere (surface and groundwater), and the biosphere (the total life on earth). As depicted in Fig. 8.1a (Deevey, 1972), each of these components is interconnected. Thus, the general dynamic processes (Chapter 7) of material and energetic introduction, transformation, translocation, sequestration, and dissipation are also interconnected. Some of these processes are the objects of study in a number of different disciplines. For example, *diagenesis* (i.e., the transformation of organic and inorganic sediments) is of interest to the geochemist (Mason, 1958, pp. 150–154) and to the limnologist and the aquatic biologist (Pamatmat *et al.*, 1973; Rheinheimer, 1974, pp. 144–145). An evaluation of potential impacts of a specific project on the diagenesis of lake sediments (and the consequences of these impacts to other components of the environment) may, therefore, require the assessment team to deal with the different disciplines that have something to say about the same process. In one situation, it may be a specialist in the biological disciplines who has the most useful information; in another, it may be a specialist who typically deals with the abiotic environment. The assessment team should not restrict information gathering by its own conception of where the line between abiotic and biotic processes should be drawn. *Once a dynamic process has been identified within the context of one specialty, the team should inquire as to how other specialties also focus on or otherwise consider that process.*

Another basic example of abiotic and biotic coupling is the concept of *habitat*. There is much argument as to whether this concept rightly belongs to biology or to other disciplines and/or subdisciplines (e.g., ecology, microclimatology, soil science, limnology). The practical problem for assessment purposes is: "Where do I go, and what specialist do I talk to in order to get good habitat information most rapidly?" The root of this problem is the fact that the term "habitat" has been utilized by numerous persons in pursuit of divergent interests. The most generally used meaning of the term is simply the place in which an organism lives (Odum, 1971, p. 234). If one gives emphasis to the idea of "place," one may tend to give emphasis to the physical and chemical parameters of a geographical location (temperature, relative humidity, surface texture, color). Thus the concept would appear to refer to an abiotic quality of a particular piece of

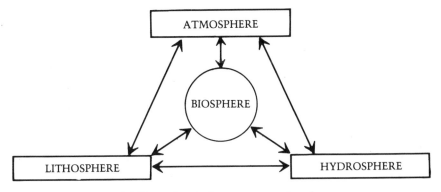

(a) Biogeochemical components of earth

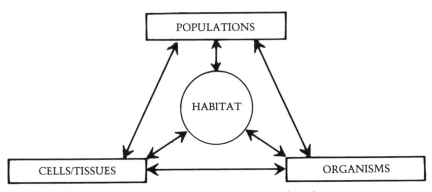

(b) Abiotic and biotic components of the biosphere

FIGURE 8.1. *Abiotic and biotic interrelationships.*

the environment. Conversely, if emphasis is given to the idea of "organism," the concept might appear to be essentially a biological concept.

As depicted in Fig. 8.1b, the idea of habitat as a "place" with physical and chemical characteristics is related to various biological components of the environment in a way that is similar to the relationship between the biosphere and the basic abiotic components of earth's environment.

In short, and for all practical considerations, you cannot have the physics and chemistry of land, water, and air without having some biology, and you cannot have a biological system (cells, tissues, organs, organisms) without having some type of physical and chemical habitat in which that system actually exists.

Where, then, are the best sources of information on habitat for a given project assessment? There is no general rule. In one situation, it may be surficial geologists, soil scientists, and hydraulic engineers. In another situation, it may be the fisheries biologist, the wildlife biologist, the herpetologist, or the ornithologist. *The assessment team should, therefore, not assume that it can safely consult only a single discipline or type of discipline for information on habitat.*

Abiotic and Biotic Compartmentalization

As has already been pointed out, the geochemist deals not only with chemical and physical phenomena, he also deals with biological processes that figure in those phenomena. Similarly, the wildlife biologist deals not only with biological phenomena, he also deals with physical and chemical processes that figure in those phenomena. It is, of course, a question of emphasis. *The best source of data on dynamic abiotic processes is the one that gives emphasis to physical and chemical processes; the best source of data on dynamic biological processes is the one that gives emphasis to biological processes.*

Not only is it a question of emphasis between these two broad categories of disciplines, it is also a question of the emphasis that individual disciplines and subdisciplines give to particular concepts within each category. For example, what is the difference between biology and ecology? If I (as a member of an assessment team) need information that is clearly biological (as opposed to being abiological), do I go to an ecologist? What is an ecological impact anyway?

Historically, ecology is a subdiscipline of biology. But its definition and scope have varied (and still do) from one professional to another. Ecology has been described by professionals as "scientific natural history [Elton, 1968, p. 11]," as the "study of organisms in relation to their environment [Oosting, 1956, p. 3]," "the scientific study of the interactions that determine the distribution and abundance of organisms [Krebs, 1972, p. 4]," "the study of interrelationships between organisms and their surroundings [Ricklefs, 1973, p. 11]," and as "environmental biology [Odum, 1971, p. 3]." Of course, in contemporary lay terms, "doing ecology" seems to include everything from "picking up trash and litter along a stream bank" to "putting bricks in the toilet." Given such a plurality of definitions, then, to whom might the assessment team reasonably expect to go in order to collect ecological information on a project area? To wildlife biologists? To agriculturalists and horticulturists? To game wardens? To responsible community leaders interested in the environment? In fact, each of these sources has been utilized at one time or another as a prime source of (so called) ecological data and information for the purposes of impact assessment. Ecology, as a formal discipline, may be over 100 years old, but people are easily (and understandably) confused as to just what it is.

For purposes of succeeding chapters, we will use the term *"ecology"* to denote that subdiscipline of biology that gives primary emphasis to the concepts of (a) population, that is, a collection of organisms of the same species or type; and (b) community, that is, a collection of different and interacting populations (Odum, 1971). Ecological impacts will be considered, therefore, to be impacts on those components and dynamic processes that the ecologist uses to talk about phenomena related to populations and communities. By way of contrast, biological impacts will be considered to be impacts on the components and processes that biologists use to describe the structure and functioning of cells, tissues, organs, and individual organisms.

Examples of Dynamic Abiotic and Biotic Processes

Although the specific site of the proposed project must determine which abiotic and biotic processes are particularly relevant for purposes of a given assessment (Chapter 6), some general processes that involve material and energetic introductions, transformations, translocations, sequestrations, and dissipations in soil, water, air, and in organisms are typically important and might be considered. Each of these may be affected by project-related activities during construction, maintenance, and operational phases.

Examples of typical abiotic and biotic mechanisms that accomplish the introduction of materials into soils include: the deposition of organic materials, such as animal wastes, detritus, and other organic debris accumulated during biological succession, precipitation of inorganic and organic materials from the atmosphere, weathering of previously covered or sealed substrates, settling or dustfall of wind-dispersed fine materials, and the impingement of photic and thermal energies from the sun. Some of these same processes are involved in the introduction of materials into water (e.g., deposition, precipitation). However, surface waters also receive inputs through the suspension and/or dissolution of contaminating materials in tributary waters, and from the overturn (or destratification) of waters that results in the introduction of nutrients from bottom muds into surface levels. In estuaries, tides are important inputs of energy and accomplish the work of fertilization (an introduction of materials) as well as waste removal (a dissipation of materials). The atmosphere receives water through the process of evapotranspiration, which is the sum total of water lost (dissipated) from the soil through evaporation and transpiration (the movement and loss of water through plants), as well as from the convective and advective transport in the atmosphere of both natural and pollutant materials. The introduction of materials and energies into organisms include such mechanisms as feeding (herbivory, carnivory, detritivory, omnivory); respiratory mechanisms; osmosis; and auditory, visual, and other sensorial stimulation.

Once materials have been introduced into soils, waters, air, and organisms in the project area, they are variously transformed, translocated, sequestered and/or dissipated, depending on a variety of individual and interrelated abiotic and biotic factors.

For example, detritus that is deposited on surface soils by overhanging vegetation may be transformed or altered by biological decomposition, chemical oxidation, disintegration, compaction, and leaching. Leaching (of dissolved materials) and eluviation (of suspended materials), and subsequent percolation of surface water, which contains both dissolved and suspended materials, also accomplish the translocation of materials through different soil horizons and, from there, the introduction of such materials into groundwater. Translocation of surface deposited materials may also be accomplished by erosion.

Sequestration (or entrapment) of materials by soils may be accomplished through adsorption onto particle surfaces, absorption into materials, chemical chelation, and other mechanisms. Chemical chelation also occurs in water, as well as physical sequestration through sedimentation. Materials may be sequestered in individual organisms by the process of biomagnification, and by metabolic transformations accompanying biological productivity. They may be dissipated from the organism through transpiration and guttation (in plants), and through movement and various other behavioral patterns (in animals).

These and other examples of dynamic processes in soil, water, air and organisms are included in Table 8.1. Additional processes may, of course, be identified and included.

The importance of the assessment team constructing a list (such as Table 8.1) of ongoing abiotic and biotic processes in the proposed project area cannot be overemphasized. Such a list provides the basis for:

1. Identifying key concepts utilized by various specialists who have disciplinary interest in different environmental components and phenomena
2. Identifying (by consulting these specialists and/or the relevant literature) those variables and controlling parameters and conditions that are generally considered important for understanding environmental phenomena in the existing project area
3. Relating specific project activities to such parameters and conditions
4. Defining basic mechanisms whereby project effects on one or more dynamic processes in one or more environmental compartments can, in turn, affect other dynamic processes in the same or other compartments

Table 8.1
Examples of Dynamic Process in Soil, Water, Air, and Organisms

General dynamic processes	Soil	Water	Air	Organism
Introduction of matter and energy into compartment	illuviation deposition precipitation weathering aggradation solar radiation succession burrowing	deposition precipitation contamination destratification migration solar radiation tidal, and other flows of water	evaporation advection convection transpiration solar radiation emission	feeding respiration osmosis absorption audio, visual, olfactory, and other sensation
Transformation of matter and energy within compartment	weathering oxidation compaction dispersion diagenesis disintegration crystalization	oxidation reduction metabolism diagenesis mineralization decomposition	photooxidation catalytic oxidation	photosynthesis respiration metabolism productivity reproduction
Translocation of matter and energy within compartment	infiltration erosion leaching eluviation burrowing	diffusion migration precipitation flow migration	precipitation inversion advection convection interception	transpiration diffusion circulation metabolism
Sequestration of matter and energy within compartment	chelation adsorption absorption covering	chelation sedimentation consolidation stratification stagnation	inversion stratification stagnation	biomagnification productivity
Dissipation of matter and energy within compartment	evapotranspiration leaching eluviation erosion	evaporation dilution overflow drawdown	advection precipitation settling condensation	excretion guttation radiation catabolism

That general dynamic processes in the physical environment may be manifest in processes specifically and differently defined by various disciplines (as in Table 8.1) necessarily gives emphasis to the need for close coordination between the assessment team and appropriate specialists and other sources of information. It also underscores a basic attitudinal issue in any impact assessment—namely, that assessment team members who persist in the notion that they can perform assessment tasks "mechanically," without bothering to learn new things about the environment, are essentially detrimental to the entire assessment process. Certainly a design engineer who is not willing to learn about soil processes and how they are related to water, and the biologist who is not willing to learn about atmospheric advection and convection and their relationship to water, soil, and organismic processes are both likely to ensure that the assessment will become a superficial and irrelevant exercise. *The fact is, that even when focusing on individual abiotic and biotic impacts of projects, the assessment team has to understand that the project affects complex yet integrated systems and that the nature of any environmental system is ultimately determined by the nature and action of very specific dynamic processes.*

Project Effects on Abiotic and Biotic Systems

Potential effects on abiotic and biotic systems may be identified by first identifying how different project activities may influence dynamic processes in the soil, water, air, and organisms in the project area. *Probable effects* can only be estimated in terms of the data that are specific to the proposed project and to the project site. Therefore, potential effects are those effects that are reasonable in light of theory or general understanding; probable effects are those effects that are reasonable in light of both theory and actual fact. The identification of both the potential and the probable effects on abiotic and biotic systems depends upon an informed understanding of specific dynamic processes.

For example, turbid particles are introduced naturally into a stream through a number of biotic and abiotic processes (Table 8.1). Construction activity may also introduce such particles. Project-induced turbidity may be different from natural turbidity in terms of (a) type (i.e., organic or inorganic); (b) amount; (c) size and other physical characteristics; and (d) timing of introduction. Regardless of source, turbid particles in water will be translocated, transformed, sequestered, and/or dissipated according to any number of abiotic and biotic processes. They may, for instance, settle out rapidly (e.g., under low hydraulic gradient and high specific gravity). Consequent siltation or downstream aggradation may, in turn, affect other processes, including the diagenesis of downstream muds. One of the consequences of diagenesis, of course, is the release of important plant nutrients. Thus, if project-induced sediment is organic, a potential abiotic consequence of its introduction into a streambed (even if it settles out rapidly after input) is an increase in downstream concentrations of nutrients. A potential biological consequence of enhanced nutrient levels may, in turn, be an increase in the productivity of plant biomass. Whether or not such a sequence of events will actually occur depends on various factors, including: ambient nutrient concentrations, type of aquatic vegetation, and the amount and timing of nutrient releases through diagenesis.

Other potential effects (abiotic and biotic) of introducing turbid particles into streams include: abrasion and coating of fish gills, and, thus, the respiration of fish; covering of developing fish spawn and the diffusion of oxygen into oxygen-requiring eggs; alteration of downstream bottom-habitat, and, thus, the biological community utilizing that habitat; the reduction of sunlight penetration through the water column, and, thus, the photosynthetic production of oxygen at lower depths and the heating of surface waters. Again, whether or not any of these potential effects on abiotic and biotic aspects of the aquatic environment are, in fact, probable effects depends upon specific data. Some of this information would include: existing fish species, timing and place of fish reproduction, abrasive and chemical characteristics of turbid particles, existing downstream bottom-habitat; ambient turbidity levels and relationship to thermal regime of the stream, etc.

As indicated by this example, project impacts in aquatic environments include various interactions between abiotic and biotic components within water. The same can be said about any major environmental compartment. But potential and probable impacts of project activity not only depend on abiotic and biotic relationships within a single environmental compartment (e.g., water, air, soil, organism), they also depend on biotic and abiotic interrelationships between all environmental compartments.

For example, a project may require clearcutting of trees (cutting down to the ground of all trees in an area). Depending upon prevailing wind vectors (i.e., velocity and direction), the removal of trees may result in a measurable increase in seasonal velocity as measured at the surface of an impoundment downwind of the project. Now the velocity of wind across the surface of water is an important factor in the vertical movement of water (translocation) in certain lakes. It may very will be, for instance, that the lake is thermally stratified, that bottom and surface waters therefore do not mix during the summer and that ambient winds are not strong enough to set up a gyrus that can overcome existing thermal stratification. However, clearcutting of trees may result in an incremental increase in the velocity of summer winds. Depending upon the topography of the area, the morphometry of the lake basin, and the degree of thermal stratification (i.e., thermal impedance), prevailing winds after clearcutting may be sufficient to overcome thermal impedance to mixing in the lake. If, in fact, this situation occurs, nutrients may be brought to the surface as bottom and surface waters mix, and may thus result in an enhanced productivity of algae at the surface of the lake.

Few people would automatically correlate clearcutting in an area distantly removed from a lake with changes in abiotic (e.g., concentration of nutrients) and biotic (e.g., plant productivity) components of that lake. But, as I have just discussed, it is possible—at least in theory. Whether or not it will happen depends, again, on site-specific and project-specific data. It may very well be, for instance, that any incremental increase in wind velocity as a result of clearcutting would, in fact, be insufficient to overcome thermal impedance to mixing. Whether this is the case or not cannot be answered theoretically. It requires specific data and information (e.g., ambient velocities; amounts and height of trees to be cut, thermal regime of the lake, topography of the area) that can be analyzed according to current understanding of atmospheric- and hydrodynamics.

In some instances, the prediction of impacts on abiotic and biotic systems can be based on quantitative paradigms; in other instances, predictive methodologies will be essentially qualitative. As a general rule, the assessment team should ensure that, wherever possible, quantitative methodologies are examined for their relevance, especially where such methodologies are generally accepted by the scientific or technical community as standard methodologies. However, the assessment team should also be aware that:

1. Standard techniques are not currently available for the quantitative analysis and prediction of all environmental phenomena.
2. Numerous quantitative models are currently in development and/or may be available, but many of them have not been generally verified.
3. The National Environmental Policy Act does not require every environmental amenity to be turned into a number for purposes of evaluating impacts, but specifically points out that assessment shall include consideration of currently nonquantifiable environmental amenities.

Predicting the Probable from the Possible

It is often difficult to identify the specific predictive tools utilized in an assessment merely by reading the formal EIS. This is, of course, a

PREDICTING THE PROBABLE FROM THE POSSIBLE   131

major failing of such EISs. *Any objective review of assessments by other agencies, experts, and the public at large requires that the basis of all judgments and evaluations pertaining to impacts be identified.*

There are probably four basic predictive techniques that are typically utilized in impact assessment. These include (a) quantitative modeling; (b) "worst-case" calculation; (c) laboratory and/or field experimentation; and (d) quantitative and qualitative thresholds.

Quantitative models have been utilized for predicting and/or evaluating both abiotic and biotic impacts of project development. Examples include air dispersion models for the prediction of rates and extent of fallout of suspended particles, hydrodynamic models for the prediction of downstream locations of aggradation and other hydraulic phenomena, water-quality models for the prediction of changes in water quality with respect to time (e.g., in impoundments) and distance from point of discharge (e.g., streams and rivers receiving chemical and thermal wastes), nutrient budget models for the prediction of effects of introducing or reducing specific nutrient inputs into surface waters, forests, agricultural lands, etc. Sometimes quantitative models have been specially designed for use in a particular project. Sometimes the models which have been used in the assessment of a particular project have been previously developed for general use.

Probably the most efficient means available to the assessment team for identifying such models are:

1. The individual specialists and experts consulted for the purpose of identifying relevant dynamic processes
2. Regulatory agencies having specific responsibility for individual environmental resources (Appendix A)
3. Literature searches by special organizations having comprehensive access to environmental data and information (e.g., the National Technical Information Service, U.S. Department of Commerce, Springfield, Virginia 22161)

The technique of using worst-case calculations is often a useful technique that can save much time and money otherwise spent in tracking down and utilizing inconclusive models and other quantitative techniques. For example, an assessment team considering the impact of road-salt runoff into a lake can spend vast amounts of time and money trying to determine the amount of salt that will actually get into the lake. This is a very complex chemical and physical question and may be modeled in a number of different ways. One way to overcome this complexity is to assume, simply, that all salt applied to the highway will get into the lake. The assessment team can then concentrate its efforts on the impacts of resulting salt concentrations in the lake and reduce to an absolute minimum that time and money otherwise wasted on technical arguments that, for all practical purposes, cannot be resolved with any certainty. Other examples of the worst-case approach may include estimating amounts of nutrients and toxic materials to be released as a result of disturbing bottom muds during construction, estimating the amounts and timing of introductions of pesticides into various biotic components during maintenance phases, and evaluating impacts (both aquatic and terrestrial) of erosion from work areas during construction. Worst-case calculations are particularly useful in identifying those abiotic and biotic impacts that are subject to environmental quality standards and regulations. In such cases, worst-case calculations can be directly compared with standards for the purpose of identifying significant impacts.

Laboratory and/or field experimentation will probably become an increasingly important means of predicting and evaluating project impacts on the physical environment. Examples of this approach include the use of bioassays (Environmental Protection Agency, 1969; American Public Health Association, American Public Works Association, and Water Pollution Control Federation, 1976) to determine the effects of project-induced nutrients, toxins, and other materials on individual biological systems; laboratory studies of decay rates for turbidity caused by construction near or in aquatic systems; monitoring of induced effects on specific environmental components in controlled experimental areas with the proposed (or a similar) project area. Important sources of information on such techniques include (a) governmental agencies having research and regulatory authority (e.g., U.S. Environmental Protection Agency, U.S. Fish and Wildlife Service, and state regulatory authorities); and (b) professional organizations. A useful volume for identifying these agencies and organizations is the *Conservation Directory* (National Wildlife Federation, 1976).

Predictive techniques based on quantitative and qualitative thresholds rely typically upon past experience with specific types of projects in specific types of environments. For example, it may be that construction noise (above certain thresholds) in previous projects is known to have affected the successful reproduction of important game species of birds, depending on the timing and duration of the noise. Under such circumstances, the assessment team for a current project may conclude that it is highly probable that similar construction noise (i.e., at or above the threshold of previous projects) will have similar impacts on individual species. Such a conclusion gives priority, of course, to an experiential understanding of the local environment and may preclude the modeling, calculation, or experimental manipulation of those variables that are specific to the current project. While previous experience is an important factor in making judgments of environmental impacts, the assessment team should not assume that previous experience is the most important factor and, therefore, that other predictive techniques should be rejected a priori. This is particularly true when impact thresholds have become more or less offically recognized and incorporated into environmental standards. As discussed in Chapter 6, meeting environmental standards does not ensure that important environmental impacts cannot occur. While the assessment team must use existing standards in order to identify and evaluate project impacts, the team should also consider the scientific and technical issues that underly those standards and that may indicate project impacts that otherwise might be overlooked merely because the project conforms to standards. Quite typically, the literature that presents environmental standards also gives excellent scientific and technical discussions of their limitations, possible uses, and their ramifications with respect to numerous abiotic and biotic components of the environment. The best source of this literature is the U.S. Environmental Protection Agency (including both regional and Washington offices).

Regardless of specific predictive tools utilized, the assessment team should ensure that all predictions of project impacts include some estimate of probability. For example, is it highly probable that infiltration and percolation of water into and through soils in the project area will be reduced, with consequent sheet erosion in the area? Or is it only possible, with little likelihood that it will be important if, in fact, it does occur? The assignment of relative probabilities, whether qualitatively or quantitatively expressed, is necessary in order to assign priorities to those impacts which must finally be considered by decision-makers.

Predicted impacts should also be identified with respect to their direct and/or indirect relationship to project activities. For example, the abrasion of fish gills by turbid particles released into water during excavation is an indirect effect of excavation. The direct effect of excavation is the turbidity itself. If the effect of turbidity on fish is considered important enough for decision-makers to consider its mitigation, the distinction between direct and indirect causation becomes important. In this particular example, mitigation could be implemented at the point where the project causes the turbidity (e.g., aprons and checkdams may keep turbid particles within the immediate work area), or where the turbidity affects the respiration of fish (e.g., the release of turbid particles may be timed to coincide with times when particularly sensitive fish are not present in the stream). In short, *the careful distinction between direct and indirect effects allows for a more comprehensive and precise identification of possible mitigation methods*, including those that may be implemented during project-planning, design, construction, operation, and maintenance phases.

Additional criteria that can be utilized for evaluating the relevance of predictions of impacts on the physical environment have already been identified in Chapter 6. These will be discussed in greater detail in Chapter 12.

References

American Public Health Association, American Water Works Association, and Water Pollution Control Federation
  1976   *Standard methods for the examination of water and wastewater* (14th ed). Washington, D.C.: American Public Health Association.
Deevey, E. S., Jr.
  1972   Biogeochemistry of lakes: Major substances. In *Nutrients and eutrophication: The limiting-nutrient controversy*, edited by G. E. Likens. Lawrence, Kansas: American Society of Limnology and Oceanography, Publications Office, Allen Press, (Special Symposia, Vol. I).
Elton, C.
  1968   *Animal ecology.* London: Methuen & Co.
Environmental Protection Agency (EPA)
  1969   *Provisional algal assay procedure.* Corvallis, Oregon: Pacific Northwest Water Laboratory. (Report No. E-14)
Krebs, C. J.
  1972   *Ecology.* New York: Harper & Row.
Mason, B.
  1958   *Principles of geochemistry* (2nd ed.). New York: John Wiley & Sons.
National Wildlife Federation
  1976   *Conservation directory.* Washington, D.C.: Conservation Education Service, National Wildlife Federation.
Odum, E. P.
  1971   *Fundamentals of ecology* (3rd ed.). Philadelphia: W. B. Saunders.
Oosting, H. J.
  1956   *The study of plant communities: An introduction to plant ecology.* San Francisco: W. H. Freeman.
Pamatat, M. M., R. S. Jones, H. Sanborn, and A. Bhagwat
  1973   *Oxidation of organic matter in sediments.* Washington, D.C.: Office of Research and Development, U.S. Environmental Protection Agency. (EPA-660/3-73-005)
Rheinheimer, G.
  1974   *Aquatic microbiology.* New York: John Wiley & Sons.
Ricklefs, R. E.
  1973   *Ecology.* Newton, Massachusetts: Chiron Press.

9

Ecological Impacts: Aquatic Environment

Introduction

As discussed in Chapter 8, ecology is a subdiscipline of biology that focuses on biological populations and communities. However, "the breadth of the subject allows for areas of specialization which scarcely overlap. Studies of root distribution in relation to soil moisture have little superficial similarity to those dealing with the fluctuation of populations of algae in a lake. Nevertheless, both are ecology, and the investigators are presumably both grounded in the same ecological principles (Oosting, 1956, p. 7).

One way to subdivide the numerous interests that together make up ecology is to differentiate *autecology* from *synecology*. Autecology focuses on an individual organism or an individual species; synecology focuses on groups of different organisms (e.g., different species) that are associated together as a unit (Odum, 1971, p. 6). Frequently, autecological information and data are most easily available from specialists who do not call themselves ecologists—as, for example, the wildlife biologist, the fisheries biologist, the ornithologist, and the herpetologist. In each case, primary emphasis is given to the life history and behavior of a specific type of organism (either an individual organism, or a population of that organism). Contrarily, the primary emphasis of synecology is the overall system of which a particular organism or population is only a single (even if a major) component.

For purposes of impact assessment, it is most often convenient to assume that autecological data and information are most readily available from biological specialists other than those specifically called "ecologist." This is not to say that there is, in fact, any difference between a wildlife biologist who is primarily interested in "game mammals" and an ecologist. It is only to point out that the best information on a particular species is usually available from those biologists who specialize in the life history and behavior of that species and that such biologists typically do not have the title "autecologist," but more mundane titles, such as bacteriologist, zoologist, and botanist. Quite typically, however, the title "ecologist" is meant to imply that its holder is interested in how whole ecosystems (comprising more than one species or type of organism) work. For purposes of this volume, we will use ecology synonymously with synecology and will include autecological impacts as biotic impacts (as discussed in Chapter 8) that may have ecological (i.e., synecological) consequences.

In order to facilitate information gathering and analysis during the assessment of ecological impacts, it is also useful to differentiate between the ecologies of basic types of environment—namely, aquatic systems (including fresh and saltwater), terrestrial systems, and wetland systems. Again, such a differentiation reflects more the compartmentalization of existing data and information than it does any inviolate individuality of these complex systems. Just as abiotic and biotic factors are interconnected (Chapter 8), so are aquatic, terrestrial, and wetland ecosystems. Yet, disciplinary specialization (as well as biological specialization) has resulted in somewhat different literatures (and even somewhat different jargon) for each of these basic types of ecosystems.

Habitat and Niche

While there are many very specific and also very general definitions of the term "ecosystem," it is generally understood to represent a functional unit that comprises a living community (i.e., a collection of different populations) and the nonliving (abiotic) components of the community's surroundings (Oosting, 1956, p. 29; Odum, 1971, p. 5; Smith, 1974, pp. 19–20). Of course, the phrase "functional unit" may be variously interpreted, and should be given more explicit meaning. For purposes of all following discussion, an ecosystem will be considered as *a living community and its abiotic surroundings that interact in such a manner as to manifest a stable pattern of material cycling and linked energy transformations.*

Similarities and differences between aquatic, terrestrial, and wetland ecosystems may be discussed in terms of the similarities and differences between their respective: (a) communities; (b) abiotic surroundings; (c) stability; (d) material cycling; and/or (e) energy transformations.

Any comparison of these basic characteristics of aquatic, terrestrial, and wetland ecosystems, however, ultimately depends on the comparison of respective habitats—the "place" where an organism lives, and (2) niches—the "functions" that organisms perform within their habitats (Odum, 1971, pp. 234–235; Smith, 1974, pp. 24–25).

HABITAT

As discussed in Chapter 8, for all practical purposes of impact assessment there is a biology to be associated with any environmental chemistry and physics; there is also a chemistry and a physics associated with any biology. An individual fish (a biotic component of water) living in an impoundment not only responds to the physics and chemistry of its habitat, it also contributes to or otherwise affects the physics and chemistry of that habitat. Similarly, a population of the same species or type of fish (an autecological component of water) also responds and contributes to its own habitat. And, in much the same way, a community of fish, benthos (bottom dwelling organisms), and microscopic plants and animals (a synecological component of water) also respond and contribute to their own habitat.

At the level of the individual organism, it is only the individual organism that takes things (e.g., nutrients, energy) from its chemical and physical habitat and returns things (e.g., wastes) to that habitat. If suitable nutrients and/or energies are not present, or if its wastes are not removed or suitably transformed within that habitat, the organism must move to another or perish. Of course, different organisms have different potentials for acclimating to stressful conditions.[1] But when these potentials are exceeded, the habitat becomes inimical to the organism.

At the level of the population, a number of similar organisms both take and contribute things from and to the population's habitat. Because the organisms within the population are similar, they have similar needs; they may therefore compete and/or cooperate for individual habitat. As with the individual organism, conditions may become inimical to the population. Yet, there is an important difference with respect to population responses to stress. In a population, no two individuals are precisely alike. Some individuals will experience an environmental stress more severely than others. Thus some individuals will be selected out by the changing physics and chemistry of their habitat as being the best suited,

---

[1] See Liebig's "law" of the minimum, and Shelford's "law" of tolerance (Odum, 1971).

within that habitat, for survival. It is a basic tenet of biology that natural selection is a phenomenon of populations and not of individuals.

At the level of community, the phenomenon of natural selection also applies—because a community is a collection of different populations. In fact, some populations in a community (e.g., predators) may enhance selective forces already acting on other populations (e.g., prey). However, in addition to antagonistic interactions between two or more populations in a community, there are also mutualistic relationships that enhance the habitat (i.e., reduce its stress) for both participating populations. Theoretically, a community (as a collection of at least two different populations in the same environment) may exhibit up to nine different types of interactions (Odum, 1971, p. 211–233).

From these examples it is evident that habitat not only varies with the physical and chemical characteristics of the environment, but also with organismic, population, and community phenomena that are directly and indirectly related to those physical and chemical characteristics. Thus, aquatic, terrestrial, and wetland ecosystems differ not only with respect to gross differences in the physics and chemistry of water, land, and water–land interfaces, but also with respect to:

1. Characteristic stresses on the acclimation potential of individual organisms
2. Selective forces that act on typical populations, and/or
3. Characteristic competitions and/or mutualisms between various populations in the community

NICHE

Habitat has been described as an organism's "address" (where it lives); and niche, as the organism's "profession" (what it is doing in its habitat) (Odum, 1963, p. 27). Historically, professional ecologists have given somewhat different emphasis to various aspects of the niche concept. Some, for example, utilize it to refer specifically to what the organism does—they give emphasis to the *functional role of the organism*. Others utilize the concept to describe different *subdivisions of the physical environment* (i.e., habitat) in which the organism acts out its role(s). The former approach typically generates descriptions of ecological niches in terms of dynamic processes that are carried on by the organism; the latter generates descriptions of the physical and chemical factors that give competitive edge to one species or another (and thus allow the successful population to carry on ecological functions). The modern view has been described (Smith, 1974) as involving both approaches—that is, the "functional role of an organism in the ecosystem as well as its position in time and space (p. 240)."

From the perspective of the impact assessment team, it is probably most useful to give priority to the functional aspects of the niche concept. This is because of the power of this concept to identify functional similarities between populations that are obviously different in terms of their biology.

For example, basic niches (in a general, functional sense) in ecosystems include (1) primary production, (2) consumption, and (3) decomposition.

Each of these niches may be defined by a certain type of work that has to be done in ecosystems. The myriad of organisms in various ecosystems, despite their apparently infinite diversity in shape, color, size, and behavior, are nothing more than different looking and different behaving machines by which such ecological work is accomplished.

Primary producers are those organisms that transform the radiant energy of sunlight and inorganic nutrients into living materials (i.e., they carry on photosynthesis). The only requisite "design specification" for such a living transformer is that it have chlorophyll. Thus, all green plants, as long as they can carry on photosynthesis, act as primary producers. It may be a pine tree, or a microscopic, single-celled plant floating in the ocean. The ecological niche for both is the same. Each is doing the work of transforming the inorganic into the organic.

Consumers are those organisms that cannot transform sunlight energy into the chemical energies of organic molecules. Therefore they must consume preformed organic materials and transform these into the living substance of their own bodies. Herbivores transform the chemical energy of plant tissue into the chemical energy of their own animal tissues; carnivores transform the chemical energy of other animal tissue into the chemical energy of their own; omnivores transform the chemical energy of both plant and animal tissue into the chemical energy of their own; detritivores transform the chemical energy locked up in dead organic remains into the chemical energy of their own tissues.

The niche of consumption, then, is a concept that allows the ecologist to see an ecological equivalence (in function) between a cow and a grasshopper, for both do the work of herbivory; between a wolf and a bass, for they both do the work of carnivory; between a carp and a man, for they both do the work of omnivory; and between a snail and an earthworm, for they both do the work of detritivory. Of course, the biological differences between any two of these organisms are immense. But ecologically, one is often the functional equivalent of the other.

Decomposers are those organisms that accomplish the disintegration and mineralization of the organic back into the inorganic, thus releasing back into the environment those inorganic nutrients previously locked up in the complex molecules of life by the primary producers. It may be a bacterium, a fungus, an earthworm, or a beetle that does it. Or it may be any number of other organisms that transform the organic back into the inorganic. The important thing, ecologically, is not so much who does the work, but rather, that the work is done.

These three basic ecological niches give emphasis to a particular perspective about organisms in general—namely, that whatever else organisms are, they are at least machines that do work. Moreover, they are particular kinds of machines. They are transformers—machines that take in energy and matter of one kind and transform it into energy and matter of another kind. And since there are three basic kinds of work to be done in an ecosystem, there are essentially three kinds of living transformers that one might expect to find in ecosystems.

Each type of transformer can be found in typical aquatic, terrestrial, and wetland ecosystems. They manifest, however, different biologies, according to the constraints of habitat that are peculiar to each.

As depicted in Fig. 9.1, one way to view these three basic types of ecosystems is to envision each as a specific combination of niche

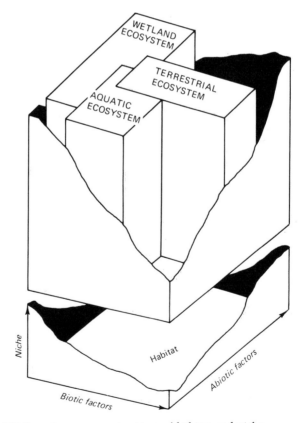

FIGURE 9.1. *Ecosystems as constructions of habitat and niche.*

functions and of the various biotic and abiotic factors of habitat. While the basic niches of primary productivity, consumption, and decomposition are common to all of the three types of ecosystems, it is habitat that varies from one to the other, thus providing the physical and temporal space wherein different populations can perform their interdependent functions. As depicted in Fig. 9.1, aquatic, terrestrial, and wetland ecosystems are not independent of one another. Many of the biotic and abiotic factors that define habitats available in one can also define the habitats available in another. Thus there is linkage or coupling between each of three types of ecosystems.

Overview of Aquatic Ecosystems

The literature on aquatic ecosystems may be variously organized according to different scientific, technical, and social issues and concerns. For example, the literature may be organized according to (a) general type of aquatic system; (b) aquatic biota; and/or (c) environmental phenomena and issues of special interest.

Various hierarchies may be employed for organizing literature with respect to general types of aquatic ecosystems. A simple hierarchy might be:

1. Saltwater systems[2]
   a. tidal
   b. nontidal
2. Freshwater systems
   a. surface waters
      (1) lentic (ponds, lakes, impoundments)
      (2) lotic (creeks, streams, rivers)
   b. groundwater
      (1) water-table aquifers
      (2) artesian aquifers

Each of these categories may, of course, be subdivided and expanded.

Literature which may be consulted by the assessment team may also be organized according to the taxonomy, life history, and population dynamics of basic types of aquatic populations, including:

1. Aquatic plants
   a. bacteria and fungi
   b. algae
   c. vascular plants

[2] Salt-water systems will be discussed in more detail in Chapter 11.

2. Aquatic animals
   a. zooplankton (microscopic, free floating)
   b. benthos (bottom dwelling)
   c. nekton (macroscopic, strong swimmers)

Again, many subdivisions of these categories may be made. Such subdivisions reflect the specialized interest of individual authors and the intended use (e.g., research, industry, education) of the literature.

Finally, there is a vast literature which is available for purposes of impact assessment and which focuses on particular issues, including (a) toxicology of aquatic pollutants; (b) eutrophication (nutrient enrichment) of recreational waters; (c) fish-pond management; (d) sanitary surveys of streams and impoundments; (e) chemical and biological modeling of lakes; and (f) water-quality indices.

While these hierarchies and categories are certainly not inclusive either of the scope or of the specialization of literature pertaining to aquatic ecosystems, they are sufficient to underscore an important point for the assessment team—specifically, that aquatic ecology is a field of many specializations and that *the nonprofessional had better have a clear idea of what he or she needs from the technical literature before going looking for it*. Literature that is consulted merely because "it has something to do with aquatic ecology" is in all probability not the least bit more useful for impact assessment than literature that is randomly selected from among a mixed collection of novels, logarithmic tables, and biographies of dead Romans.

The assessment team can ensure that its members will have a clear idea of what is needed if it emphasizes the importance of an ecological overview based on niche functions and habitat constraints within aquatic ecosystems. Such an overview should highlight:

- the functions (niches) that are present in a particular aquatic system and how they may be interrelated with one another
- The kinds of habitats that determine which types of aquatic populations will be able to carry out niche functions
- those dynamic abiotic and biotic process (Chapter 8) that typically influence (and in turn are influenced by) the response of the aquatic community both to available niches and to available habitats
- Types of project actions that can directly and indirectly affect ecosystem structure and dynamics

NICHE FUNCTIONS

An example of how the three basic niches may be interrelated in aquatic ecosystems is depicted in Fig. 9.2.

In this figure, the rectangles represent the ecological work to be done

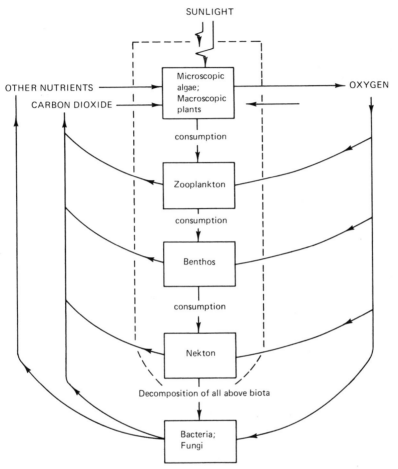

FIGURE 9.2. *An example of niche interrelationships and material cycling in aquatic ecosystems.* [Adapted from U.S. Department of Transportation. 1975. Ecological Impacts of Proposed Highway Improvements. Washington, D.C. Federal Highway Administration]

(i.e., primary productivity, consumption, and decomposition) by living transformers that are adapted to the available habitats. As already discussed, the work of primary production can only be accomplished by plants that have chlorophyl. Thus, in a generalized aquatic ecosystem, one might expect to find (a) aquatic macrophytes (submerged, emergent, and/or floating vascular or "higher" plants); (b) phytoplankton (microscopic, free-floating algae); (c) periphyton (microscopic algae which grow attached to some substrate such as a stone, leaf); and (d) pigmented bacteria that can carry on photosynthesis.

Regardless of the biological diversity of aquatic populations that carry out primary production, the ecological importance of these populations is that they transform radiant sunlight energy into the chemical energy of living plant tissue. This chemical energy of plant tissue then becomes the energy input into consumers that subsequently transform it into the chemical energy of animal tissue.

An example of such an herbivorous consumer (also called a *primary consumer*) is the zooplankton (microscopic, free-floating animals). There are many different species of zooplankton, and not all of them consume living plant cells (e.g., phytoplankton, periphyton). But those that do may be said to act as primary consumers and to perform the same kind of ecological function as a cow does when it grazes on grass. They transform the substance of living plant material into the substance of living animal material, and thus they provide energy inputs into those organisms that, in turn, prey upon them (i.e., the *secondary consumers*). As depicted in Fig. 9.2, the benthos (bottom-dwelling animals such as clams, insect larvae, and mud worms) may be important secondary consumers in aquatic ecosystems. Such secondary consumers transform the energies and materials of primary consumers into the chemical energy of their own flesh, and similarly serve as energy inputs into the next series of living transformers, the *tertiary consumers*. Tertiary consumers are represented in Fig. 9.2 by the nekton (animals that can swim against the currents of water, for example, fish).

It is important to note that the examples of organisms included as primary, secondary, and tertiary consumers in Fig. 9.2 may be inappropriate for a given aquatic ecosystem. It has already been mentioned, for example, that some zooplankters eat plants and that others do not. Similarly, not all fish are tertiary consumers. Many organisms also have a broad range of possible foodstuffs they can utilize; they can, therefore, function as primary, secondary, or other higher-order consumers at one and the same time, or at different periods in their life cycles.

All primary producers and all the various consumers are not, of course, devoured by other organisms. Many producers and consumers live out their lives, die, and contribute their dead bodies to the organic matter in bottom muds. This dead organic matter (detritus) undergoes decomposition by various aquatic organisms, including aquatic bacteria and fungi, that transform detrital materials and energies into the living biomass of their own bodies.

As indicated in Fig. 9.2, the sequential consumption of one organism by another and the subsequent transformation of one type of matter and energy into another type involves the cycling of oxygen, carbon dioxide, and other inorganic nutrients throughout the aquatic ecosystem. While large amounts of oxygen become dissolved in water through the mechanical agitation of waves, riffle areas, and waterfalls, the dissolved oxygen in many quiescent waters is primarily derived through the photosynthesis of primary producers. However, not only is there a photosynthetic production of oxygen by primary producers, but also a consumption of that oxygen by the same primary producers through the process of respiration. The total photosynthetic production of oxygen in a unit period of time (e.g., 12 hours) is referred to as *gross oxygen production*. The total respiratory consumption of oxygen in the same unit period of time and by the same plants can be measured and subtracted from the gross oxygen production. The difference between gross oxygen production and respiratory consumption is referred to as the *net production of oxygen*. If the net production of oxygen in water is zero (the gross production and respiratory consumption are equal), there is no photosynthetically derived oxygen available for the respiratory needs of the zooplankton, benthos, and nekton.[3] In such a situation, the respiratory needs of these organisms must be met by other sources of oxygen, such as mechanical agitation of water and diffusion of oxygen from the atmosphere through the water column.

As indicated in Fig. 9.2, the respiration of consumers and decomposers (as well as of primary producers) results in the production of carbon dioxide that is required by photosynthesizing plants. Also, the decomposition of organic materials[4] results in the release of inorganic nutrients, such as nitrates and phosphates, that are also required for photosynthesis. Thus, the biological processes of photosynthesis and respiration are coupled in the aquatic ecosystem and essentially serve to link the chemical inputs and wastes of one population to the inputs and wastes of other populations. The specific manner in which niche functions, populations that perform those functions, and consequent material cycling are all interconnected, linked, or otherwise coupled in an aquatic ecosystem constitute the *trophic structure* of that system. *There can be no realistic assessment of project impacts on any aquatic ecosystem without a basic understanding of the trophic structure of that ecosystem.* This understanding is, in turn, based upon knowledge of specific *food chains* in aquatic systems in the project area and its environs.

A food chain is the pattern of simple, sequential consumption of one organism by another. There are two basic types of food chains—the *grazing food chain* (i.e., the sequential consumption, which begins with the consumption of living plant material), and the *detritus food chain* (i.e., the sequential consumption, which begins with the consumption of detritus). An important characteristic of both types of food chains is the vast amount of energy that is lost as heat when the biomass of one organism is transformed into the biomass of another organism. This lost energy accounts for about 80–90% of all food energy transformed at each

---

[3] See discussion of "compensation point" (Russell-Hunter, 1970).

[4] See discussion of "biochemical oxygen demand" (American Public Health Association, American Water Works Association, and Water Pollution Control Federation, 1976).

step in the food chain (Odum, 1971, p. 63). Thus, in a grazing food chain which includes the transformation of phytoplanktonic food–energy by zooplankton and then by benthic organisms (i.e., two sequential consumptions), only about 10% of the phytoplanktonic food–energy consumed by zooplankton actually becomes zooplanktonic biomass. In turn, only about 10% of zooplanktonic food–energy consumed by the benthos actually becomes benthic biomass. Thus it takes about 100 units of phytoplanktonic food–energy to provide only 1 unit of benthic biomass.

Food chains (both grazing and detritus) are typically interconnected in natural ecosystems. Interconnected food chains are referred to as a *food web*. Food webs exist because organisms typically have more than one type of food supply.

Examples of grazing and detritus food webs are included in Fig. 9.3. Note that an individual population (e.g., nekton) may have alternative food supplies (e.g., zooplankton and benthos). Also note that each population may participate simultaneously in a grazing and a detritus food web. Finally, the depicted food webs also include predatory birds and other animals that link aquatic food webs with terrestrial food webs (Chapter 10 and 11).

HABITAT CONSTRAINTS

The different habitats of different aquatic systems determine what types of biological populations can, in fact, carry out niche functions.

For example, in deep lakes and ponds that have steep shorelines, there is relatively little probability that rooted macrophytes will be able to establish themselves to the point that they become the major primary producers in the lake. In such a situation, it is more likely that the major primary producers will be the phytoplankton, which are specifically adapted for floating at surface depths (where the light for photosynthesis is). Similarly, in ponds and streams that have suitable substrate, the major primary producers may be the periphyton, which are specifically adapted to utilize different types of substrates. In shallow ponds having appropriate hydrological regimes, one might expect submerged, emergent, and floating vascular (or "higher") plants to fill the niche of primary production. Of course, in many aquatic systems, various combinations of different plant populations (both macro- and microscopic) can be found.

In general, aquatic populations are adapted to such general attributes of the water medium as buoyancy, oxygen concentration, temperature, substrate, light transmission, and hydrodynamic flow. Projects may directly and indirectly affect these (and other) attributes and thus affect the ability of organisms to utilize aquatic habitats. Also, project activities may directly and indirectly affect specific populations by preventing or

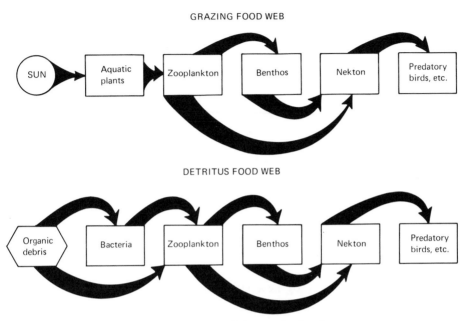

FIGURE 9.3. *An example of grazing and detritus food webs.*

enhancing a population's successful utilization of its adaptive capabilities.

Examples of direct and indirect effects on these attributes of the water medium and/or on the utilization of biological adaptations include:

1. *The introduction of organic silts into water.* These silts undergo decomposition and can result in reduced oxygen concentrations. Although the consequent oxygen stress may be of quite a short duration, it can be a significant change in aquatic habitat with respect to certain population dynamics, such as fish spawning, egg development, and migration. Also, the turbid particles not only can reduce oxygen concentrations through their decomposition, but they can (if they stay in suspension for periods of days to weeks) reduce light transmission through the water column, thus reducing the photosynthetic production of oxygen in lower depths. The silt can also sediment out to change bottom substrate, can abrade fish gills, and clog the gills of fish and benthic organisms. In summary, then, turbid particles can directly alter the aquatic habitat by the deposition of new substrate or by changing light transmission, and indirectly alter it by reducing oxygen concentrations. They can also affect the successful utilization of biological adaptations to aquatic habitats (e.g., by clogging fish gills).

2. *The introduction of paints and other materials having a low*

specific gravity. Because of their low specific gravity, paints and other materials come into direct contact with those organisms that are specifically adapted to float at the water surface, for example, certain algae, vascular plants, and larvae. Affected plants may be major primary producers, and thus their loss may trigger other changes in the trophic structure of the aquatic ecosystem. Contrarily, certain project activities may enhance the successful utilization of biological adaptations.

For example, the deposition of rocks and boulders (construction wastes) in streams having little hard substrate can provide habitat to plants (e.g., periphyton, mosses) and animals (e.g., insect larvae) specifically adapted to attach themselves to such substrate. Depending on the manner of deposition of construction wastes, riffle areas may also be created. These riffles may increase oxygen concentrations through mechanical agitation of water and, thus, indirectly enhance the habitat of aerobic biota of various kinds. The same riffle areas may also provide rest areas for fish migrating upstream. Ecological assessment requires the balancing of adverse impacts (coating of floating plants with paints) with positive impacts (providing new substrate) from a systems perspective.

As discussed, projects may have direct and indirect effects on (*a*) the physical, chemical, and biological attributes of habitats; and/or (*b*) the successful utilization of biological adaptations to those habitats. The assessment team should, therefore, make a concerted effort to identify key attributes of aquatic habitats in the project area and the basic adaptations of aquatic biota that utilize those habitats. Relevant data and information can be solicited by means of the following kinds of questions which can be addressed either to biologists or to the specialized literature:

1. Which plants in the project area are adapted (and how) for floating, growing at submerged depth, and for growing emergent from water?
2. Which plants and animals are particularly adapted (and how) for living on relatively persistent substrate (e.g., muds at the bottom of a lake), and which (and how) for living on relatively unstable substrate (e.g., wave-washed shoreline)?
3. Which organisms are particularly adapted (and how) to fluctuations in water flow, and in oxygen and nutrient concentrations?

It is important for the assessment team to realize that individual species of generally similar populations (e.g., algae) can be extremely diverse with respect to their adaptations to aquatic habitats. It is therefore desirable to make these questions as species-specific as possible.

Impacts on Aquatic Ecosystems

All impacts on individual populations of aquatic biota have the potential for altering the aquatic ecosystem of which they are a part. Thus, all dynamic abiotic and biotic processes discussed in Chapter 8 should be examined for their direct and indirect influence on the niche functions and habitats that together determine ongoing food webs in the project area and its environs. For example, the assessment team may identify sheet erosion as a probable, short-term consequence of excavation during the construction phase. The team should also consider how sheet erosion may affect aquatic food webs. Such effects may be related to (a) habitat modification that precludes the presence and thus the functioning of particular populations (e.g., benthic species); and/or (b) influences on niche functioning by other important populations which nevertheless remain in the system (e.g., phytoplankton).

A practical way in which the assessment team can ensure comprehensive consideration of potential impacts on aquatic food webs is to continually ask the question, "So what?" with respect to each and every abiotic and biotic impact it has previously identified—that is, *the assessment team must consider how each abiotic and biotic impact can directly and indirectly affect those food webs that characterize the aquatic ecosystem under consideration*. Abiotic and biotic impacts that are identified but never integrated into an overview based on total system dynamics are always indicators of an incomplete assessment.

The process by which the assessment team may undertake the integration of individual impacts on abiotic and biotic components of the environment into an ecological systems overview of project impacts can be described in terms of three phases.

In the first phase (Fig. 9.4), probable impacts on abiotic components (impacts on the physical and chemical attributes of soil, water, and air) and biotic components (impacts on cells, tissues, organs, and individual organisms or species) are identified by relating project impacts on general dynamic processes, including material and energetic introduction, transformation, translocation, sequestration, and dissipation (Chapter 8).

In the second phase, probable abiotic and biotic impacts are evaluated for their direct and indirect influence on habitat (including vertical, horizontal, and temporal variations) and niche functions (including primary production, consumption, and decomposition).

In the third phase, probable impacts on habitat and niche functioning are evaluated for their influence on the population dynamics[5] (autecolog-

---

[5] Useful analytical categories for the study of population and community dynamics are discussed by Odum (1971), and Smith (1974).

ical dynamics) of individual populations (including population density, natality, mortality, age distribution, etc.), and on synecological dynamics and components (including food webs, competition, and mutualism).

As indicated in Fig. 9.4, an important consideration at each phase of this integrative effort is the concept of dynamic interaction—interaction between abiotic and biotic compartments, between habitat and niche, and between autecological and synecological components and processes. Key concepts for evaluating these interactions include: material and energetic dynamics (Phase I), biological adaptation (Phase II), and trophic structure and dynamics (Phase III).

In evaluating the influence of project-induced abiotic and biotic impacts on aquatic ecosystems, the assessment team should be guided by two basic considerations:

1. The geographical extent of an aquatic ecosystem typically extends beyond its own shoreline or banks, and
2. An aquatic ecosystem is typically in some state of change or development and should not be viewed as intrinsically static.

GEOGRAPHICAL EXTENT

Individuals with assessment responsibility have often mistakenly assumed that projects have to be implemented in or next to a body of water in order to have impacts on its ecology. Thus, the evaluation of impacts on groundwater has often been linked with the evaluation of impacts on surface water only at the editing phase of the written EIS, and not in the analytical and integrative phases of the overall assessment. The fallacy of this approach can be easily demonstrated by reviewing some of the basic abiotic and biotic attributes of ground and surface waters, and how these attributes can be coupled in an actual situation.

Groundwater is that water which is found below the water table. Above the water table, in the *zone of aeration,* any water is held to the surfaces of rock and soil particles by capillary forces and will not flow into a well. Below the water table, in the *zone of saturation,* all openings—including crevices, crannies and pores—are completely filled with water (Environmental Protection Agency, 1977).

The flow of groundwater below the water table depends on the porosity and permeability of stone–soil substrate. Substrates having high porosities and permeability (e.g., sand and gravel) facilitate the flow of groundwater. Substrates that have relatively small and disconnected spaces (e.g., clay–silt), retard the flow of groundwater and constitute poor aquifers.

Groundwater may be found in two basic types of aquifers: (a) an *artesian aquifer,* containing groundwater that is confined under pressure

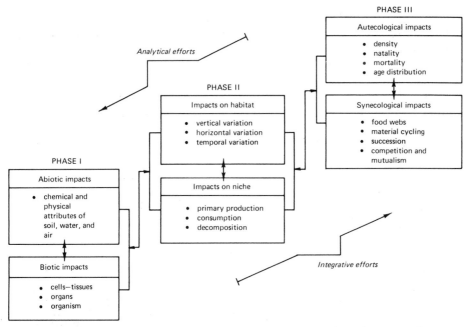

FIGURE 9.4. *Three phases in the integration of impacts into an ecological overview of project impact.*

between layers of impermeable rock; and (b) a *water-table aquifer*, containing groundwater that is not confined under pressure.

Because of geological faults and fissures, water can escape from artesian aquifers in the form of springs. Groundwater may also flow from a water-table aquifer into surface waters, including impoundments and rivers. In fact, groundwater can constitute the major volume of riverine and other surface waters.

A basic difference between artesian and water-table aquifers is that the artesian aquifer is generally continuous over great distances, and thus its area of *recharge* (replenishment) may be located far away from where it might be tapped by a well or a spring. Groundwater contained in water-table aquifers is replenished locally and is more immediately responsive to the local infiltration of soils by precipitation.

Among the most important biotic, abiotic and ecological processes in impoundments (discussed in previous sections) are photosynthesis and respiration, stratification and overturn, nutrient cycling, and energy distribution through aquatic food webs. These processes can be influenced by the chemical, physical, and biological components of groundwater inputs, as well as by the amounts of these inputs and their periodicity.

For example, bacterial respiration in bottom sediments releases nutrients into water. However, when impoundments stratify (because of cold-water inflows, depth, etc.), nutrient-rich waters are trapped at the bottom of the impoundment and cannot contribute to primary production until overturn. Similarly, during stratification, little if any oxygen from surface waters (epilimnion) diffuses to bottom waters (hypolimnion), with the result that bottom waters can become anaerobic. This is especially true when previously inputted groundwaters (perhaps interrupted or altered by project development) are important sources of oxygenated water.

One of the basic differences between running (lotic) and impounded (lentic) waters is that the dimension of time (which is so central to physical, chemical, and biological processes in impoundments) becomes the dimension of length in running water. Thus, while organic molecules are decomposed over days and months in a lake, those same molecules are decomposed over miles in a river. As organic molecules (from groundwater sources as well as from surface runoff) are inputted into running waters, a process of *natural purification* is initiated by aquatic bacteria and fungi. This process involves the decomposition (and, therefore, transformation) of organic materials, the distribution of decomposition products over some linear distance, and the sedimentation of these products according to hydraulic gradients along the bed. Riffles provide important oxygen inputs, speed up decomposition processes, and, therefore, shorten the zone of active decomposition. These also aid in the evaporation (or dissipation) of volatile materials. Pool areas, as well as meander-banks, are important sediment traps, and serve to sequester potentially toxic materials and food supplies that, in turn, enhance biological productivity.

Depending on the rate of decomposition and on hydraulic flow (which can be altered by interrupting contributory groundwater flow), the mineral products of decomposition are made available to downstream primary producers that, in their turn, can replenish oxygen supplies previously reduced by the respiration of upstream bacterial and fungal populations.

With respect to these basic features of ground and surface waters, it is clearly possible that:

1. Groundwater recharge areas (for both artesian and water-table aquifers) can be physically and/or chemically altered by project activity (e.g., as a result of excavation, putting in borrow, on-site storage of leachable materials, soil compaction, runoff from completed and operation projects).
2. Hydrological and/or chemical alteration of groundwater aquifers can result in changes in the hydrological, thermal, nutrient, and toxic regime of receiver lentic waters (including lakes, ponds, and man-made impoundments).

3. Hydrological and/or chemical alteration of ground-water aquifers can result in changes in hydrological, thermal, nutrient, and toxic regime of receiver lotic waters (including creeks, streams, and rivers).

Just as impacts on groundwaters may have abiotic and biotic (and thus ecological) ramifications in both lentic and lotic surface waters, so may project impacts on the local atmosphere and terrestrial environment. Thus the ecosystem dynamics of lentic and lotic surface waters even far removed from a project can be influenced by project-induced changes in groundwater, air, and terrestrial components located in the immediate project area.

AQUATIC SUCCESSION

In order to identify the impacts of projects on an aquatic ecosystem, it is necessary to know what the ecosystem is doing in the absence of the project (Chapter 7). In general, the assessment team should consider that an aquatic ecosystem is at some particular point of its overall ecological development, and that a proposed project may enhance, retard, change the direction of, or be essentially neutral with respect to that development.

An important example of ecological development or succession of lakes and ponds is the phenomenon of *eutrophication*.

Eutrophication is the process by which nutrient-poor lakes and ponds (i.e., *oligotrophic* waters) become nutrient rich (i.e., eutrophic waters). The progress or succession from oligotrophy to eutrophy is a natural phenomenon, but it can be greatly enhanced by human activity. The human enhancement of eutrophication is called cultural eutrophication.[6]

Various characteristics of lakes and ponds have been examined for their usefulness as indicators of the succession from oligotrophy to eutrophy (Welch, 1963; National Academy of Sciences, 1969). Many of these characteristics can only be determined by laboratory analyses or field measurement.[7] Some characteristics of eutrophic waters, however, are discernible by gross physical features of a pond or lake, including (a) shallowness of basin; (b) organic bottom muds; (c) abundance of macrophytes; (d) abundance of microscopic algae; and (e) poor transparency due to suspended algae (Welch, 1963, p. 344; Lee, 1970, p. 3; Hynes, 1971, p. 141).

There are numerous chemical and biological processes in lakes and ponds that are interrelated and susceptible to influence by human activi-

---

[6] The following discussion of natural and cultural eutrophication is adapted from a previously authored report (Massachusetts Division of Environmental Quality Engineering, 1977).

[7] See Environmental Protection Agency (1974).

ties. Among those processes that are particularly relevant to the eutrophy of surface waters are photosynthesis and respiration, stratification, and diagenesis (or chemical alteration) of bottom muds.

Many factors govern the actual rate of photosynthesis and the production of plant biomass (i.e., primary production) in water (Goldman, 1974). However, those factors that are generally considered to be key factors are the availability of light energy and the availability of certain nutrients (Russell-Hunter, 1970, p. 2).

Once produced, the biomass of aquatic plants performs a number of important functions in the aquatic ecosystem, including:

• Production of oxygen
• Shading and cooling of sediments
• Slowing water movements; thus providing habitat for some benthic organisms
• Regeneration of substances from sediments
• Provision of surfaces for attachment by bacteria and other organisms
• Provision of food, nest-building material, and sites for egg attachment for insects and fish
• Provision of nesting sites for fish
• Protection of small fish from predation
• Conversion of inorganic material to organic material
• Provision of food and habitat for game birds and other animals
• Anchoring soil by means of root systems

While these and other functions of plant biomass (Mulligan, 1969, p. 465; Gaudet, 1974, pp. 24–37) are key functions in aquatic ecosystems, dense growth of aquatic vegetation (i.e., "blooms") often produce stressful conditions for a variety of aquatic biota.

For example, the respiratory requirements of vegetation can be sufficient to cause the net productivity of oxygen to approach zero. As discussed previously, this can result in oxygen stress on other aerobic organisms in aquatic food webs. The creation of anaerobic zones in eutrophic waters can also be mediated by the shading of underlying plants by dense growth at the surface. Such shading prevents the photosynthetic production of oxygen in bottom waters. The respiration of bottom-dwelling organisms is, nevertheless, unaffected. Meanwhile, nutrient chemicals at the surface are rapidly taken up by the luxuriant surface growth. If this condition persists, bottom waters may undergo oxygen depletion, and surface waters may undergo nutrient depletion.

Consequences of oxygen depletion in shallow eutrophic ponds include the production of obnoxious smells through the anaerobic decomposition of bottom muds, loss of habitat for fish and other aquatic organisms, and the raising of the pond bottom due to the heavy accumulation of dead vegetative biomass. In deeper ponds and lakes, which undergo thermal stratification, nutrient-depleted surface waters are not easily replenished by nutrient-rich bottom waters. In such lakes, bottom waters and sediments act as nutrient traps (Hendricks and Silvey, 1973) that release plant nutrients from muds to surface vegetation only during those periods (e.g., spring and fall) when the thermal barrier to mixing between bottom and surface water disappears.

Bottom muds are composed of materials produced within the pond or lake (autochthonous materials), as well as materials produced outside and transported by wind and runoff into the pond or lake (allochthonous materials). The chemical composition of allochthonous materials varies greatly, depending on the nature of the watershed and geological conditions of the basin (Ruttner, 1969, p. 191). Autochthonous materials are typically organic.

The diagenesis (or chemical transformation) of allochthonous and autochthonous muds is mediated by a number of physical, chemical, and biological factors. In eutrophic waters, aquatic bacteria vigorously decompose organic materials (Pamatmat et al., 1973) with frequent formation of anaerobic zones. Such zones facilitate the loss of hydrogen sulfide that not only produces obnoxious odors, but that also (in the presence of iron) forms offensive, black, iron-sulfide deposits. Also, the microbial release of bubbles of hydrogen, methane, and carbon dioxide causes bottom sediments to be transported repeatedly through the water column where they can be further transformed (Rheinheimer, 1974, pp. 144–145). One of the important consequences of these transformations, of course, is the release of plant nutrients into the water column. The release of materials, such as phosphorus and nitrogen, from organic sediments has been studied extensively (Wetzel, 1975; Williams and Meyer, 1972; Wildung and Schmidt, 1973; Austin and Lee, 1973; Hendricks and Silvey, 1973), and is generally considered to be sufficient to support large populations of aquatic vegetation.

Project design and development, as well as completed and operational projects, can result in the enhancement or in the retardation of eutrophication and its consequences. Project-mediated enhancements may include direct inputs of nutrients (e.g., runoff of nutrient-rich soils from construction sites, seepage from field septic systems), and indirect inputs (e.g., secondary development in project area as a result of project completion, recreational usage due to improved access or secondary development). Project-mediated retardation of the eutrophication process may include the direct removal of nutrients (e.g., dredging out of bottom muds), as well as the reduction in nutrient availability (e.g., covering nutrient-rich muds with "clean fill"). Actual project contributions to the eutrophy of a lake or pond cannot, therefore, be determined theoretically. They must be evaluated on a site and project-specific basis.

Concluding Remarks

Impacts on aquatic resources include impacts on abiotic, biotic, and ecological attributes of water resources. While these attributes of water resources are interrelated, they are not synonymous. The difference is a question of emphasis. Abiotic and biotic impacts are typically impacts on individual components and interrelationships. Ecological impacts are impacts on the overall potential of an aquatic resource to do the work of linking together various abiotic and biotic transformations of matter and energy. This important difference can be highlighted by means of an analogy.

For example, one may consider a random assortment of automobiles, buildings, roads, people, and power plants as a collection of abiotic and biotic items. One may also study the potential impacts of a proposed project on each of these individual items. Yet, automobiles, buildings, roads, people, and power plants, as well as any number of other abiotic and biotic items can, in fact, be arranged in various patterns of interrelationships. They may be parts of a rural or an urban community; they may also be a part of a collage hanging in a museum. It is the task of ecology to study the interrelationships of such abiotic and biotic items in the form of a real community and not from the perspective of a pretty picture.

The science of ecology, therefore, assigns no intrinsic artistic, philosophical, or intellectual worth to individual components of the environment, but rather, it assigns systemic functions—functions that describe the sequential passage of energy through the environment and the recycling of materials from one component to another. Ecological impacts are, accordingly, primarily impacts on these systemic functions and, only secondarily, on those components that perform the functions.

With respect to aquatic ecosystems, the assessment team should, therefore, not be so concerned about game fish that it fails to consider the ecological roles of individual fish species; not so concerned about water quality standards for swimming that it fails to consider the effects of different parameters of water quality on benthos; and not so concerned about the macrocomponents of human aesthetics that it fails to consider project impacts on microbial processes. This is not to say that the human use of aquatic ecosystems is unimportant in impact assessment, but rather, that the assessment team should not equate the human interest in water with scientific understanding of its ecological dynamics.

By focusing on systemic functions in aquatic resources, aquatic ecology provides the assessment team with an important tool for meeting the decision-making goals of NEPA (Chapter 4).

For example, ecological considerations allow the assessment team to identify the significance of innumerable, individual abiotic and biotic impacts with respect to an integrated system (i.e., the aquatic ecosystem).

The importance of individual abiotic and biotic changes with respect to a whole system aids the decision-maker in setting priorities. For instance, is a fish kill important enough to warrant redesign? One way to measure its importance is to look at the fish kill in terms of its ramifications to ongoing food webs. Sometimes a fish kill can have little or no impact on aquatic food webs, and the ecosystem will remain essentially unaltered. Sometimes a fish kill can dramatically change an aquatic ecosystem. Of course, the ecological ramifications of a fish kill are not necessarily as important as its social ramifications!

Second, ecological considerations force the assessment team to consider long-term versus short-term environmental costs and benefits. For example, an increase in turbidity may be a short-term abiotic consequence of construction. But what of various long-term effects on population and trophic dynamics?

Third, ecological considerations absolutely require careful delineation of direct and indirect consequences of project actions. Many examples of this requirement have been discussed throughout this chapter.

References

American Public Health Association, American Water Works Association, and Water Pollution Control Federation
    1976   *Standard methods for the examination of water and wastewater* (14th ed.). Washington, D.C.: American Public Health Association.

Austin, E. R., and G. F. Lee.
    1973   Nitrogen release from lake sediments. *Journal of Water Pollution Control Federation, 45*(5): 870–879.

Environmental Protection Agency
    1974   *An approach to a relative trophic index system for classifying lakes and reservoirs (Working Paper No. 24).* Corvallis, Washington: National Eutrophication Survey, Pacific Northwest Environmental Research Laboratory.

    1977   *Conducting sanitary surveys of water supply systems.* Chicago: U.S. Environmental Protection Agency, Region V.

Gaudet, J. J.
    1974   The normal role of vegetation in water. In *Aquatic vegetation and its use and control,* edited by D. S. Mitchel. Paris: UNESCO.

Goldman, C. R. (Editor)
    1974   *Primary productivity in aquatic environments.* Berkeley, California: University of California Press.

Hendricks, A. C., and J. K. G. Silvey
    1973   Nutrient ratio variation in reservoir sediments. *Journal of the Water Pollution Control Federation, 45*(3).

Hynes, H. B. N.
    1971   *The biology of polluted waters.* Toronto: University of Toronto Press.

Lee, G. F.
    1970   *Eutrophication.* Springfield, Virginia: The University of Wisconsin Water Resources Center, National Technical Information Service. (PB 197-697)

Massachusetts Division of Environmental Quality Engineering
1977 *Draft environmental impact report, control of aquatic vegetation in the Commonwealth of Massachusetts*. Boston: Division of Waterways, Commonwealth of Massachusetts.
Mulligan, H. F.
1969 Management of aquatic vascular plants and algae. In: *Eutrophication: Causes, consequences, correctives*. Washington, D.C.: National Academy of Sciences, Printing and Publishing Office.
National Academy of Sciences
1969 *Euthrophication: Causes, consequences, correctives*. Washington, D.C.: National Academy of Sciences, Printing and Publishing Office.
Odum, E. P.
1963 *Ecology*. New York: Holt, Rinehart and Winston.
Odum, E. P.
1971 *Fundamentals of ecology* (3rd ed.). Philadelphia: W. B. Saunders.
Oosting, H. J.
1956 *The study of plant communities*. San Francisco: W. H. Freeman.
Pamatmat, M. M., R. S. Jones, H. Sanborn, and A. Poltagwat
1973 *Oxidation of organic matter in sediments*. Office of Research and Development, Washington, D.C.: U.S. Environmental Protection Agency. (EPA-660/3-73-005)
Rheinheimer, G.
1974 *Aquatic microbiology*. New York: John Wiley & Sons.
Russell-Hunter, W. D.
1970 *Aquatic productivity: An introduction to some basic aspects of biological oceanography and limnology*. London: The Macmillan Company, Collier–Macmillan Limited.
Ruttner, F.
1969 *Fundamentals of limnology* (3rd ed.). Toronto: University of Toronto Press.
Smith, R. Leo
1974 *Ecology and field biology* (2nd ed.). New York: Harper and Row.
Welch, P. S.
1963 *Limnology* (2nd ed.). New York: McGraw-Hill.
Wetzel, R. G.
1975 *Limnology*. Philadelphia: W. B. Saunders.
Wildung, R. E., and R. L. Schmidt
1973 *Phosphorus release from lake sediments*. Washington, D.C.: Office of Research and Monitoring, U.S. Environmental Protection Agency. (EPA-R3-73-024)
Williams, J. D. H., and T. Mayer
1972 Effects of sediment diagenesis and regeneration of phosphorus with special reference to Lakes Erie and Ontario. In *Nutrients in Natural Waters* edited by H. E. Allen and J. Kramer. New York: John Wiley & Sons.

10

Ecological Impacts: Terrestrial Environment

Introduction

Terrestrial environments, like aquatic environments, are comprised of interrelated abiotic and biotic components. And, as in aquatic environments, material and energetic transformations in the terrestrial environment are sequentially linked by grazing and detritus food webs which accomplish the ecological work of primary production, consumption, and decomposition. As depicted in Fig. 10.1, primary producers include the trees, shrubs, forbs (wildflowers), and grasses; primary consumers include a large diversity of mammals, insects, birds, reptiles, and amphibians; secondary and tertiary consumers include the predators; and, finally, decomposers include the bacteria, fungi, and other biota that mineralize the organic molecules of detritus and release essential plant nutrients back into the environment. Thus, while the ecological work carried on in terrestrial ecosystems is the same as that carried on in aquatic ecosystems (Fig. 10 Page 143), the biota that accomplish that work are different.

Differences in the biology of terrestrial and aquatic environments are directly related to differences in habitat constraints in these two environments.

Biological adaptations in terrestrial and aquatic environments reflect habitat constraints, including constraints related to (a) gravity; (b) oxygen concentrations; (c) temperature variations; (d) availability of water; and

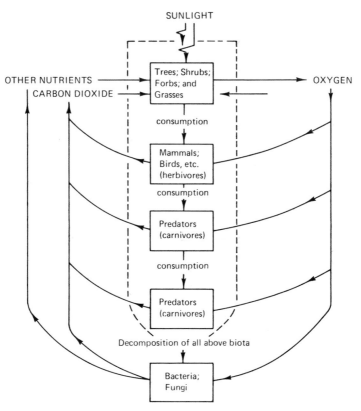

FIGURE 10.1 *An example of niche interrelationships and material cycling in terrestrial ecosystems.* [Adapted from U.S. Department of Transportation. 1975. Ecological Impacts of Proposed Highway Improvements. Washington, D.C. Federal Highway Administration.]

(e) the nature of substrate. For example, water is itself a supporting medium and imparts a buoyancy to living things that counteracts the effects of gravity. But in the terrestrial environment, there is essentially no counterforce to gravity. Thus terrestrial biota must be adapted to withstand gravitational stress. Adaptations include supportive structures, such as the woody tissue of trees and shrubs, the bone–tissue of vertebrates, and the chitinous exoskeleton of insects. Water also acts as a buffer to abrupt changes in temperature. Yet, air does not; thus the daily temperature of the terrestrial environment generally varies greatly. Many terrestrial biota must, accordingly, be adapted to these changes. Finally, the spatial and temporal availability of water in the terrestrial environment can also vary greatly, and terrestrial biota must be adapted to these variations if they are to survive.

These and other examples of biological adaptations to the constraints of terrestrial and aquatic habitats are included in Table 10.1. It is impor-

TABLE 10.1

Comparison of Some Important Habitat Constraints in Terrestrial and Aquatic Environments[a]

Important habitat constraints	Some comments on terrestrial environment and biota	Some comments on aquatic environment and biota
Gravity	Exo- and endoskeletons to support biomass. Supportive structures can be substantial percentage of total biomass of organism.	Water itself acts to support biomass. Adaptations primarily to maximize or utilize buoyancy.
Oxygen concentration	Concentration in air is generally the same over space and time. Can vary greatly in soils, depending on water content. Some specific plant adaptations to wet soils.	Can vary greatly with location and time. Aquatic biota often adapted to spatial and temporal variations.
Temperature variation	Can vary greatly with location and time. Terrestrial biota often adapted to spatial and temporal variation.	Heat capacity of water tends to prevent abrupt change. Aquatic biota generally not adapted to abrupt and large changes.
Availability of water	Can vary greatly with location and time. Terrestrial biota often adapted to spatial and temporal variation.	Generally not a constraint to aquatic biota.
Nature of substrate	Generally stable and therefore not usually a major constraint to terrestrial biota.	Generally unstable and transitory. Many types of adaptation to temporal changes in substrate.

[a] Based on materials by George Camougis, New England Research, Inc. Unpublished Lecture.

tant to note that the comments included in this table are very general and do not take account of many local factors in both terrestrial and aquatic environments. However, from the perspective of a broad overview of the world's terrestrial and aquatic environments, it is possible to state that:

1. Terrestrial organisms (flora and fauna) generally have structural and/or behavioral adaptations to the unrelieved force of gravity, abrupt and wide variation in temperature, and to spatial and temporal changes in the availability of water.
2. Aquatic organisms (flora and fauna) generally have structural and/or behavioral adaptations to abrupt and wide variation in the concentration of dissolved oxygen and to the transitory nature of substrates.

For purposes of assessing impacts on terrestrial ecosystems, it is essential that the assessment team:

1. Identify important habitat constraints within the proposed project area
2. Relate geographical and temporal variations of these constraints to the structural and/or behavioral adaptations of current populations
3. Predict project impacts on habitat constraints, and consequent effects on current populations
4. Evaluate overall consequences on the trophic dynamics of terrestrial communities in both project and nonproject areas

Each of these tasks will be discussed in the following sections.

Identifying Habitat Constraints

As discussed in Chapter 9, habitat is the place in which organisms can be found—a place that is characterized by various abiotic attributes, (such as temperature, relative humidity) that can be influenced by various biotic factors, including organismic, population, and community factors. The ecological importance of habitat is that it gives spatial and temporal location to "living transformers" that accomplish the work of primary production, consumption, and/or decomposition. Projects that alter habitat, therefore, have the potential for changing the types of organisms that will be located at a specific location at a specific time and, therefore, for changing the trophic structure of terrestrial ecosystems. For purposes of assessing ecological impacts it is therefore necessary that the assessment team identify ongoing habitat constraints within the project area. Such constraints may be described in terms of horizontal, vertical, and temporal variations in habitat.

Two basic approaches may be utilized in the evaluation of geographic and temporal aspects of habitat—a reconnaissance-type evaluation and a vegetative analysis (DeVos and Mosby, 1971, pp. 135–172).

A *reconnaissance-type evaluation of habitat* in the project area is typically employed for the evaluation of a large-area habitat and includes various techniques for evaluating the suitability of the project area for individual animal species. Such techniques include:

1. Cover mapping—mapping of the project area by symbols that represent vegetative cover, including overstory (e.g., trees), understory (e.g., shrubs), and ground cover (e.g., mosses)
2. Cover density evaluation—mapping of the project area by symbols

that represent both qualitative and quantitative aspects of terrestrial cover
3. Soil evaluation—mapping the project area with respect to soil types, including physical and chemical characteristics
4. Evaluation of food production, availability, and consumption—mapping of the project area with respect to the total yearly crop (of food for specific organisms), the quantity of food available at a particular time, and the proportion of available food actually consumed
5. Identification of plant indicators—mapping the project area in terms of plant species that indicate conditions of climate, soils, moisture, and previous disturbance.

A *vegetative analysis of habitat* in the project area is the typical approach employed for a detailed examination of the vegetative aspects of wildlife habitat. A variety of techniques may be employed; these techniques measure one or more of the following aspects of vegetation:

1. Presence–absence (a simple statement of the presence or absence of a particular vegetative species)
2. Basal area (the proportion of ground surface occupied by a species)
3. Cover (the vertical projection of aboveground parts on to the ground)
4. Frequency (the percentage of sample plots in which a species occurs)
5. Density (the number of individuals per unit area)
6. Dominance (some estimate of the comparative position of a particular species in an area)
7. Importance (the value of a given plant for a given purpose)

With respect to these two basic approaches to evaluating habitat constraints over the project area, several points should be emphasized for the purpose of impact assessment.

First, *the assessment team should consider itself primarily qualified to compile existing literature* on reconnaissance-type and vegetative evaluations of habitat. Under no circumstances should the assessment team assume that it can undertake actual field studies based on these techniques without expert guidance. Field studies undertaken for the purpose of habitat evaluation absolutely require professional inputs in design, operational, and analytic phases of such studies.

Second, *the assessment team should assume that some literature (including reconnaissance-type and vegetative evaluations) does exist and is relevant to the general project area.* While this assumption may ultimately prove to be incorrect, at least the team will first have exhausted possible sources of information before committing itself to expensive and time-consuming field studies.

Appropriate agencies and organizations that should be contacted for possible information include federal and state fish and wildlife and geological survey agencies, environmental quality and/or natural resource agencies, local and regional universities, and local and regional wildlife or conservation organizations. It is also important to consult other EISs that have been completed for previous and/or ongoing projects in the general project area. Such EISs may have directly relevant information or may identify additional sources of information. The assessment team should also be aware that while such agencies and organizations may not have actual habitat maps of the general project area, they may have tabulations of soil and vegetation types, aerial photographs (black and white, color, and/or color–infrared) that can be utilized and transferred to map overlays. Particular attention should be given to data and information derived from the utilization of remote-sensing techniques (National Academy of Sciences, 1970; Corps of Engineers, 1974).

Two key concepts are useful for soliciting appropriate data and information from these agencies and organizations—plant associations, and biomes.

A *plant association* is a collection of plant species that have a high probability of occurring together in a local area. A plant association may include relatively few species, or it may include dozens of species. Individual plant associations (also referred to as *plant formations* or *vegetative types*) are typically named after the predominant species in the association, or after two or more prevalent species. In some instances, plant associations may also be assigned a number (Society of American Foresters, 1954).

Examples of the general information required by the assessment team are:

1. What plant associations are present in the project area?
2. How many acres does each association cover in the project area?
3. What individual species of each association are likely to occur in the project area?
4. Do individual associations and species in the project area occur elsewhere in the general region? Are they common or uncommon in the region? How many acres?

A *biome* is a community of plants and animals that is characteristic of a general climatic area (Odum, 1971, p. 378; Walter, 1973; Cloudsley-Thompson, 1975). Biomes are often named after the dominant type of vegetation, for example, coniferous forest biome, the deciduous forest biome, the grassland biome. However, the concept of biome is a broad, community concept that includes consideration of the interrelationships of plants and animals (Smith, 1974, p. 256). Thus, in soliciting information on the biome(s) in which a project is located, the assessment team is soliciting information on plant associations (including individual plant species and animal species typically found in habitats characterized by the presence of these plant associations.

In some instances, of course, the assessment team will find little specific information that can be utilized to describe the general habitat constraints of the project area. In such cases where no reconnaissance or vegetative evaluations of habitat have been performed, the assessment team should (a) compile whatever information is available on soils, climate, hydrology, vegetation, etc.; (b) consult local botanists, zoologists, and ecologists (and/or the general literature) in order to identify possible plant associations and animal populations; and (c) conduct such aerial and field reconnaissance as may be required in order to identify the type, location, and extent of actual plant and animal associations in the project area.

It is improbable that the assessment team will have to undertake detailed, long-term field studies of habitat in the project area in order to complete a reasonable assessment of project impacts. However, such field studies may be required in order to fully evaluate project impacts on special issues, including threatened and endangered species. In such instances, it is requisite to utilize professional guidance. This is particularly important for evaluating project impacts on "critical habitat" which is provided for by Section 7 of the Endangered Species Act of 1973 (Schreiner, 1976).

Relating Habitat to Biological Adaptations

The assessment team should assume that plants and animals found within the project area are specifically adapted to those habitats—adapted in terms of their bodily structure and characteristics, their behavior, and their life cycles. While this assumption may not be true in specific cases, it is only for the professional biologist to point out these cases.

Biological adaptations may be described in very general or in very specific terms. The assessment team should become familiar with some of the basic concepts that underlie an evaluation of both general and specific biological adaptations. Such concepts facilitate the information-gathering phase of impact assessment by focusing the assessment team's attention on key considerations.

An example of an important concept that gives perspective to the whole question of biological adaptation is Shelford's law of tolerance (Odum, 1971, pp. 107–109). This law essentially states that the deficiency or excess of any physical or chemical parameter can influence or directly determine the presence and success of an organism in a particular habitat.

The implications of Shelford's law are depicted in Fig. 10.2. In this example, the range TA1–TA2 represents the ambient temperature range in a given habitat during a particular portion of the season. The minimum temperature in this period is TA1, and the maximum is TA2. The Law of Tolerance states that deficiencies in this temperature range (temperatures less than TA1) and excesses (temperatures greater than TA2) can determine both the presence and the success of an organism (e.g., Species A). As indicated in the figure, TA2 represents the near extreme temperature at which there is a growth of Population A. At temperatures just above TA2, Population A stops increasing, and the percentage of the population that cannot reproduce increases dramatically. At still higher temperatures (greater than T1), Population A disappears.

Figure 10.2 also depicts the succession of Species A by Species B at temperatures greater than T1. In such a case, Species B is considered to be better adapted to elevated temperature than is Species A.

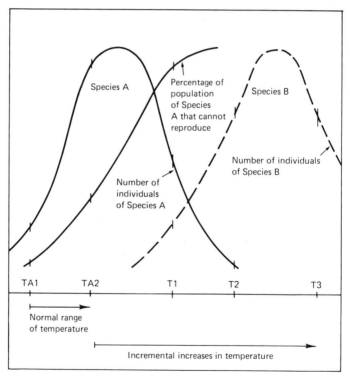

FIGURE 10.2 *An example of population changes due to incremental increases in te:n-perature.*

Several points should be emphasized with respect to Shelford's law and its depiction in Fig. 10.2:

1. The assumption of this law is that there is an optimum range for each habitat parameter for each type of organism. When a parameter is less than or greater than this optimum, the presence and success of the organism in that habitat are in jeopardy. This is merely another way of saying that the organism is adapted to limited variations in habitat and that when structural, and/or behavioral adaptations are insufficient to deal with actual variations (mediated, perhaps, by project activities), the organism disappears from that habitat.

2. It is highly unlikely that the disappearance of a particular kind of biological population will result in an empty habitat. Nature abhors biological as well as physical vacuums. The disappearance of one population is, in fact, the usual first step in a sequential colonization of the altered habitat by another population specifically adapted to the altered habitat.

In light of these two considerations, the assessment team should understand that project activities that can affect the successful utilization of adaptations by existing populations not only have the potential for causing the demise of those particular populations—but also for initiating a biological succession in the project area. Frequently, assessment teams have focused only on populations that might be excluded from the project area and have given little if any attention to those populations that may be introduced. It is necessary in any comprehensive assessment to consider both exclusions and introductions.

Shelford's law has general applicability to aquatic, terrestrial, and wetland ecosystems and underlies any understanding of biological adaptation to these environments.

A concept with more limited application, and particularly relevant to the adaptations of terrestrial vegetation, is the root-to-shoot concept. While the *root-to-shoot ratio* (Odum, 1971, p. 375) is a numerical ratio between below-ground biomass (root) and aboveground biomass (shoot) of vegetation, it is referred to here as a concept—because the numerical quantity is as much a tool for understanding complex and interrelated phenomena as it is for measuring a particular attribute of vegetation.

For example, a typical terrestrial plant is a living machine that does the ecological work of primary production. Yet, most terrestrial plants are basically two machines in one. The aboveground portion carries on photosynthesis and, therefore, requires water and nutrients that are supplied by the below-ground portion. The below-ground portion, in turn, requires food–energy (ultimately supplied by the aboveground portion) in order to pull water away from the surfaces of soil particles where it is tightly bound. If the roots do not get the energy from photosynthesizing leaves, water and nutrients will not be obtained. But if water and nutrients are not obtained, the leaves will not be able to carry on photosynthesis. The root-to-shoot ratio is a numerical representation of the particular balance reached between the energy-producing leaves and water–nutrient-supplying roots. This balance may be assumed to be optimum for a particular plant in a particular habitat. This balance may, therefore, be influenced by project-mediated changes in that habitat. For example, alteration of groundwater hydrology or dust-fall of chemicals and materials on leaves can result in a readjustment of root-to-shoot ratios. One consequence of such a readjustment may be, of course, a wilting of the plant. If a new and adequate balance of roots and shoots cannot be achieved, the plant will die.

With respect to root-to-shoot ratios, the assessment team should consider that:

1. The root-to-shoot ratio represents a vegetative growth pattern which is adaptive to above- and below-ground constraints of terrestrial habitats.
2. The assessment team should, therefore, consider the root-to-shoot ratio of vegetation in the project area in precisely the same manner as it would consider any biological adaptation—that is, the teams should identify key environmental parameters (e.g., groundwater hydrology, sunlight, nutrient concentrations) that actually determine the adaptive growth pattern.

Another concept that is important for relating terrestrial habitats in project areas to biological adaptation is the general biological concept of *population dispersal* (Odum, 1971, p. 200). Population dispersal includes *emigration* (the one-way outward movement of a population from an area), *immigration* (the one-way inward movement of a population into an area), and *migration* (the periodic departure and return of a population).

These forms of population dispersal should be viewed by the assessment team as behavioral adaptations to habitat constraints on actual biological needs. Such an approach precludes the cursory consideration of animal-behavioral patterns that often emanate from the human attitude that animals, like people, do whatever occurs to them to do. *Assessment of project impacts on animal behavior should be predicted on the assumption that behavior is a mechanism for achieving fulfillment of real needs.*

In the process of relating behavioral adaptations such as population dispersal, the assessment team should distinguish between *nocturnal* (nighttime), *diurnal* (daytime), *diel* (24-hour), and *seasonal* patterns. Projects may affect each of these behavioral patterns. Security lighting, for example, may affect nocturnal migrations; equipment operation during the day may affect diurnal migrations; fencing may affect diel migrations; and, finally, the location of the entire project may be such as to interfere with normal seasonal migrations.

Project Effects on Habitat Constraints and Biological Adaptations

As soon as habitat constraints and related biological adaptations are identified, the assessment team can begin to predict potential project impacts on autecological components of the terrestrial environment (Chapter 9). There are two basic steps to be taken: (a) identifying the actual uses of existing vegetation by other terrestrial biota; and (b) identifying project impacts (direct and indirect) on existing vegetation and on the use of this vegetation by other terrestrial biota.

Existing plant associations and species in the project area perform numerous important functions for both local and migratory biological populations, including:

- Food (both living and dead plant material)
- Shelter (short- and long-term)
- Sites for breeding and rearing of offspring
- Materials for nest-building

An important concept that serves to integrate these functions is the concept of carrying capacity.

The *carrying capacity* of a land area[1] is the potential of that area for supporting an animal population. Carrying capacity is typically expressed as a number (i.e., the number of organisms of a particular species per unit area of land). While the determination of carrying capacity requires the professional judgment of wildlife biologists, it is generally possible for the assessment team to collect important information on carrying capacity in the project area without recourse to professional field studies. Basic approaches include:

1. Estimating acreage in project area for each type of plant association
2. Estimating acreage of each plant association to be clearcut during construction
3. Estimating acreage to be replanted (if any) with what type of association

---

[1] Carrying capacity can also refer to the potential for water to support aquatic biota. This is typically done in fisheries biology. However, the most common, general usage of this concept is with respect to terrestrial and wetland ecosystems.

4. Using existing literature and/or local expertise to develop a list of individual species in each association
5. Using information on existing associations and individual plant species, and on proposed plantations, to identify potential uses of vegetation by local and migratory animal species

Once plant associations and representative species are known, the assessment team can utilize existing guides to wildlife and their food habits (e.g., Martin *et al.*, 1951) to identify animal species likely to utilize each vegetative type for food. Once associated animal species have been identified, the assessment team should consult with local agencies, organizations, and individuals in order to ensure that the compiled list of animals is, in fact, representative of resident and migratory species in the projected area. The reviewed compilation can then serve as a guide to more specialized literature on the structure, behavior, and life cycles of individual animal species. The assessment team should consult this literature in order to identify particular habitat requirements, tolerance to various environmental factors, and other characteristics that might be affected by the proposed project. *Of special importance is the need for the assessment team to utilize all compiled information for the purpose of eliciting estimates (from literature and/or practicing professionals) of the current carrying capacity of the area for different animal species and the likely carrying capacity for these same species after the proposed project has been implemented.* One useful guide to the carrying capacity of habitats that differ with respect to overall quality is available through the United States Department of Agriculture (1968). Others can be obtained from various wildlife and conservation organizations, state government agencies, and local colleges and universities that have curricula in wildlife biology and management. These guides are very useful for providing quantitative answers to the following types of questions:

1. If 20 acres of a particular kind of association is to be clearcut, and no subsequent replanting will occur, what will be the consequence in terms of the carrying capacity of this area for existing animal populations?
2. If landscaping will result in the establishment of 10 acres of new plantation, how many and of what kind of animal populations might be expected to immigrate into the area?

The assessment team should not assume that projects always result in a decrease in carrying capacity. As already discussed, changes in habitat can preclude the existence of one type of organism in the project area for one species and increase it for another. Projects can also increase the carrying capacity of the project area for populations which are already in the area, but which are constrained by the availability of food, shelter etc. For example, deer may currently exist in a project area that is largely characterized by dense, mature forests. It may very well be that the existing deer population is relatively small because of the limited browse material available in mature forests. Clearcutting of forested areas typically allows for an enhanced growth of low shrubs that are an important food supply for deer. Thus clearcutting can actually increase the carrying capacity of the area for deer populations. However, the same project can result in circumstances that tend to decrease the overall carrying capacity for deer. A possible set of circumstances and their effect on carrying capacity are indicated in Fig. 10.3.

It is of paramount impoitance that the assessment process include an evaluation of the total effect of all such circumstances on the carrying capacity of the project area for all major groups of terrestrial biota. If the assessment team focuses on only one aspect of project development (e.g., clearcutting) or on only one type of organism (e.g., game mammals), potential impacts on the total terrestrial ecosystem will be overlooked. *A comprehensive assessment will, therefore, include an assessment of project impacts on the carrying capacity of the project area for all major biological populations (i.e., primary producers, consumers, decomposers) and will include consideration of impacts to be expected in all phases of project development, from early construction to operational phases of the project.*

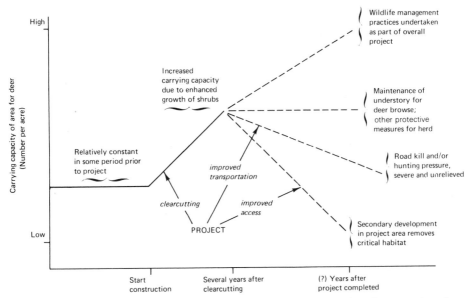

FIGURE 10.3 *Some possible changes in carrying capacity for deer due to project development.*

In each phase of project development there can be project-mediated changes in habitat and, thus, changes in carrying capacity. For example, changes in hydrology and soil compaction can affect root-to-shoot ratios of remaining vegetation. One possible consequence of such alterations can be the wilting of certain plant species and the consequent loss of food and/or shelter for those biota that are dependent on that vegetation. Air pollution during the construction and operational phases can directly affect sensitive vegetation (Naegele, 1973) and also result in the loss of vegetative food supplies and shelter. Security lighting, depending on the frequency of light and its intensity can also injure sensitive plant species (U.S. Department of Agriculture, 1973). Of course, project-mediated changes in local habitats can directly affect animals and their use of local habitats. For example, changes in the frequency, duration, and periodicity of ambient noise levels during construction and operational phases can preclude the successful reproduction of various species and/or interrupt diel and/or seasonal migrations.

As a general guide, the assessment team should consider that:

1. Project-mediated changes in any ongoing process of energetic and/or material introduction, transformation, translocation, sequestration, and dissipation (Chapter 8) can directly affect plants and animals.
2. Any effect of an individual plant or animal population can have indirect consequences on other plant and animal populations.
3. Direct and indirect impacts on plants and animals in the project area may affect plant and animal populations far removed from the immediate project area.

Particular note should be made of a certain type of argument often found in EISs. The argument proceeds as follows: The project is likely to result in some temporary and/or long-term displacement of animals in the project area, but these displaced populations will be able to migrate to other habitats in the surrounding region, and there will, therefore, be no significant long-term impact on populations other than a change in their geographical distribution.

Such an argument is quite logical—as long as (a) suitable habitat is, in fact, available in the general region; and (b) the current carrying capacity of this alternative habitat for the displaced population is not already saturated. However, if in fact either one of these conditions is not met, the term "displaced population" is very misleading. While these organisms may very well migrate to other available habitats, they can, in fact, supersaturate the carrying capacity of those habitats for that species. When the supersaturation of carrying capacity occurs, it is possible that losses due to increased competition, disease, and starvation can actually exceed the total number of organisms originally displaced.

Just as it cannot be assumed a priori that displaced organisms will, in fact, be able to succeed in alternative habitats, so it cannot be demonstrated that all alternative habitats are always saturated with respect to their carrying capacity for individual populations. While policy memoranda of various government officials can declare that the carrying capacity of all habitats is assumed to be at a maximum, it is possible that actual field studies of alternative habitats may prove the contrary.

Evaluating Consequences on Trophic Dynamics

As discussed in Chapter 9, all ecosystems should be considered to be at some particular point in an overall ecological development. The point is, of course, that communities of biological populations and interrelated abiotic factors are, in fact, more often characterized by constant change than by stability (Oosting, 1956, p. 236).

The term *succession* typically refers to predictable ecological changes. An example of succession in the aquatic environment is the eutrophication of lakes and ponds (Chapter 9). Examples of succession in the terrestrial environment include *primary succession* and *secondary succession* (Oosting, 1956, p. 240; Odum, 1971, p. 258).

Both primary and secondary successions (also called *ecological succession*) are characterized by predictable changes in species and in community processes that (a) result from modifications of the physical environment by the community; and (b) culminate in an ecosystem that is stable with respect to its overall climate (Odum, 1971, p. 251).

Primary succession begins on a bare area where no vegetation has previously grown (e.g., bare rock, new islands); secondary succession begins in areas that have been previously vegetated (Oosting, 1956, p. 240). The evaluation of ecological impacts of most projects typically includes consideration of impacts on secondary succession.

The total sequence of biological communities that replace one another during secondary succession is referred to as the *sere* (Odum, 1971, p. 251). The sere may be subdivided into three general types of *seral stages:* the pioneer stage, developmental stages, and the climax stage.

In the pioneer stage, plant and animal organisms may be generally characterized as opportunistic. They are typically small, have short and simple life histories, and reproduce rapidly. The community as a whole tends to be characterized by simple, grazing food chains, rates of photosynthetic production that are greater than rates of respiratory consumption, and poorly organized zones of vertical and horizontal patterns of distribution.

In the climax stage, organisms are generally large, have long and complex life cycles, and are less apt to reproduce rapidly. The community as a whole tends to be characterized by detritus food webs, rates of photosynthetic production that are roughly equal to rates of respiratory consumption, and highly organized zones of vertical and horizontal patterns of distribution.

Organismic and community characteristics of developmental stages are intermediate to the characteristics of pioneer and climax stages.

As a general rule, the best information on the stage of succession in a project area is available from local sources, including regional offices of federal agencies, state agencies, local colleges and universities, and local botanical and wildlife associations. It is important to contact these or other appropriate agencies and organizations in order to determine (a) the present seral stage in the project area; (b) probable future seral stages if the proposed project is not undertaken; and (c) probable future seral stages if the proposed project is undertaken.

The importance of these determinations lies in the fact that ecological succession is a natural process and, even in the absence of human activity, results in the sequential replacement of one biological community by another until some steady state between the community and its abiotic environment is reached (i.e., the climax stage is established). Thus it is necessary to isolate project-mediated changes in populations from natural, successional changes in populations. For example, as plant succession proceeds from pioneer stages to developmental stages, habitat changes will result in the loss of certain animals. A particular project may speed up the loss of such animals by removing habitat more quickly than it would otherwise disappear through succession. A project may also serve to prevent the loss of such animals by keeping the project area in an early phase of succession.

It is unlikely, of course, that a typical project will alter the ecological succession in an entire region. Most often, projects impact highly localized successions. This typically results in the establishment of an *ecotone* (or *edge*), a zone of transition between different types of communities. Examples of ecotones include the zone of shrub growth between a forest and an open field, and the demarcation line between tall, native grasses and short, planted grasses. Habitats within an ecotone are different from the habitats on either side of it. Thus organisms that utilize ecotones for food and shelter are not typical of the biological communities found outside of the ecotone.

Some types of ecotones are often viewed as positive consequences of project development. For example, the shrubby ecotone between a mature forest and mowed-grass enbankments along a highway can be an important food supply for deer. It is also possible that the ecotone will serve to bring deer too near the highway, and thus increase the roadkill of deer. It is also possible, of course, "to edge a forest to death" (e.g., turn it all into an open field) by piecemeal introduction of ecotones in previously stable mature forests. The assessment team should evaluate the overall consequences of ecotones with respect to regional ecology. *The assessment team should definitely not assume that ecotones will always have desirable consequences for terrestrial biota and ecosystems.*

The importance of a regional overview in assessing impacts on terrestrial ecosystems cannot be overemphasized. Professional ecologists tend to deal with whole forests, not just a few acres; and with whole grasslands, not just a few square miles—that is, they tend to take a biome perspective (Reichle, 1970; Ohmann *et al.*, 1973). Project impacts along an alignment, say, through a forest, should therefore be viewed in terms of the forest ecology of the overall region and in terms of the interrelationships between the forest and other biomes in that region. How large, then, is a region?

There is no general rule that the assessment team can utilize in order to delineate the acreage of its concern. It depends on particular interrelationships among specific primary producers, consumers, and decomposers. However wide the area required to identify basic interrelationships among these living energy transformers is precisely the area required for a comprehensive understanding of potential impacts on terrestrial ecosystems characterized by those transformers. Whatever the size of that area might be, it is quite certain that it will extend beyond the property lines and rights-of-way of the proposed project.

Concluding Remarks

The vast and highly diverse literature on the abiotic and biotic components of the terrestrial environment presents obvious difficulties to an assessment team that is not professionally trained in all of the relevant disciplines and subdisciplines. There are always those who would argue that only persons who have such training belong on the assessment team. Yet, such an argument is quite inappropriate when one considers that all of the disciplines that can be brought to bear on terrestrial environments are still but a fraction of the disciplines that are relevant to assessing impacts on the total human environment. From a practical point of view, then, assessment teams will include people who must rely less on their individual knowledge of innumerable terrestrial components and processes than on basic organizational skills. Previous sections of this chapter have, therefore, focused on a limited number of concepts that meet basic organizational needs of an assessment team faced with the complexity of terrestrial ecosystems. The most general and pervasive of these needs is, of course, to ask the right questions. The most common problem is, of course, to know when one is getting the right answers! With respect to this problem there is a vast difference between the following two questions that might be asked in the progress of impact assessment:

1. What can happen during blasting in the project area?
2. What biological populations will be specifically affected, and how, by blasting during June?

It is very probable that the first question will elicit an opinion from just about anyone and that the team member who poses the question will be left with the problem of deciding which answers are founded in fact, and which are pure speculation.

The second question is much more specific and indicates that its framer fully expects that:

1. Specific reasons can be given for identifying some populations as being more susceptible than others to blasting.
2. Different populations might be affected differently.
3. Seasonality is at least one factor to be considered.

It is necessary for the assessment team to expect this kind of specificity in all responses to its proferred questions. *When specificity is lacking in answers supplied by any informational source, it is an inappropriate informational source for assessing impacts on terrestrial ecosystems.*

The importance of such concepts as plant associations, biomes, carrying capacity, biological adaptation, plant and ecological succession, ecotone, etc., lies, therefore, in the fact that questions that utilize these concepts provide some basis for evaluating the relevance of answers given in return. Such concepts are certainly tools for unlocking storehouses of technical knowledge (which also includes knowledge that some things are as yet unknown). But they are also measures of the depth and relevance of the knowledge of those who would offer their advice and guidance to the assessment team.

References

Cloudsley-Thompson, J. L.
 1975 *Terrestrial environments.* New York: John Wiley & Sons.
DeVos, A., and H. S. Mosby
 1971 Habitat analysis and evaluation. In *Wildlife management techniques* (3rd ed., revised), edited by R. H. Giles. Washington, D.C.: The Wildlife Society.
Martin, A. C., H. S. Zim, and A. L. Nelson
 1951 *American wildlife and plants.* New York: Dover.
Naegele, J. A. (Editor)
 1973 *Air pollution damage to vegetation.* Washington, D.C.: American Chemical Society.

National Academy of Sciences (NAS)
  1970   *Remote sensing, with special reference to agriculture and forestry.* Washington,
         D.C.: National Academy of Sciences.
Odum, E. P.
  1971   *Fundamentals of ecology* (3rd ed.). Philadelphia: W. B. Saunders.
Ohmann, L. F., C. T. Coshwa, and R. E. Lake
  1973   *Wilderness ecology: The upland plant communities, woody browse production,
         and small mammals of two adjacent 33-year-old wildfire areas of northeastern
         Minnesota.* St. Paul, Minnesota: North Central Forest Experiment Station, U. S.
         Department of Agriculture.
Oosting, H. J.
  1956   *The study of plant communities.* W. H. Freeman, San Francisco.
Reichle, D. E. (Editor)
  1970   *Analysis of temperate forest ecosystems.* Springer-Verlag, New York.
Schreiner, K. M.
  1976   Critical habitat: What it is—and is not. *Endangered Species Technical Bulletin, 1*
         (2).
Smith, R. L.
  1974   *Ecology and field biology* (2nd ed.). New York: Harper and Row.
Society of American Foresters
  1954   *Forest cover types of North America (exclusive of Mexico).* Bethesda, Maryland:
         Society of American Foresters.
Corps of Engineers
  1974   *An assessment of remote sensing applications in hydrologic engineering.* Davis,
         California: The Hydrologic Engineering Center, Corps of Engineers.
U.S. Department of Agriculture
  1968   *Technical Note, RE: How Much Wildlife?* Upper Darby, Pennsylvania: Regional
         Technical Service Center, Soil Conservation Service, U.S. Department of Agricul-
         ture.
  1973   *Security lighting and its impact on the landscape.* Beltsville, Maryland: Agricul-
         tural Research Service, Plant Genetics and Germplasm Institute, Agricultural
         Environmental Quality Institute. (Publication CA-NE-7)
Walter, H.
  1973   *Vegetation of the earth, in relation to climate and the eco-physiological condi-
         tions.* New York: Springer-Verlag.

11

Ecological Impacts:
Wetland Environments

Introduction

It has been estimated (Shaw and Fredine, 1956) that a mimimum of 45 million acres of wetland out of a national total of 127 million original acres have been drained and/or filled. While these figures mean that over a third of the natural wetland habitat in the nation has been converted to human use, the known reductions in natural wetland habitat in individual states are considerably higher. For example, in a period of about 100 years (from 1850), the following percent reductions in wetland acreage have been recorded:

- Arkansas   −51%
- California   −79%
- Illinois   −95%
- Indiana   −79%
- Iowa   −90%
- Missouri   −91%

The absolute loss of any natural resource is, of course, of central concern to anyone who holds a preservationist or conservationist ethic. Yet, one does not have to be a preservationist or a conservationist to recognize that wetland ecosystems are intimately interconnected with aquatic and terrestrial ecosystems and to realize that the loss or alteration of wetlands can, therefore, have important consequences on the total physical environment. As depicted in Fig. 11.1 (and as discussed in previous chapters), aquatic and terrestrial ecosystems may be directly interconnected. Each may also receive or contribute material and energetic inputs and/or products to coastal- and inland-wetland ecosystems. Thus the assessment team should view wetlands as potentially important links between aquatic and terrestrial communities.

Some important abiotic effects of project development within wetland environments have been identified (Darnell *et al.*, 1976, p. 179) and include loss of wetland habitat, reduction of habitat diversity, modification of normal seasonal flow patterns, fluctuation in water levels and flow rates, reduction in flow volume, increased downstream flooding, creation of canals in swamps and marshes, increase in turbidity, increase in sedimentation, clogging of stream riffles, filling of pool areas, alteration of stream-bottom topography, reduction in light penetration, elevation in temperature, modification of natural chemical composition, increased oxygen demand, addition of chemical pollutants, buildup of bottom pollutants, and increase in salinity (in coastal estuaries, marshes, and swamps).

While the assessment team may have to focus its analytic effort on the probability, duration, and magnitude (Chapter 6) of any one or all of these potential impacts on wetlands, the assessment team must understand that the ecological significance of individual impacts ultimately requires integrative effort on the part of team members—an effort characterized by the attempt to relate individual abiotic impacts on wetland environments to the total aquatic, terrestrial, and wetland ecosystems in the region. For example, an increase in downstream turbidity as the result of construction activity in a wetland may be of primary concern to those members of the assessment team who are interested in project impacts on

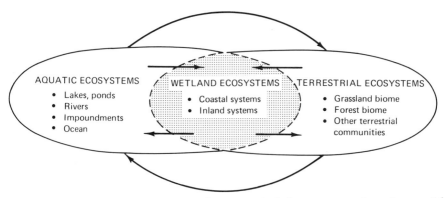

FIGURE 11.1. *Wetland ecosystems as dynamic links between aquatic and terrestrial ecosystems.*

the recreational use of downstream waters. Thus, the amount, timing, and duration of sediment loading of downstream waters will likely be evaluated in light of water-quality standards for primary and secondary recreational uses of downstream waters. Yet, as already discussed in Chapter 9, turbid particles can also have a variety of autecological and synecological impacts on aquatic ecosystems. An ecological overview of the consequences of construction in a wetland area would, therefore, require information on the trophic structure and mineral cycling of the receiving riparian ecosystem. In a similar fashion, the reduction of habitat diversity due to filling in a wetland area may be of interest to those primarily interested in aesthetic impacts in the project area. But what of the importance of those habitats to animal populations (both indigenous and migratory) that fulfill important niche functions in both local and distant terrestrial ecosystems?

Historically, wetlands have typically been viewed as "useless areas"—as "wastelands" to be drained and filled and converted into economically productive land, thus, the high rate of their disappearance over a past century that is generally characterized by economic growth. While it is generally true that this perspective is changing—that wetlands are today more often seen as having ecological values that have to be considered and weighed in the decision-making processes of project development—it is all too often also true that wetlands are viewed as being primarily important to ducks and "nature freaks." Both the narrow economic perspective and the myopic environmental perspective of wetlands are false. Wetlands are not useless wastelands; nor are they important only for hunting and nature walks.

Classification of Wetlands

Coastal and inland wetlands have been previously subdivided into 20 classificatory categories (Shaw and Fredine, 1956, p. 15). These categories were based on consideration of various parameters, including geographic location, depth, influence by ocean water, and importance to waterfowl. A major effort to update this classification system has been undertaken by the U.S. Fish and Wildlife Service and has resulted in an interim classification of wetlands and aquatic habitats of the United States (Cowardin et al., 1976). This classification system is an important informational input into the process of assessing project impacts on wetlands.

According to the Interim Classification System,[1] a wetland is "land where an excess of water is the dominant factor determining the nature of soil development and the types of plant and animal communities living at

---

[1] A revised version of this system is available in Cowardin et al., 1977.

the soil surface. It [i.e., wetland] spans a continuum of environments where terrestrial and aquatic ecosystems intergrade [Cowardin *et al.*, 1976, p. 6]." All wetlands categorized in the Interim Classification System are included in a hierarchical scheme that includes the following taxa:

1. Provinces
2. Ecological systems and subsystems
3. Habitat classes, subclasses, and orders
4. Habitat types

The *province* is the most general component of the hierarchical classification scheme and refers to the climax vegetation (Chapter 10) of a particular geographic region. Provinces included in the classification scheme include all inland "sections" described by Bailey (1975), as well as the following coastal provinces:

- The *Arctic Province* (southern tip of Newfoundland, northward around Canada to the west coasts of the Arctic Ocean, Bering Sea, and Baffin and Labrador Basins)
- The *Acadian Province* (northeast Atlantic coast from the Avalon Peninsula to Cape Cod)
- The *Virginian Province* (middle Atlantic coast from Cape Cod to Cape Hatteras)
- The *Carolinian Province* (south Atlantic coast from Cape Hatteras to Cape Kennedy)
- The *West Indian Province* (Cape Kennedy to Cedar Key, Florida, and the southern Gulf of Mexico, the Yucatan Peninsula, Central America, and the Caribbean Islands)
- The *Louisianan Province* (northern coast of the Gulf of Mexico from Cedar Key to Port Aransas, Texas)
- The *Californian Province* (Pacific coast from Mexico northward to Cape Mendocino)
- The *Columbian Province* (northern Pacific coast from Cape Mendocino to Vancouver Island)
- The *Fjord Province* (Pacific coast from Vancouver Island to the southern tip of the Aleutian Islands)
- The *Pacific Insular Province* (surrounding all of the Hawaiian Islands)

The *ecological system* refers to wetland and aquatic habitats that share the influence of one or more dominant hydrological, geomorphological, or chemical factors. Five major ecological systems are recognized in the Interim Classification System:

(1) the *Marine System* (coastal land and water systems with un-

obstructed access to the open ocean), including intertidal and
subtidal subsystems

(2) the *Estuarine System* (coastal land and water systems which are
semienclosed by land, with open, partially obstructed, or sporadic
access to the ocean, and with a measurable quantity of ocean-
derived salt in the water), including intertidal and subtidal subsystems

(3) the *Riverine System* (between the channel bank and extending to
and including vegetation channelward of the bank), including (*a*)
the high gradient subsystem (area of fast flow, and where the
stream bed consists of rock, cobbles, or gravel with occasional
patches of sand); (*b*) the low gradient subsystem (area of slow
flow, and where the stream bed is mainly sand, silt, and clay); and
(*c*) the tidal subsystem (area where stream bed gradient is low and
velocity fluctuates under tidal influence)

(4) the *Lacustrine System* (all nontidal habitats situated in depres-
sions that are bound by wave-formed or bedrock shoreline-
features, or that are at least 20 acres in extent at the deepest
portion of the catchment and are devoid of trees, shrubs, or
persistent emergent vegetation), including (*a*) the littoral subsystem (the area from the shoreward boundary to the maximum
depth of effective light penetration; and (*b*) the profundal subsystem (the area where there is insufficient light penetration for the
growth of rooted or adnate macrophytes)

(5) The *Palustrine System* (all nontidal wetland and aquatic habitats
where water is not restricted to a definable channel, wave-formed
or bedrock shoreline-features are absent, and persistently nonvegetated and deepest portions of the catchment are less than 20
acres in extent); no subsystems of the palustrine ecosystem are
recognized in the Interim Classification System

Habitat classes, subclasses, and *orders* are based on dominant vegetation, the form of substrate for nonvegetated areas, persistence of leaves
(evergreen and deciduous vegetation), soil type, substrate texture, substrate origin, and dominant sedimentary-animal communities. An example of the hierarchical relationship of habitat order, subclasses, and
classes for the palustrine ecological system is shown in Fig. 11.2. It
should be noted that each of the five major ecological systems utilized in
the Interim Classification System may be found in one or more provinces.

The fourth major component of the hierarchical Interim Classification System is *habitat type.* Habitat type is composed of habitat order
plus appropriate modifiers for *water regime,* and *water chemistry.*

Modifiers for water regime include modifiers for both nontidal and
tidal water regimes as follows:

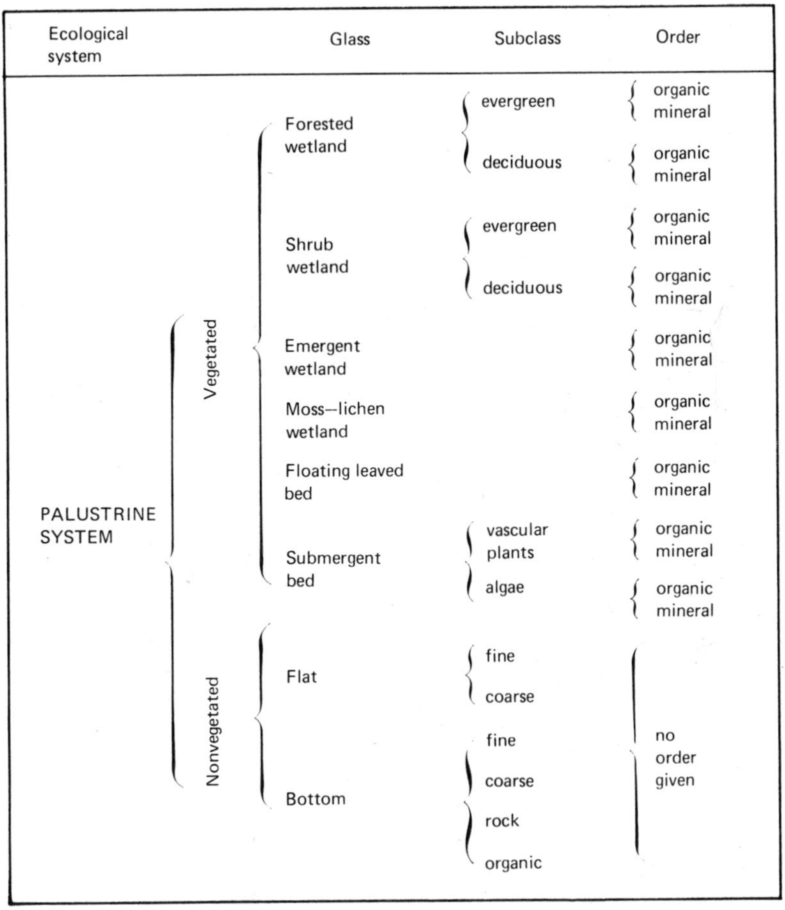

Ecological system	Glass	Subclass	Order
PALUSTRINE SYSTEM — Vegetated	Forested wetland	evergreen	organic / mineral
		deciduous	organic / mineral
	Shrub wetland	evergreen	organic / mineral
		deciduous	organic / mineral
	Emergent wetland		organic / mineral
	Moss–lichen wetland		organic / mineral
	Floating leaved bed		organic / mineral
	Submergent bed	vascular plants	organic / mineral
		algae	organic / mineral
PALUSTRINE SYSTEM — Nonvegetated	Flat	fine	
		coarse	
	Bottom	fine	
		coarse	no order given
		rock	
		organic	

FIGURE 11.2 *An example of a wetland classification hierarchy for the palustrine ecological system. Note that no subsystem of the palustrine ecosystem is recognized.* [*After Cowardin* et al., *1976, p. 16.*]

NONTIDAL WATER REGIME
- Saturated (no surface water, and only slight annual fluctuation of water table)
- Temporarily flooded (surface water seldom covers area for more than 10 consecutive days)
- Seasonally flooded (surface water covers area, but for less than half a year)
- Semipermanently flooded (surface water covers area for more than half a year, but not permanently)
- Permanently flooded (surface water covers area at all times of the year, every year)

- Intermittently flooded (surface water covers area during irregular intervals)

TIDAL WATER REGIME
- Irregularly flooded (tidal waters cover land surface less frequently than once a day)
- Regularly flooded (normal tides alternately flood and expose land surface at least once every day)
- Subtidal (land surface permanently flooded, but depth fluctuates with the tides)

Modifiers for water chemistry include modifiers for salinity and pH. Salinity is classified as hypersaline (> 40‰ salinity), saline (40–30‰ salinity), subsaline (30–.5‰ salinity), and fresh (<0.5‰ salinity). The classification for pH includes acid (<5.5 pH units), circumneutral (5.5–7.4 pH units), and alkaline (>7.4 pH units).

In addition to these modifiers for water regime and water chemistry, special modifiers have been proposed for the determination of habitat type, including:

- Impoundment (habitat created or modified by manmade barrier to gravitational flow of water)
- Dugout (habitat within a basin excavated by man)
- Canal (habitat within a channel excavated by man where no previous channel existed)
- Irrigated (where water table has been raised above its natural level by mechanical means and where dominant vegetation is used as agricultural crops or forage)
- Farmed (where soil surface has been mechanically altered for the production of crops)

The Interim Classification System is obviously complex—especially to those who have assessment responsibility and who are also not professional ecologists. It is, therefore, necessary to highlight several points about the Interim Classification System (or other similar classification systems now in use in various states).

First, the complexity of these subsystems reflects the abiotic and biotic complexity and diversity of wetland ecosystems. The assessment team should, therefore, immediately reject any simplistic notion that a wetland is "just another swamp" that deserves only cursory consideration by the assessment team. Rather, the assessment team should attempt to classify wetlands in the project area and its environs, using both federal and state classification systems. Appropriate federal and state agencies should be contacted for specific advice as to which classification scheme is desirable. It may be necessary to utilize more than one scheme in order to ensure that both local and national concerns will be addressed by the assessment process.

Second, any classification scheme for wetlands should be viewed by the assessment team as an important tool for facilitating the overall assessment effort, and not as "one more thing we have to do." The Interim Classification System identifies abiotic and biotic factors that have to be considered in evaluating the ecological importance of a particular wetland. The precise identification of these factors is a prerequisite for soliciting relevant expert information and helps to identify expert informational sources from nonexpert sources. For example, there is a tremendous difference in precision between the following two questions:

1. What are potential ecological impacts of dredging in a slightly saline, temporarily flooded, mineral rich, emergent, palustrine wetland in the Acadian Province?
2. What are potential ecological impacts of dredging in wetlands near the coast?

The first question is phrased in a manner that gives professionals basic data (i.e., on province, ecological system, habitat class, order, etc.) which they require in order to begin their evaluation. The second question is phrased in so general a manner as to elicit opinions from most anyone, regardless of their knowledge and scientific understanding of wetland dynamics. In short, a classification system such as the Interim Classification System is a means by which the assessment team can begin to maximize scientific input into the evaluation process and, at the same time, minimize irrelevant emotionalism and opinion.

Third, a classification system (especially one which is utilized nationally) provides the assessment team with a basis for identifying the possibilities and probabilities of specific impacts. For example, it is quite ridiculous to suggest that, because a certain impact occurred as a result of one type of project in one wetland, the same impact can be expected as a result of similar projects in other wetland areas. As indicated by the Interim Classification System, wetlands are quite diverse. Thus any extrapolation from one project assessment to another should be made with full consideration of this diversity. A classification system such as the Interim Classification System is an important tool for identifying those wetlands that are similar and/or dissimilar from an ecological perspective, and thus those cases where extrapolation from previous experience might reasonably be attempted.

In spite of these advantages of the Interim Classification System, it is likely that the nonecologist who nevertheless has assessment responsibility will be somewhat frustrated by its complexity and by the diversity of wetland types described. It is, therefore, important that members of the assessment team focus on general attributes of wetland ecosystems. This approach will facilitate the team's understanding and appreciation of the overall classification system.

General Attributes of Wetland Ecosystems

Intertidal and subtidal subsystems of marine and estuarine ecological systems include a variety of habitat classes, including forested, shrub, and emergent wetlands, rock shores, beaches/bars, flats, vegetated and non-vegetated reefs, floating leaved and submergent beds, and nonvegetated bottoms (Cowardin et al., 1976, p. 12–13). Each of these habitats provides a variety of materials and energies to biological communities. These communities, depending on the nature and amount of specific materials and energies, may exhibit both adaptive and stressful responses.

Characteristic energies and sources of biological stress in coastal wetlands have been identified (Odum et al., 1974, pp. 25–28) and include breaking waves, tidal currents, temperature variation, high rate of sedimentation, salinity, dissolved oxygen depression, light, organic and inorganic enrichment, and industrial wastes.

Which of these energies and sources of biological stress are most important in a given wetland depends, of course, on the specific type of wetland. For example, rock shores receive the energy imparted by breaking waves; emergent wetlands receive the energy of tidal currents. *The assessment team should use appropriate wetland classification schemes to identify (through experts and/or scientific literature) those energies and materials that are most important for understanding the ecology of a particular wetland.*

Once appropriate energies and materials have been identified (i.e., once a homorphic model of the particular wetland has been identified), the assessment team should attempt to identify the basic interrelationships between these energies and materials and the biological community (i.e., by constructing an isomorphic model of the wetland).

An example of an isomorphic model for an estuarine, emergent wetland is depicted in Fig. 11.3. Two points about this model should be emphasized.

First, the major energetic and material inputs into this wetland ecosystem are (a) sun energy (for photosynthesis); (b) tidal energy (for periodic input of water, flushing out of wastes, and distribution of biomass); and (c) organic matter and nutrients (from terrestrial sources).

These energies and materials accomplish the fertilization, irrigation, and essential energizing of phytoplanktonic and vascular photosynthesizers and thus they directly influence primary production.

Second, phytoplankton and vascular plants (upon decomposition by

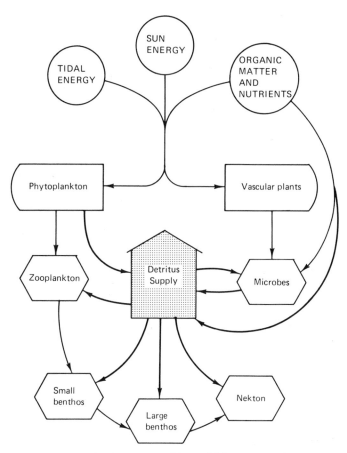

FIGURE 11.3 *An example of an isomorphic model for an estuarine, emergent wetland.* [*Adapted from Odum, 1971, p. 64.*]

microbes), as well as inputted organic matter contribute to the detritus reservoir within the emergent wetland ecosystem. This detritus reservoir, as in climax communities in terrestrial ecosystems (Chapter 10), plays an essential role in the trophic dynamics of the wetland ecosystem. It is typically the major food supply for a variety of salt-marsh biota, including zooplankton, benthos, and nekton (Gosselink *et al.*, 1974, Nixon and Oviatt, 1973; Odum and Smalley, 1959; Teal, 1962; Redfield, 1972).

Of course, some marine and estuarine ecosystems are not vegetated and/or do not depend on large reservoirs of detritus (e.g., reefs, rocky shore areas, and flats). *The assessment team should be careful that it does not assume that projects can have no ecological impacts merely because of the absence of vegetation in a wetland area within a right-of-way.* Nonvegetated wetland areas can be important, for example, for their sequestration in sediments of potentially toxic materials. They can also be important for decomposing complex organic wastes and exporting released nutrients to other vegetated ecosystems. In general, the assessment team should assume that minimal primary productivity in a particular coastal wetland is *not* a measure of the ecological significance of that wetland. Thus, *the assessment team should always strive to look at a particular coastal wetland in the context of interconnected aquatic, terrestrial, and other wetland ecosystems.*

Inland wetlands (including riverine, lacustrine, and palustrine ecosystems) include habitat classes that are similar to habitat classes found in coastal wetlands (e.g., forested, shrub, and emergent wetlands, beaches/bars, rocky shores, and flats). Some inland wetlands, such as vegetated, low-gradient reaches (of riverine systems), and emergent wetlands in littoral areas (of lacustrine systems) also depend on detritus, just as do salt-water marshes. Other inland wetland habitats, however, are quite different from coastal habitats, as, for example, the riffles and pools of riverine systems and intermittently flooded flats in lacustrine systems. In general, the assessment team should consider the following potential functions of inland wetlands:

1. Source of detritus for downstream communities
2. Habitat for migrating (diurnal, nocturnal, diel, and seasonal) terrestrial animals (Chapter 10)
3. Recharge areas for water table and artesian aquifers (Chapter 9)
4. Nesting and reproductive habitat for terrestrial and aquatic biota
5. Sequestration (long- and/or short-term) and/or dissipation (e.g., through evaporation) of potentially toxic materials
6. Grazing food supply for indigenous and migratory animals
7. Interception and mineralization of organic runoff
8. Hydraulic baffling and mediation of downstream currents
9. Critical habitat for threatened or endangered or other protected species
10. Interception and sedimentation of turbid runoff waters

In some cases, a particular wetland may perform a number of these functions. For example, a vegetated wetland may provide both grazing and detritus food through its primary productivity. Emergent vascular plants may also influence current and, thus, influence downstream riparian habitat at the same time as they provide nesting habitat for indigenous and/or migratory terrestrial species. Finally, reduced flow rates through the wetland (because of the baffling effect of vascular plants) can increase sedimentation within the wetland and thereby enhance the potential of the wetland to intercept and sequester potentially toxic materials (e.g., from runoff in industrial areas) in bottom muds. The sequestration of these materials in wetlands can prevent toxic effects on downstream communities.

In some cases, of course, a particular wetland may be devoid of vegetation, or sustain vegetation of such a kind or in such small amounts that it is of little value as graze or detritus to birds and animals. Such a wetland may, nonetheless, be important in terms of regional ecology because of its hydrologic potential as a recharge area for those aquifers that provide water to local plant associations. Or the wetland may serve primarily as a natural sedimentation basin that reduces sediment loading of other surface waters in the general area, thus reducing potential biological stress associated with the sediment loading of those waters.

In light of these examples, it is clear that the assessment must take a systems overview of both coastal and inland wetlands. Such an overview should include the following considerations:

1. All wetlands, regardless of location and type, perform some type(s) of work.
2. The work performed by a particular wetland may directly influence both local and distant ecosystems, including other wetland ecosystems as well as aquatic and terrestrial ecosystems.
3. The work performed by a particular wetland, and thus its interrelationships with other ecosystems, can be determined by identifying (a) energetic and material inputs into the wetland; and (b) energetic and material outputs from the wetland.

Assessing Impacts

There are some excellent, practical guides for indentifying potential project impacts on various types of wetlands, including those by Silberhorn et al. (1974), Darnell et al.. (1976), and Clewell et al. (1976). The assessment team should utilize such guides whenever possible. However, the assessment team should realize that there is no finalized, universally accepted set of guidelines for the comprehensive identification and assessment of project impacts on wetland ecosystems. Each compilation of guidelines has its limitations. As with the assessment of project impacts on aquatic and terrestrial ecosystems, *the assessment team must strive to utilize available guidelines within an already defined assessment strategy, rather than base that assessment strategy entirely on available guidelines.*

An assessment strategy that is appropriate for the comprehensive evaluation of project impacts on wetland ecosystems should include the following task objectives for members of the assessment team:

1. Characterize wetlands in the immediate project area in terms of basic energetic and material inputs and outputs (using, when available, federal, state and other classification systems which may be found in the scientific literature)
2. Characterize wetlands in the general region in a comparable fashion
3. Identify key abiotic and biotic factors in each wetland as determined by the appropriate classification system or by other pertinent scientific literature
4. Relate each key abiotic and biotic factor to physical, chemical, and biological processes (including autecological and synecological processes) in aquatic and terrestrial environments in the region—for example, processes that determine energetic and material introductions, transformations, translocations
5. Identify potential direct impacts of project activities on these aquatic and terrestrial processes and possible subsequent (indirect) impacts on project area and regional wetland ecosystems
6. Identify potential direct impacts of project activities on project-area and regional-wetland ecosystems
7. Evaluate cumulative direct and indirect impacts on regional wetlands
8. Identify possible consequent impacts on project area and regional aquatic and terrestrial ecosystems

The basic assumptions in this strategy are (a) that projects have the potential for affecting wetlands both in the immediate project area and in the general region outside of the project area; and (b) that project impacts on wetlands may be both direct and indirect. As depicted in Fig. 11.4, the determination of direct and indirect impacts on wetlands in a project area or in the general region requires consideration of how regional wetland ecosystems are interrelated with regional aquatic and terrestrial ecosystems.

It is clear that the indirect impacts on these ecosystems are often difficult to assess in quantitative terms, because they depend on so many intervening factors between the source of the impact and the impact itself. Some direct impacts on wetlands (e.g., total sediment loading during construction) are also difficult to assess quantitatively. In such cases where it is impossible to evaluate an impact quantitatively, the assessment team should, nevertheless, consider the impact qualitatively. Wherever possible, of course, the assessment team should assess impacts quantitatively, as in:

1. Loss of (so many) acres of emergent vegetation that typically produces (so many) tons of biomass per acre per year

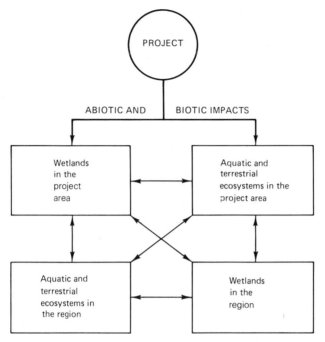

FIGURE 11.4 *Overview of project impacts on wetlands in the project area and in the region.*

2. A (some) percentage of reduction in wetland carrying capacity for waterfowl (or other) populations
3. A (some) percentage of reduction in detritus that is exported to estuaries and that supports (some) pounds of detritus feeding fish per year
4. The permanent loss of tertiary treatment of land runoff in a watershed of (so many) acres

While it may be advisable in specific instances to conduct quantitative studies of wetlands, most assessments can probably utilize quantitative data that are already available. Some of these data (e.g., biomass of different trophic levels, rates of productivity) are available from scientific studies of a specific wetland that may be similar to a wetland under consideration by the assessment team. Data may also be obtained from literature that deals with wetlands of an entire state (e.g., Silberhorn *et al.*, 1974) or in large geographic regions (e.g., Odum *et al.*, 1974; Shaw and Fredine, 1956). The assessment team should contact appropriate federal and state agencies, universities, and other organizations that have an interest in wetlands in order to identify data that may be relevant to the proposed project. Expert opinion should be solicited as to the reasonableness or unreasonableness of using data obtained from one wetland ecosystem for the assessment of project impacts on another wetland.

References

Bailey, R. G.
  1975 *Ecoregions of the United States.* Ogden, Utah: U.S. Forest Service.
Clewell, A. F., L. F. Gainey, Jr., D. P. Harlos, and E. R. Tobi
  1976 *Biological effects of fill-roads across salt marshes.* Tallahassee, Florida: Florida Department of Transportation.
Cowardin, L. M., V. Carter, F. C. Golet, and E. T. LaRoe
  1976 Interim classification of wetland and aquatic habitats of the United States. In *Proceedings of the National Wetland Classification and Inventory Workshop,* edited by J. Henry. Washington, D.C.: Office of Biological Services, Fish and Wildlife Service, U.S. Department of the Interior.
Cowardin, L. M., F. C. Golet, and E. T. LaRoe
  1977 *Classification of wetlands and deep-water habitats of the United States: An operational draft.* Washington, D.C.: Office of Biological Services, Fish and Wildlife Service, U.S. Department of the Interior.
Darnell, R. M., W. E. Pequegnat, B. M. James, F. J. Benson, and R. A. Defenbaugh
  1976 *Impacts of construction activities in wetlands of the United States.* Corvallis, Oregon: U.S. Environmental Protection Agency, Office of Research and Development, Corvallis Environmental Research Laboratory. (EPA-600/3-76-045)
Gosselink, J. G., E. P. Odum, and R. M. Pope
  1974 *The value of the tidal marsh.* Baton Rouge, Louisiana: Center for Wetlands Resources, Louisiana State University. (LSU-SG-74-03)
Nixon, S. W., and C. A. Oviatt
  1973 Ecology of a New England salt marsh. *Ecological Monographs, 43*(4):463–498.
Odum, H. T.
  1971 *Environment, power, and society.* New York: Wiley–Interscience.
Odum, H. T., B. J. Copeland, and E. A. McMahan (Editors)
  1974 *Coastal ecological systems of the United States* (4 vols.). Washington, D.C.: The Conservation Foundation.
Odum, E. P., and A. E. Smalley
  1959 Comparison of population energy flow of a herbivorous and a deposit feeding invertebrate in a salt marsh ecosystem. *Proceedings of the National Academy of Sciences, 45*(4):617–622.
Redfield, A. C.
  1972 Development of a New England salt marsh. *Ecological Monographs, 42* (2):201–237.
Shaw, S. P., and C. G. Fredine
  1956 *Wetlands of the United States.* Washington, D.C.: Fish and Wildlife Service, U.S. Department of the Interior. (Circular No. 39)
Silberhorn, G. M., G. M. Dawes, and T. A. Barnard, Jr.
  1974 *Coastal wetlands of Virginia: Guidelines for activities affecting Virginia wetlands* (Interim Report No. 3). Gloucester Point, Virginia: Virginia Institute of Marine Science.
Teal, J. M.
  1962 Energy flow in the salt marsh ecosystem of Georgia. *Ecology, 43*(4):614–624.

12

Criteria of Adequacy of Assessment: Physical Environment

Introduction

There appear to be at least several schools of thought with respect to criteria of adequacy of assessment of project impacts on the physical environment.

One argues that adequacy is ultimately determined by court decisions. From this perspective, an inadequate assessment is one that results in court injunctions against project development.

Another maintains that adequacy should be determined by those criteria historically employed by professional scientists in the conduct of field and laboratory research. Thus, an adequate assessment is one that can meet the stringent measures of scientific publication.

Still another school of thought argues that the only possible measure of adequacy is the precise match between predictions made in the assessment document and actual future events. Thus, an inadequate assessment is one that fails to forecast the future accurately.

In the best of all possible worlds one might suggest that these individual perspectives are not, after all, mutually exclusive—that judges, scientists, and anyone else can reach agreement as to what is a good assessment. But in a less than the best of all possible worlds we know that this is not the case. "Good" legal decisions can flout scientific fact as easily as not. "Good" scientific studies can have as much immediate utility for decision-makers as a wet tea bag. And "good" predictions can often be made as well by soothsayers as by practitioners of either the law or science.

It is a fact, nonetheless, that persons having assessment responsibility have been threatened with the loss of their or others' jobs if their assessment "ends up in court." It is also a fact that persons having assessment responsibility have been threatened with scientific inquisition by Ph.D.s too numerous to contemplate except in nightmares. And it is also a fact that most everyone who has to make recommendations for action based on his or her own assessment of what currently is and what is likely to become in the future is constantly threatened by the prospect that the future will prove the person an absolute ass. In full cognizance of the practical, day-to-day reality of these considerations, this general rule is offered—if the assessment of projects is primarily guided by concern to avoid legal action, scientific review, or any form of future criticism, the assessment is probably a waste of time.

The reason that such an assessment is probably a waste of time is that impact assessment under NEPA is intended to present decision-makers with information that can facilitate the making of *better* decisions—not just to facilitate decision-making, but to facilitate decision-making that results in better decisions. Saving one's job or one's "face," or otherwise preserving harmonious relations among diverse professionals are not (regrettably) practical measures of the adequacy of meeting this difficult task of improving decisions.

Some General Criteria

It is necessary to distinguish the assessment process from the Environmental Impact Statement. The assessment process is the sum total of analytical and integrative tasks (Chapter 6) undertaken by the assessment team. The EIS is a specific document that is required by NEPA and that meets the guidelines of the agency designated as the "lead" agency. Thus, the EIS derives from the assessment process, but does not necessarily depict that process in its entirety. Criteria for evaluating the adequacy of the EIS will be discussed in Chapters 21 and 22. This chapter focuses on criteria that may be used for evaluating the adequacy of the assessment process as it relates to the assessment of impacts on the physical environment. In general, these criteria pertain to (a) the comprehensiveness of the overall assessment approach; (b) technical and scientific inputs; and (c) recommendations by the assessment team.

Impacts on the physical environment can occur at all phases of project development, from early planning to operational and maintenance phases. Some of these impacts are quite direct—such as changes in hydrology as a result of excavation during the construction phase, or the mortality of organisms due to the toxic effects of pesticides applied during the maintenance phase. Other impacts are indirect, involving a chain of chemical, physical, biological, and even social events. For example, during the early planning phase it may happen that local residents will alter their own land use because of their perception of the inevitability of the proposed project. Some lands may subsequently be "developed" in an attempt to enhance their economic value prior to project development and right-of-way acquisition. Other lands, previously tended for recreational and aesthetic value, may be abandoned. Such changes in ongoing land-use, whether they occur because of economic speculation or because of a sense of desperation, can affect existing wildlife habitat and, therefore, the trophic dynamics within plant and animal communities.

In addition to potential direct and indirect impacts of a particular project on ecosystems, there are the so-called *incremental impacts* to consider. These impacts derive from multiple projects undertaken in a region over a period of time. While an individual project may result in a relatively small and unimportant impact on the physical environment, numerous projects having the same type of impacts can have an important additive effect. For example, land clearing for one project may result in an immeasurably small reduction (e.g., 2%) in the regional carrying capacity for a particular population. However, 5 or 10 additional projects, each having a similarly small impact on carrying capacity, can collectively result in a measurable and important reduction.

In light of these considerations, it is important that the assessment team take a comprehensive overview of both the proposed project and other actual, proposed, and potential projects in the region. Specific guidelines which might be used for evaluating the comprehensiveness of the team's approach include:

1. All phases of the proposed project should be considered, including early systems planning, design, location, acquisition, construction, and operation and maintenance phases.
2. All project activities in each phase of project development (e.g., blasting, clearcutting, mowing, relocation of residents) should be identified and evaluated for potential impacts on the physical environment.
3. The timing and duration of each project activity should be related to other important events and activities in the general project area and its environs, including seasonal changes in meteorology and hydrology, animal migrations, and patterns of recreational and other uses of natural resources.
4. Cumulative impacts of the proposed project and all other ongoing and potential projects in the general region should be considered.

A great deal of emphasis was given in previous chapters to the interrelatedness of the various components of the physical environment. No assessment can be said to be adequate that does not address interrelationships. What is required is a systems approach. An assessment that utilizes a systems overview of the physical environment will typically ensure that:

1. Individual abiotic and biotic components of the physical environment in the project area are identified so as to give a comprehensive overview of the current physical environment.
2. Dynamic interrelationships between individual and biotic components are defined on the basis of scientific and technical principles and concepts, and on the basis of previous local experience.
3. Key factors (including physical, chemical, biological, and social factors) that can influence these dynamic interrelationships are identified and specifically related to the project area and its environs.
4. The past history of the physical environment in the project area and its environs is examined and short- and long-term projections of the likely future development of the physical environment, assuming the proposed project is not undertaken, are made.
5. All potential changes in the physical environment as a result of the proposed project are distinguished from changes that would occur in the absence of the proposed project and are discussed in terms of their probability, magnitude, and duration.
6. Mitigation and/or enhancement alternatives (including design and management alternatives) are identified as early in project development as possible and are examined for their practical feasibility and efficacy and for their overall impact on the physical environment.
7. All individual project impacts on the physical environment are examined for consequent effects on physical, chemical, organismic, autecological, and synecological components of distantly located ecosystems.

The probability, magnitude, and duration of project impacts on the physical environment are determined by complex abiotic and biotic environmental factors *and* by the technical details of project development. In assessing project impacts on the physical environment, the assessment team must focus precisely on numerous technical and scientific issues. Opinions about project impacts that are not founded on technical and scientific knowledge and understanding have no place in the assessment of project impacts on the physical environment.

The assessment team must constantly strive to document its findings in terms of the relevant professional literature. Some of this literature may be obtained from academic sources; some of it is most easily available through appropriate government agencies. Generally, adequate documentation can be ensured if the following steps are taken:

1. All federal (Appendix A), state, and local government agencies, universities and other organizations having regulatory, scientific and/or technical interest in the physical environment of the general project area are identified and utilized as important informational sources.
2. Individual local residents having special expertise and experience with respect to the local physical environment are identified and utilized for data gathering and/or evaluation.
3. Literature surveys are conducted for the purpose of identifying issues and findings about which there is important scientific disagreement, as well as those about which there is general agreement.
4. Only standard, professionally recognized methods are utilized in those cases where field and/or laboratory studies are undertaken as part of the assessment process.
5. Major conclusions are based on previously tested principles and hypotheses, and not on developing, state-of-the-art research.
6. Quantitative techniques are employed wherever possible and generally accepted by professionals in relevant disciplines.
7. Clearly subjective interpretations, valuations, etc. are specifically identified as such.

The assessment team should continually remind itself that the single most important reason for taking a comprehensive, well-documented approach to assessment is to make recommendations to decision-makers. Moreover, a well-designed assessment process should result in recommendations *before* the development of the EIS itself—or, in other words, the assessment process, once undertaken, should be able to influence all subsequent project development. For example, the assessment team can prioritize its efforts so as to ensure the identification of critical habitats for threatened or endangered species in the general region—thus the team can be in the position to influence early corridor location and/or design phases so as to avoid possible impacts on those populations. Of course, an assessment team will never find itself in such a position if it fails to take certain organizational steps, including (a) coordination of key assessment tasks with key decision-points in overall project development; (b) enforcement of strict compliance with assessment timetables by all team members; and (c) maintenance of all assessment files and findings in such a manner as to facilitate their rapid utilization for memoranda, informal meetings, and conferences that can influence project development.

All recommendations for alternative project location, design, and/or management that actually emanate from the assessment of the physical environment should be based on (a) knowledge of regional diversity in aquatic, terrestrial, and wetland environments; (b) a clear understanding of how the proposed project can affect any or all of these environments; and (c) documented evidence that the recommended alternative action can in fact achieve a desired environmental objective.

Special Issues: Analytical Tasks

Analytical tasks of impact assessment were discussed in Chapter 6, and included (a) the identification of relevant standards, guidelines, concepts, principles, and experiences; (b) the generation of technical and environmental data and information; and (c) the identification of possible impacts and their interactions. While many of the general criteria of adequacy outlined above are relevant to these tasks, some attention should be given to special issues that are directly related to the efficient and successful completion of these tasks. These issues include the use of technical jargon, the relevance of data and information, and the conduct of an interdisciplinary evaluation of project impacts on the physical environment.

TECHNICAL JARGON

The EIS is a public document and should be written in a style so as to be understandable by the layman (Chapter 21). However, the process of assessing project impacts on the physical environment requires the collection and evaluation of highly technical and scientific data and information. In its day-to-day assessment activities, the assessment team must use specialized technical and scientific jargon. In fact, the assessment team that persists in using nontechnical language and concepts during data collection and analysis is, in effect, cutting itself off from the very sources of professional information and guidance that it must depend on in order to conduct an adequate assessment. For example, the phrases "balance of nature," "environmental integrity," "environmental quality," and "healthy ecology," may be useful when writing a newspaper or magazine article, or making a speech, or for cocktail conversation. But they are not formal disciplinary concepts that professionals actually utilize as analytical tools for probing the complexity of the physical environment. The assessment team should seek information on "trophic dynamics" and "food chains" rather than on "the balance of nature"; on "homeostatic mechanisms" in a particular community, rather than on the "environmental integrity" of that community; on "population dynamics," rather than on "environmental quality"; and on "mineral cycling" and "species diversity," rather than on "healthy ecology."

This is not to say, of course, that all professionals in any particular discipline use the same concepts in precisely the same way. There is often disagreement among experts with respect to technical terminology and its application. The assessment team must come to recognize different meanings and uses of technical terms. An understanding of differences in the professional use of technical jargon facilitates both the gathering of information and the eventual interpretation of data.

The proper use of technical and scientific jargon in the assessment of project impacts on the physical environment can be enhanced by the assessment team's adherence to the following rules:

1. Precise meanings of technical and scientific terminology should be documented.
2. Professional technical and scientific glossaries, dictionaries, and monographs should be utilized as prime documentation.
3. Individual terms should have mutliple documentation, rather than singular documentation.
4. Regional variants of standard technical terms should be clearly identified.

RELEVANCE OF DATA AND INFORMATION

In some assessments it would appear that the collection of voluminous data and information has been substituted for the careful analysis of data as the chief measure of adequacy of assessment. This approach cannot be tolerated. File drawers full of data are not the objective of impact assessment under NEPA—but the careful evaluation of data for purposes of decision-making is. Thus, the assessment team must ensure that only those data that can and will be interpreted and evaluated are collected. Some basic guidelines for identifying and using relevant data include:

1. All data and information to be collected should conform to appropriate regulatory requirements of federal, state, and/or local government agencies.
2. All other data and information to be collected should be shown to be directly relevant to the construction and analytical use of homomorphic and isomorphic models of the general project area.
3. Requirements for the amount, precision, and geographic setting or location of all data to be collected should be documented in the professional literature.
4. Field and/or laboratory studies should only be undertaken in order

to generate required data that are otherwise unavailable and should utilize only recognized, standard methodologies and techniques.

5. All sources of collected data and information should be clearly and precisely identified.
6. Careful attention should be given to the proofreading of transcriptions, tabulations, and summaries of collected and generated data.
7. All interpretations of collected and generated data should be documented in the appropriate regulations and/or in technical and scientific literature.

INTERDISCIPLINARY EVALUATION:

The National Environmental Policy Act clearly calls for interdisciplinary effort in the assessment process. In many instances, interdisciplinary efforts are mistakenly confused with multidisciplinary efforts. The difference in important. For example, most universities have a multidisciplinary structure—the university houses individual specialists, such as biologists, political scientists, and geographers, who individually pursue general educational objectives. They do this largely by training students in their respective disciplines. Few universities have an interdisciplinary structure—a structure that requires that individual specialists collectively and *interdependently* pursue educational objectives. At the risk of even further oversimplification, one might describe a multidisciplinary study as one during which participants meet infrequently and tell other participants what they have individually accomplished, and an interdisciplinary study as one during which its participants have to meet frequently, if only to get some idea as to what they should have been doing since the last meeting!

The importance of interdisciplinary assessment (as opposed to multidisciplinary assessment) under NEPA cannot be overemphasized. Interdisciplinary assessment requires each participating specialist to look at his or her special area of interest from the perspective of other and very different areas of interest. The consequence of this enforced broadening of disciplinary perspectives is an enhanced probability that the assessment team will take a comprehensive approach to project assessment. For example, a biologist may be a poor hydraulics engineer; and a hydraulics engineer a poor biologist. But these two specialists, if they work together, will find various aspects of the physical environment that are of common professional interest (e.g., effects of vascular plants on flow and sedimentation rate, effects of stream bed gradient on downstream benthic habitat).

While there are no universal guidelines for ensuring that interdisciplinary studies do not devolve to multidisciplinary studies during the conduct of the assessment process, several general guidelines might be useful, including:

1. The assessment team should conduct frequent in-house meetings and conferences in which the entire team can discuss findings and future directions.
2. Summaries of technical interpretations of different types of collected and generated data (e.g., soil surveys, habitat and vegetation surveys, water quality data) should be distributed to all team members, regardless of their major area of expertise and responsibility.
3. Ad-hoc subgroups of the assessment team, comprised of individuals having diverse expertise and responsibility, should be periodically organized for the purpose of reviewing the progress and direction of the overall assessment.
4. Work schedules, the location of assessment-related files, and the general physical surroundings of the assessment team should be conducive to informal interpersonal interaction among team members.

Special Issues: Integrative Tasks

Integrative tasks identified in Chapter 6 included (a) the evaluation of the significance of impacts; (b) identification and evaluation of mitigation and/or enhancement techniques; and (c) the making of recommendations.

There is no universally accepted measure of the significance or insignificance of project impacts on the physical environment. This is because some subjective valuation is always involved in the estimate of significance. For example, what is more significant—the loss of a deer herd, or the loss of a cold-water fishery? Both, of course, are significant or insignificant, depending upon subjective values. *The assessment team has to come to grips with the fact that physical, chemical, and biological sciences cannot tell us what is more or less desirable.* Technically, there is as real an ecology to a piece of human feces floating in a stream as the ecology of "blue water." And there was, technically, as "good" an ecology among the communities and habitats of the Age of Reptiles as there is among the communities and habitats of contemporary parklands. The disciplines of hydrology, hydrodynamics, soil mechanics, chemistry, aerodynamics, biology, and ecology, and all other disciplines and subdisciplines that probe the physical environment *do not* define absolute preferences for one set of environmental circumstances over another set of circumstances. They merely tell us what is happening, what is likely to happen, and how—it is for people to decide what shall be preferred, not nature. While no known physical laws describe environmental preferences, it is nevertheless the task of the assessment process to separate "significant" from "insignificant" impacts so that decision-makers can know how to prioritize their efforts. Various criteria for evaluating the significance of impacts on the physical environment within the context of the total human environment will be discussed in Chapter 22. In this section, it is appropriate to focus on selected criteria which should be addressed in the overall assessment of significance

While the following criteria may be approached individually in an objective manner, it should be stressed that the manner in which these criteria are ultimately prioritized is finally influenced by subjective valuation. These include:

1. The probability, magnitude, and duration of individual impacts (Chapter 5)
2. The geographical extent of each impact (e.g., highly localized, regional, supraregional)
3. The degree of coupling between each impact and other subsequent and dependent impacts (e.g., causes numerous—or few— secondary impacts)
4. The potential for technical control and/or reversal (e.g., not currently reversible by any practical means, reversible at relatively high—or low—cost)

The identification of appropriate mitigation and/or enhancement techniques is an essential component of impact assessment. The earlier that appropriate techniques can be identified in project development, the greater the probability that project location and design can be altered so as to achieve the national environmental goals of NEPA. However, the assessment team should understand that each act of mitigation and/or enhancement must be evaluated for its total impact on the environment. It is one thing to suggest a means of curing or avoiding one impact. It is quite another task to ensure that the means of mitigating or enhancing one impact does not itself have other and even more significant impacts. It is the usual problem of curing an illness by killing the patient. Thus the assessment team must consider each act of mitigation and/or enhancement as any other feature of project design and development and evaluate all possible impacts on the physical environment. The following questions are appropriately addressed during the evaluation of any technique for mitigating and/or enhancing project impacts:

1. What are alternative location, design, and management means available for mitigation and/or enhancement?

2. What are the advantages and disadvantages of each alternative means?

3. What are the environmental consequences if the recommended alternative fails in the achievement of its objective?

4. Has the recommended alternative been previously tested, and (if so) with what results?

5. What monitoring (if any) of results might best demonstrate its efficacy or failure?

6. What steps can be taken if the recommended alternative fails?

With respect to the adequacy of recommendations made during the assessment process, the following general rule generally applies—*an assessment process that does not result in a greater number of recommendations to decision-makers than are ultimately made in the final EIS is an inadequate assessment process.*

While there is no doubt much cause for consternation among both decision-makers and assessment teams in this rule, the basis for it is the very intent of NEPA itself (Section 1)—namely, that the assessment process itself (and not just the EIS) will serve as a primary decision-making tool and will, therefore, influence the earliest planning and developmental phases of projects. The alternative to this rule is to follow the example of those who designed a famous building in Boston and who apparently seem content to have substituted sound design with makeshift protocols for guessing which panes of glass will fall to the streets below. Impact assessment under NEPA was never intended as a device for the public offering of blatantly poor, environmental design-alternatives. Under NEPA, the public should not have to waste its time considering half-baked proposals.

The expenditure of public funds through the federal EIS process obligates each assessment team to meet congressional commitment to better decision-making by federal agencies. The public, therefore, is fully entitled to ask (whether by legislative command or by the dictates of common sense):

1. What specific findings and proposals by the assessment team have been incorporated into the alternative proposals for the proposed project?

2. Was the assessment team presented with a final list of alternatives, or did the assessment team have the option (and did it take it) of offering new alternatives within general guidelines?

3. What is the actual operational linkage between the assessment team and location and design functions of the proposing agency?

With respect to actual recommendations made by the assessment team, the public and appropriate decision-makers should fully expect that recommendations be made with appropriate respect for (a) timeliness with regard to decision points in project development; (b) jurisdictional and technical feasibility; (c) long- and short-term costs; and (d) probable efficacy and overall impacts on both the physical and the social environment.

Concluding Remarks

There is no consensual measure for the adequacy of assessment of project impacts on the physical environment. This is caused, in part, by the complexity of the various and diverse components of the physical environment. It is also due to the different priority given by different individuals and groups to the diverse components of the total human environment.

It is not the function of the assessment team to generalize or otherwise simplify the real complexity of the physical environment. Rather, the assessment team must present as comprehensive an overview as possible of potential project impacts and offer decision-makers the kind of information they need in order to make better decisions.

In pursuing this objective in the real world, the assessment team will often find itself faced with technical, scientific, social, and even personal dilemmas. There is, regrettably, no proven rule of thumb for extricating one's self from such a situation. Nor is there any reason to believe that such a rule would enhance the quality of environmental impact assessment.

III

Assessing Impacts: The Social Environment

13

Introduction to Assessment of Impacts on the Social Environment

Introduction

As pointed out in Chapter 1, there are four general processes that are common to ecosystems and social systems: (a) the transformation of energy from one form to another; (b) the modulation (or regulation) of energy flow; (c) environmental adaptation; and (d) environmental modification.

In Part II, various abiotic, organismic, population, and community components of the physical environment that play essential roles in these four dynamic processes were discussed, including

1. Primary producers, consumers, and decomposers (as examples of energy transformers)
2. Food webs, nutrient cycling, and various abiotic factors and phenomena governing material and energetic dynamics (as examples of energy modulation);
3. Structural and behavioral adaptation of biological populations to habitat (as examples of environmental adaptation)
4. Eutrophication and primary and secondary succession (as examples of environmental modification)

In the following chapters, the focus will be on the social environment and on those components and characteristics of social systems that play similarly important roles in the transformation and modulation of energy and in environmental adaptation and modification.

Before proceeding with these discussions, however, it is necessary to discuss two issues that are particularly relevant to the assessment of impacts on the social environment. Perhaps these issues are best defined in terms of basic attitudes that have often been expressed by members of assessment teams who are typically trained in one or more of the physical or engineering sciences, including:

1. That the National Environmental Policy Act is clearly concerned with ecology and not sociology—so why get involved in something we do not have to get involved in?
2. That the subjective, nonquantitative aspect of much of social science minimizes the potential for an objective assessment of project impacts and maximizes the probability for insoluble controversy—so why get involved in something we would rather not get involved in?

These attitudes must be countered as quickly and as directly as possible if the assessment team is to undertake its full responsibility under NEPA.

First, NEPA states that all agencies of the federal government will "utilize a systematic, interdisciplinary approach *which will ensure the integrated use of the natural and social sciences and the environmental design arts in planning and in decision-making* [emphasis added]." In addition to the required use of social sciences in the assessment process, NEPA also defines the federal commitment "to preserve important historic and cultural, as well as natural aspects of the National heritage," and therefore directs federal agencies that, "to the fullest extent possible . . . the policies, regulations, and the public laws of the United States shall be interpreted and administered in accordance with the policies set forth [in NEPA]." As interpreted by the Department of Interior, "cultural" resources include historic, archaeological, and architectural resources and require consideration, during the assessment process, of pertinent Federal legislation and executive orders, including *The Antiquities Act of 1906, The Historic Sites Act of 1935, The Reservoir Salvage Act of 1960, The National Historic Preservation Act of 1966,* and Executive Order 11593 ("Protection and Enhancement of the Cultural Environment") (U.S. Department of the Interior, 1974).

While it is true that the term *"environment"* is not defined by NEPA, or in CEQ guidelines, the term is generally considered to "include physical, social, cultural, and aesthetic dimensions [U.S. Department of Housing and Urban Development, 1974, § 58.3]. This meaning of the term "environment" is specifically noted by a member of the Council on Environmental Quality in a definition of environment as "the sum total of the physical and chemical elements in which organisms live; *for man it also includes economic, political, social, and cultural elements* [Willard, 1975, emphasis added]."

Second, the evaluation of impacts on the physical environment involves subjective as well as objective analyses. For example, there is no objective principle or scientific criterion by which one can say that deer are more important than earthworms (Chapter 12). Thus social phenomena are not the sole source of problems with respect to subjective evaluations.

The basic criterion of objective analysis in both physical and social assessment is that different people, if they use the same analytical methodology, will obtain similar data and reach similar conclusions. The limits of objective analysis in both physical and social sciences should be left to appropriate professionals to define and should not be left to the uninformed opinions of pretentious savants who do not have the requisite professional training. Finally, whatever those limits may, in fact, be, the absence of objective and/or quantitative data with respect to any element of the human environment (whether physical or social) does not exempt the assessment team from serious, methodical consideration of project impacts on that element.

For purposes of impact assessment, human values, hopes, and aspirations are as real as trees, fish, and birds. Relationships among people are as real as relationships between trophic levels. Project impacts on these social components of the human environment are as real as project impacts on physical components of the human environment, and must be considered, regardless of methodological or other (real or imagined) difficulties.

As in the assessment of impacts on physical systems, the assessment of impacts on social systems requires the identification of (a) individual components; and (b) interactions between those components. Project impacts on social systems are, therefore, impacts on social components *and* on interactions between components, or on both. The first practical problem in identifying components of social systems is precisely the problem faced in identifying components of physical systems—namely, how to differentiate from among a seemingly infinite number of components those particular components that are generally considered to be important for understanding dynamic interactions and processes.

In Chapter 7 it was suggested that one way in which the assessment team can begin to focus on relevant components of the physical environment is first to focus on general processes (by which materials and energies are introduced, transformed, translocated, sequestered and/or dissipated in soil, water, air, and organisms) and then to focus on particular manifestations of these processes (e.g., eutrophication, succession).

With respect to social systems, a similar approach may be taken— that is, the assessment team should first focus on general components and processes that are typically considered to characterize social systems and then proceed to the identification and evaluation of particular phenomena.

While there is much disagreement as to how individual social systems specifically function, there is general agreement as to important components of social systems. These components (Parsons, 1951, p. 5) include:

1. Interacting "actors" (i.e., individuals who interact with one another *within a system of expectations of one another's behavior*)
2. A physical or environmental situation in which actors interact with one another
3. Motivations of individual actors
4. A cultural context (or base) in which expectations, motivations, and physical environment are defined by means of *socially shared symbols*

While the descriptions of these components make use of common English terms, it is necessary to point out that social scientists attach special meanings to these ordinary terms. For example, the word "culture" has a meaning to social scientists that is quite different from its usual lay meaning as "pertaining to the arts." To the social anthropologist, the culture of a people has historically meant a set of behavior patterns, as in "customs," "traditions," "habits," etc. More recently, the meaning of "culture" has often been modified so as to refer specifically to rules for the governing of behavior (Geertz, 1973, p. 44).

The concept of culture as a set of rules that govern behavior has been largely influenced by the development of computer technology (Simon, 1966). This technological development has permitted comparison between the computer (as an active information processor) and the human brain. Of course, no computer, regardless of the sophistication of its design, can process any information whatsoever without a program—that is, a set of directions that are fed into the computer and tell it what to do and when and how to do it. By analogy, the human brain is conceived as a highly sophisticated information-processing machine that also requires programming in order to function. The set of programs given to the human computer is culture; and the process whereby that task is accomplished may be referred to as the *enculturation* (in anthropology) or the *socialization* (in sociology) *process*.

The underlying assumption in the concept of culture as a set of programs is that the human being is born in essentially an absolute state of ignorance about what it should do in this world, or how to do it, or why. The enculturation or socialization process is the process whereby the biological being of humankind becomes a social being, knowing who it is, who others are, and what the relationship between itself, others, and the rest of the world is and ought to be.

It is interesting to note that much of the legislation and/or regulatory literature pertaining to impact assessment uses the term "culture" in a very different sense than is typically understood by professional social scientists. In this literature, cultural resources "are sites, structures, objects, and districts significant in history, architecture, archaeology, or culture, and include archaeological resources, historic resources, and architectural resources [U.S. Department of the Interior, 1974, pp. 1–3]." Cultural resources are, in short, physical features. While the assessment team must fit its assessment to such regulatory usages of the term "culture," it is absolutely necessary that the assessment team also recognizes the disciplinary meaning of the concept of culture in diverse social sciences. For whether the concept of culture is utilized to refer to particular customs and rituals, or to particular personality types (Benedict, 1934), or to rules governing human behavior, the concept of culture is a fundamental concept in social science. And, of course, social science, as a collection of dynamic, individual disciplines, is not controlled by legislated or bureaucratized meanings assigned to the simple word "culture." *Insofar as the assessment team must deal with professionals and/or professional literature in the social sciences, the assessment team should not restrict its deliberations of cultural impacts to deliberations of project impacts on physical structures. Impacts on culture are impacts on the very mechanisms by which people learn to be the people they are.*

To the non-social scientist, it must seem to be a wild and reckless flight of imagination to imply that highways, power plants, dams, and all the other types of projects that typically require impact assessment under NEPA can actually impact on mechanisms by which people learn to be the people they are. In a similar fashion, it is a wild and reckless flight of imagination to imply that local projects can affect regional biomes or the structure and dynamics of distant wetland systems. What makes such social and physical impacts appear far-fetched is, of course, our own ignorance of the mechanisms involved. Just as the assessment team must assume that it does not understand everything about the physical environment, so it must also assume that it does not understand everything about the social environment. After all, the fact that we all are physical, chemical, and biological systems does not in any way ensure our understanding of these systems; and the fact that we all live in social systems does not in any way ensure our understanding of social systems.

Overview

As already discussed, the concept of culture generally refers to specific behavioral patterns and/or to the social programs or rules that are inculcated into the developing human being and which, in turn, influence his or her behavior. Essentially, culture is therefore what we learn as opposed to what we inherit genetically (Wilson, 1966, pp. 50–51). It is what we learn from our experience with the social world about us and from our experience with the physical world about us (as explained to us by others). It is also what we learn by our perceptions of the usefulness of knowledge that is socially given us to meeting the real necessities of our daily lives. Culture is *experiential* because it is derived from and/or mediated by our and others' experience; it is *transmittable* because the experiential knowledge of one group or generation may be passed on to another group or generation. Culture is also *directive,* because the transmitted, experiential knowledge of one generation influences the experiences of a succeeding generation. Finally, culture is *variable,* because there is imprecise transmission of cultural knowledge from one group or generation to another.

As depicted in Fig. 13.1, one way in which we may envision the generalized concept of culture is to imagine culture as a set of "social dies" that are composed of previously tested behavioral patterns and that continually give "social shape" to the biological and genetic potential of

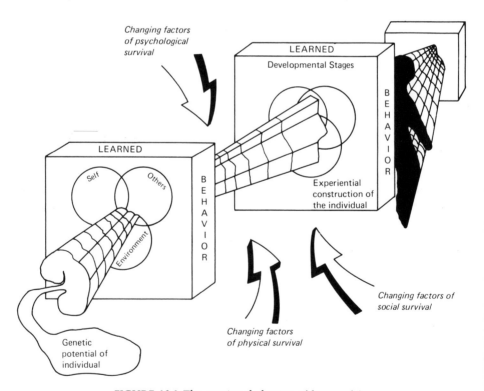

FIGURE 13.1 *The continual shaping of human life.*

each new individual. Essential components of these dies include human experience of *self, others,* and *the physical environment.* It is important to understand that psychological, social, and environmental factors that influence cultural rules can change. Thus cultures can be dynamic, changing sets of rules, as well as static sets of rules.

It is also important to understand that the concept of a culture is different from the concept of a society. The difference between these two concepts may be likened to the difference between content and form (Himes, 1968). The cultural content of social life includes those rules governing the acting (*behavioral patterns*), and the thinking and believing (*ideological patterns*) of individuals. The social form is the manner in which these rules are actually manifest in the organization of social life. Thus a society is the means whereby cultural ends become realized.

The relationships between cultural rules governing the behavioral and ideological components of social life and the physical environment in which they occur have long been the object of anthropological inquiry. While there is much disagreement as to specific relationships between what people think and do and the physical conditions in which they live, it is generally agreed that the constraints of environmental resources are basic factors in the determination of culture (Meggers, 1968, p. 20). As depicted in Fig. 13.2, interrelationships between the physical environment in which people live and the behavioral and ideological rules with which people are enculturated may be said to comprise the cultural base of a people. Just how that cultural base is actually manifested in an orderly and patterned scheme of human activities reflects the *social organization* of a particular society. While there appears to be an infinite number of ways in which a general cultural base can be translated into specific patterns of social organization, it can be assumed that these patterns all derive from the functional needs of any social system, including; (*a*) the maintenance of health and well-being of individuals; (*b*) division and coordination of labor; (*c*) communication; (*d*) training of new members; (*e*) regulation of conflicts and disruptions; and (*f*) distribution of power and wealth (Himes, 1968).

It must be emphasized that different social science disciplines give different emphasis and meaning to such terms and ideas as "culture," "social organization," and "functional needs." Even within a single discipline (e.g., sociology) or subdiscipline (e.g., community sociology), one may find meanings and usages of terms that are not generally shared among practicing professionals. Thus, the importance of the overview of social organization discussed here and shown in Fig. 13.2 is not in specific definitions of terminology, but rather in the identification of the basic kinds of components and interrelationships generally considered to be important in discussing social systems.

For purposes of impact assessment, it is useful to recognize two basic

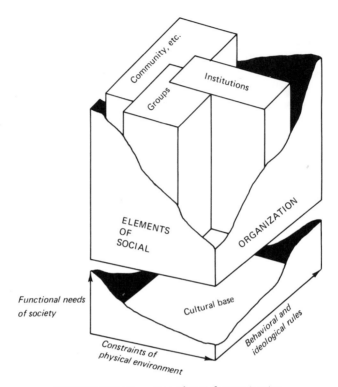

FIGURE 13.2 *Overview of social organization.*

kinds of components and interactions: (*a*) personal and interpersonal; and (*b*) institutional.

PERSONAL AND INTERPERSONAL CONTEXT

Just as the two sides of a coin cannot be separated and the integrity of the coin itself maintained, so the personal (or individual) side of human existence cannot be cleanly separated from the interpersonal (or social) side. While a person has individuality, he or she is not independent of others, if only by virtue of the enculturation or socialization process which attempts to program him or her with the collective wisdom (or ignorance) of others who have gone before. Even when the socialization process apparently fails in directing the thinking and behavior of an individual into preformed patterns, the individual learns by experience of society's attempt—and thus learns (either conformity or nonconformity) through others.

This learning of culture—which takes place in an individual and which occurs in a social context—should not be seen as primarily a kind of vocational learning that is required in order to perform a task—as in the learning of a manual skill. It is also (and probably more importantly) the learning of prescribed associations between sensory and other informational signals and mental images. For example, the words "cat," "god," "good" have innumerable meanings, depending on what we have individually learned to associate with them. Surely we were not born with mental images of their meanings. But we "know" what they mean because of the mental programming of our social experience. In the most general sense, how we individually "translate" the world about us into meanings that are interpersonally shared and provided is the object of studies in *social perception* (Secord and Backman, 1964, pp. 13–14).

That we do individually perceive the world about us in just those ways in which others also perceive it has an important consequence for the organized pattern of social life—namely, the sharing of perceptions promotes social expectations. For example, the perception of "Monday's sunrise" that is shared among thousands as "the start of a workday" is clearly the basis for those same thousands expecting certain things to happen, including traffic jams and the rush for parking space, the arrival and departure of trains and buses according to prescribed schedules, etc. But what if "Monday's sunrise" was perceived by bus drivers and conductors as "a holiday" or "a time to change schedule randomly," or by parking-lot attendants as "a day to go fishing"? The so-called pattern or organization of social life would be reduced to a shambles.

An important concept that is related to the socially shared expectations of behavior is the concept of role (Goffman, 1959). The *role* of an individual is the expectations of the person's behavior that others have because of that person's position (i.e., *status*) within society or some social group (e.g., president, mother, teacher, and priest are all statuses for which one has certain expectations as to the behavior of individuals who occupy those statuses). Of course, an individual also has his or her own perceptions of that role, and this perception may conflict or coincide with others' perceptions.

The concept of role is important because it integrates individuals into a matrix of interpersonal expectations of behavior. It is the contingency of these expectations that introduces the normative quality or predictability of social life (Secord and Backman, 1964, p. 455). For example, my getting into a plane is very much contingent on my perception of the crew (especially the pilot) as being qualified—that is, I expect the crew to act in a certain way which I have identified, either through direct experience or secondhand knowledge, as indicative of their knowing what they are doing. But, of course, the crew are hardly likely to fulfill their roles (i.e., meet my expectations of their behavior) if they do not, in turn, perceive me as a passenger intent on traveling from one place to another. Should I act suspiciously while undergoing the security check or otherwise be perceived as not being a serious passenger but as someone intent on hijacking or other mayhem, the crew most definitely will not go about their ordinary business. It is by satisfying the mutual expectations of different and interdependent role players (e.g., crew members and passengers) that social perceptions organize social systems and give it the predictability that day-to-day life requires.

As indicated in Fig. 13.2, one of the elements of social organization which both promotes and benefits from the predictability of daily life is the *social group*. While all social groups may be characterized by (a) some orientation of individual members toward one another; (b) some sharing of values and beliefs; (c) some pattern to relations among group members; and (d) statuses that are occupied by specific individuals (Himes, 1968), social scientists recognize a variety of different types of social groups.

For example, some social groups are primarily concerned with the promotion of interpersonal relationships within the group membership. In such groups, the identification between members (i.e., reference) is of paramount importance. The group may be a criminal gang, a group of housewives, or a bowling team. Regardless of its specific composition, if group members and actions give priority to the intimacy of interpersonal relationships within the group, social scientists would refer to the group as a *primary group*.

In contrast, a *secondary group* is a group in which priority is not given to the intimacy of interpersonal relationships, but to some interest held in common by group members. The key characteristic of a secondary group is the formality of its organization with respect to various group activities, including those pertaining to membership application and acceptance, communication within the group, the exercise of power and decision-making, and the resolution of conflicts. Typical examples of secondary groups might include a school committee and a working shift in a factory. While individual members of these groups may indeed expend a great deal of energy and emotion in group activities, the point is that secondary groups are primarily concerned with the achievement of objectives that are common to their memberships, and not with providing identity and life-meaning to each individual member.

INSTITUTIONAL CONTEXT

A common-sense view of institutions (which is amusing in the classroom, but misleading in the real world) is that they are names of buildings—for example, legal institutions (as evidenced by courthouses), economic institutions (as evidenced by banks), educational institutions (as evidenced by school buildings), and religious institutions (as evidenced by churches, synagogues, etc.). While the significance of institutions as important structural components of society may be rightly portrayed by the metaphorical description of institutions as buildings, the social nature of institutions is not. After all, institutions are not inanimate things. They are composed of people. The structure that they give to society is in their influence on interrelationships between people.

According to Wilson (1966, p. 406), institutions are the primary means for (a) specifying goals; (b) fixing means for achieving those goals; (c) maintaining the integrity of the social group; and (d) resolving the disrupting tendencies within social groups. Himes (1968) states that these "primary social means" may be referred to in two ways: (a) as complex, bureaucratic organizations; and/or (b) as rules that govern social action.

One way to depict an educational institution is to identify the persons who participate in that institution, their particular functions, their relative positions within the organizational hierarchy, and the roles that they play. Another approach is to identify the rules of behavior and thinking that are inculcated into those who subject themselves to socialization by the institution (i.e., the students).

These two approaches to the concept of institution are not mutually exclusive; rather, they complement one another. For instance, religious institutions may be seen both as a hierarchically organized group of people (e.g., clergy) and as a collection of rules (e.g., doctrines, values), that are promulgated within society at large by the functioning of the hierarchy. The combined view of an institution as a mechanism (bureaucracy) and as a mission (a set of rules) yields a more comprehensive and realistic appreciation of an institution as a functioning social entity than does either view alone.

As collections of rules for governing overt behavior and ideology, different social institutions may promulgate similar behavior and values. Thus the rule "thou shalt not kill" may be found as a basic rule in the socialization efforts of religious as well as of legal institutions. It is also possible, of course, that the socialization undertaken by one institution can conflict with the socialization undertaken by another. For instance, the religious code of behavior exemplified in the rule "do unto others as you would have them do unto you" can be expected to be in direct conflict with a variety of behaviors promulgated by those economic institutions historically based on monopolistic competition.

In general, the consistency of rules promulgated by different institutions varies with the general type of society. One type of society has been variously described as "pluralistic," or "associational," or as having "organic solidarity." A *pluralistic society* (such as contemporary American society) is typically large, depends upon numerous specialized and diverse groups and interests, and has a highly formalized bureaucracy. In pluralistic societies, rules (behavioral and ideological) promulgated by different institutions are often inconsistent with one another. Contrarily, a *monis-*

tic, or *communal society* (also described as a society having "mechanical solidarity") is typically small, isolated, homogeneous and is largely governed by unwritten traditions. In such societies, rules promulgated by different social institutions tend to be congruent and mutually supportive.

The size and internal diversity and complexity of contemporary pluralistic societies present historically unique challenges with respect to the role of traditional institutions as the primary programmers of the human computer. Whereas marriage and family were previously important institutionalized means of receiving and processing information about the outside world, now other institutions have become important, including governmental and educational institutions. Because of the coexistence of different institutions that have overlapping jurisdictions but different values and interests, pluralistic societies have a high potential for creating confusion among people with respect to what they should do and think—and why. This condition is referred to as *anomie.* While the concept of anomie has historically been used to refer to the normlessness of society as a property of social structure (i.e., *social disorganization*), modern usage treats anomie also as a property of individuals who must confront social structure (i.e., *personal disorganization*) (Merton, 1957, p. 161). This approach reflects the ongoing trend in social science to integrate institutional, personal, and interpersonal components of social systems.

Some Key Considerations

If one were faced with the task of describing a human being, perhaps the easiest way in which to proceed would be to begin with a simple physical description. If this were not sufficient, one might go on to discuss behavioral and mental attributes. With respect to the assessment of impacts on the social environment of man, it would seem that descriptions of the social being of humankind have too often been limited to the level of gross social anatomy, and have hardly ever progressed to the level of social dynamics. For example, it would seem that, to many assessment teams, the social being of humankind is essentially composed of bodies, buildings and money—that is, the numbers of persons to be relocated or otherwise inconvenienced; buildings to be relocated, demolished, or constructed; and money to be spent, realized, and redistributed to local, regional, and even national coffers. As evidenced by the preceding overview, the social being of humankind is more complex than mere numbers of bodies, or the physical and economic wherewithal of human life might indicate.

The social being of humankind is built of individual emotions as well as group aspirations and institutional goals; it is built of values and beliefs; of human expectations of what will happen in the future, as well as previous cultural experience; it is built of diversity of interests, as well as on consensus. Of course, one cannot easily measure a belief, count aspirations, or give a value to interpersonal relationships. Nevertheless they are real. And the mechanisms by which these largely unquantifiable elements of social life are integrated and actually expressed in the patterns of social life are the very basis of social dynamics.

Certainly there are good reasons why assessment teams have generally overlooked the impacts of projects on personal, interpersonal, and institutional dynamics and focused instead on project impacts on "social things," including the following facts:

1. The evaluation of significance with respect to social impacts (which is, after all, an important task of impact assessment) smacks of purposeful "social engineering"—a particularly politically sensitive issue for many decision-makers.
2. Assessment teams have most often been dominated (at least in the first few years after NEPA) by individuals trained primarily in the engineering and natural science disciplines.
3. Social sciences are widely viewed as a potpourri of unsubtantiated opinions.

With respect to these facts, it is necessary to identify some basic principles that should underly each assessment of project impacts on the social environment.

First, *the assessment team should understand that all projects, regardless of their manifest purpose, their location, or their physical engineering-design, do in fact, accomplish some degree of "social engineering"*—that is, all projects maintain or change, enhance or decrease, introduce or remove, restrict or distribute, direct or redirect, or otherwise affect patterns of social life. It is absolutely inconceivable that any proposed project could ever be undertaken if it could be demonstrated that the project would not sometime affect someone somewhere.

Because projects must somehow affect social life if they are to be undertaken at all, there are essentially only two types of projects—those which result in foreseen consequences and those which result in unforeseen consequences. Thus the problem of assessing impacts on the social dynamics of social systems is not that projects involve social engineering, but that the assessment process can fail to recognize or consider relationships between project- and social engineering.

Second, *the assessment team should include individuals who, either by their training or their experience, are sensitive to the concerns, con-*

cepts, and analytic techniques of professional social scientists. As discussed in Part II, the assessment team need not include professional biologists and ecologists as long as the team has access to such professionals and to the relevant literature. Similarly, the assessment team need not include professional sociologists, social psychologists, political scientists, etc., as long as the team has access to these professionals and their literature. But access alone is not sufficient. It is necessary that the access be utilized by team members who are both capable and serious in pursuance of information and data pertaining to social dynamics.

Too often those who are experientially qualified to gather and process data and information on the social environment have been just the persons excluded from the assessment process. For example, right-of-way professionals are those personnel who undertake the appraisal and acquisition of real property for project development. Much of the success of the right-of-way professional in his or her work can be attributed to his or her sensitivity to people and social issues and concerns, and to his or her ability to communicate with people having diverse backgrounds, interests, and training. Yet, in spite of these qualities and skills, right-of-way personnel are typically not included in impact-assessment teams. This particular situation may reflect bureaucratic realities in individual agencies and companies; it may also reflect the historic tendency to view impact assessment as primarily a task for engineers. Whatever the cause, any management approach to impact assessment that results in superficial consideration of social impacts is counterproductive to the goals of NEPA and should be reconsidered.

Third, *the assessment team must disregard its own preconceptions of the various social science disciplines, and deal with social scientists and their literature in the same manner as it deals with natural scientists and their literature.*

The National Environmental Protection Act calls for interdisciplinary evaluation of project impacts—*not* an evaluation by persons who represent just those disciplines that meet criteria established by the assessment team. The assessment team has no authority under NEPA whatsoever for deciding that some disciplines should not be consulted merely because they do not meet the team's conception of "objectivity" or "scientific rigor."

The statement of this principle should not be viewed as an apologia in behalf of social science disciplines. In fact, it reflects less the author's appreciation of social science than it does my observations of those members of assessment teams who persist in the misconception that impact assessment is the private preserve of natural science. Observations of these team members, though they are necessarily tainted with my own subjectivity, include:

1. Such team members are basically ignorant of the diversity, historical development, and the current technical concepts and methodologies of social science.
2. Such team members typically base their perceptions of social science on one or two undergraduate courses they have taken, or on the pronouncements of the mass media.
3. Such team members invariably manage to demonstrate their conviction that what they already know (either of physical or of social sciences), they know very well indeed; and that what they do not know is quite irrelevant.

The presence of such individuals on assessment teams should not be tolerated. Where they are tolerated, they should be ignored. Impact assessment is too complex and serious a business to be subjected to the influence of even high-grade morons.

A Basic Approach to Assessment

A comprehensive assessment of project impacts begins with an understanding that *any social system is a dynamic integration of personal (or individual), interpersonal, and institutional components.* As depicted in Fig. 13.3, projects can affect each of these components, as well as their dynamic interrelationships. For example, the taking of right-of-way may require the relocation of a particular individual. While relocation may have various consequences for the individual (both desirable and undesirable), his or her relocation may also affect interpersonal dynamics in those groups in which the individual plays important roles and to which he or she contributes inputs not easily provided by others. For instance, relocation can result in the person's being less available to other group members or may result in a complete cessation of his or her participation in group activities.

As discussed in the previous overview, personal, interpersonal, and institutional perspectives of social organization include specific consideration of:

1. Primary elements at each level (the individual, primary and secondary groups, bureaucracies, and cultural rules and sanctions)
2. Dynamic processes that characterize personal, interpersonal, and institutional elements (e.g., perception and cognition, role playing, the process of socialization)
3. Factors that influence dynamic processes (e.g., values, attitudes, and beliefs; definition of roles and statuses; conflicting pluralistic interests and concerns)

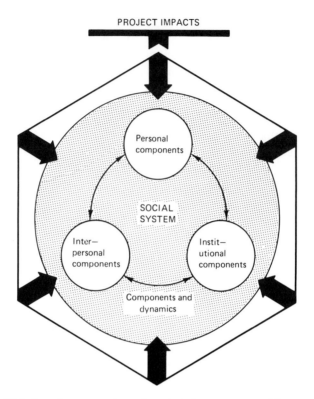

FIGURE 13.3 *A systems overview of project impacts on social components and dynamics.*

A summary of primary elements, some important dynamic processes, and some factors that influence dynamic processes at personal, interpersonal, and institutional levels of social life is included in Table 13.1. The table is not intended to be complete with respect to all potential entries, but rather, to be indicative of the type of conceptual approach that can help to ensure a comprehensive assessment process. This approach is based on the following two considerations:

1. The assessment team must have a clear idea of precisely what will be looked at in the process of assessing impacts (individuals, groups, and institutions; dynamic processes; and key factors that influence those processes).
2. The assessment team must have some understanding of the meaning and use of technical jargon employed by social science professionals.

While the tripartite partitioning of the social system into personal, interpersonal, and institutional components and dynamics can be a use-

TABLE 13.1
Some Components and Dynamic Processes of Social Systems

	Some perspectives of social organization		
	Personal	Interpersonal	Institutional
Primary element	Individual	Groups (e.g., primary and secondary groups)	Bureaucracies Cultural rules and sanctions
Some important dynamic processes	Perception and cognition Role playing Personal disorganization	Promotion of common interest and group identity Interrelationships within Group	Socialization process Identification, promulgation, and regulation of means–ends Decision-making Social disorganization
Some important factors that influence dynamic processes	Aspirations Values Beliefs Emotions Previous experience Socioeconomic class	Definition of status and role, Role strain, conflict and resolution, Demographic factors Sanctions	Diversity and conflict of pluralistic interests Rate of technological change Sanctions

ful tool for organizing the overall assessment effort, it should not be viewed as a model of the social system. The real social system, after all, does not allow for the clean separation of three types of components and dynamics. In the real social system, personal, interpersonal and institutional components and dynamics are tightly linked.

Because of the real-world coupling between social components and dynamics it is important that the assessment team conduct its work in such a manner as to maximize the integration of data and information on personal, interpersonal, and institutional components and processes. It may be quite useful, of course, if individual members focus on different components and processes. Such division of labor can greatly enhance the efficiency of data collection. However, the goal of integrating data and information requires that the team as a whole give primary attention to evaluating interrelationships between personal, interpersonal, and institutional impacts. *As a general rule, the point at which individual team members see themselves as being concerned solely with impacts on institutions, or with impacts on groups, or with impacts on individuals, is precisely the point at which the assessment team has lost sight of the real social environment of man.*

Some Examples of Social Interactions

An example of an important concept for assessing project impacts on interrelated social components and dynamics is the concept of *community cohesion*.

While the term "community" has been variously defined (Brokensha and Hodge, 1969, pp. 5–10; Zollschan and Hirsch, 1964, pp. 63–71), a definition commonly utilized for purposes of impact assessment is:

> A distinctive, homogeneous, stable, and self-contained unit of a larger spatial area defined by geographic boundaries, ethnic, or cultural characteristics of the inhabitants; a psychological unity among the residents; and the concentrated use of the area's facilities [Federal Highway Administration, undated, pp. 85–86].

The idea of "cohesion" as it might be applied to "community" is defined as "behavioral and perceptual relationships that are shared among residents of a community that cause the community to be identifiable as a discrete, distinctive geographic entity within the urban pattern."

As evidenced by these definitions, the concept of community cohesion is inclusive of personal, interpersonal, and institutional components and dynamics within a specified geographical area. For example, individuals are identified as persons perceiving their own identification and interrelatedness with one another, with groups, and with institutions; groups, as interpersonal mechanisms by which the "discreteness and distinctiveness" of the community are maintained; and institutions as "the bureaucratic basis of community self-sufficiency and identity, and providing for necessary production, consumption, distribution, socialization, social participation, social control, and mutual support [Federal Highway Administration, undated, p. 92]." Thus, in a cohesive community it is impossible to deal with individuals without dealing with specific community groups and institutions. Similarly, it is impossible to deal with either groups or institutions without dealing with specific individuals important to community life.

Another important concept that demonstrates interactions between various social components and dynamics is the concept of *mobility*. This concept may be utilized with respect to physical, social, and psychological phenomena (Wilson, 1966, pp. 643–644). For example, one is physically mobile if one can change geographic location; socially mobile, if one can change status (in the sense of public esteem); and psychologically mobile, if one can easily adapt to new places, or to new statuses and/or roles.

From a sociological perspective, physical mobility is not determined solely by either the mere possibility (e.g., having a job available elsewhere) or economic feasibility of moving (e.g., meeting the immediate costs of moving). The real potential for physical mobility is also influenced by the probability of being accepted into a new neighborhood and community (i.e., by interpersonal factors). Such acceptance (or rejection) is, in turn, influenced by definitions of roles that may be differently defined by the institutional socializations of the person who is contemplating moving and the persons among whom he or she will have to live if he does move. Finally, individual psychological adaptation to new social and physical conditions can be made either difficult or simple, depending on the degree of support by special groups in both the old and the new community, including groups of former "transplants" who are interested in smoothing things over for new arrivals.

While these are but a few of the considerations involved in evaluating the actual mobility of individuals within a social system, they are sufficient to demonstrate that individual mobility is influenced by interpersonal and institutional components and dynamics. Thus, as with the concept of community cohesion, a project may not have direct effects on a specific individual, but may significantly impact that individual by altering the interpersonal and institutional environment with which the individual has to deal. Similarly, significant project impacts on interpersonal and institutional components and dynamics may derive from impacts on personal components and dynamics of social systems.

In the following chapter, primary attention will be given to examples of both direct and indirect impacts on personal, interpersonal, and institutional components and processes. Chapters 15 and 16 will focus on potential economic and public health impacts. While both economic and public health impacts fall within the general scheme of social impacts to be developed in Chapter 14, they will be accorded individual treatment because of the preeminence typically given them in formal EISs.

Concluding Remarks

The assessment of project impacts on the social environment is generally considered to have received less attention than has the assessment of project impacts on the physical environment. Some of the blame for this might reasonably be attributed to a general misconception that the word "environment" should be interpreted as referring to the physical, chemical, biological, and ecological surroundings of man. Yet, other explanations have been offered.

It has been suggested (Singh and Wilkinson, 1974) that the preeminence given to physical considerations in the assessment process is due, in part, to the as yet poorly developed condition of social science theory and methodology. It has also been suggested in this chapter that the insensitivity of team members trained in engineering and natural-science disciplines, as well as the timidity of decision-makers with respect to the identification of projects as exercises in social engineering, have also contributed to a certain superficiality of social impact assessment.

Regardless of the cause (or causes) of the under-utilization of social science concepts and concerns in the assessment process, such a situation cannot be tolerated if impact assessment is to be conducted in accordance with the goals and objectives of NEPA. Each assessment team is, therefore, best advised to concentrate its efforts on the utilization of concepts, information and understanding as they currently exist (in both physical and social science disciplines) rather than on the construction of plausible but irrelevant rationalizations of its failure to do just that.

References

Benedict, R.
    1934   *Patterns of culture.* Boston: Houghton Mifflin.
Brokensha, D., and P. Hodge
    1969   *Community development: An interpretation.* San Francisco: Chandler.
Geertz, C.
    1973   *The interpretation of cultures: Selected essays.* New York: Basic Books.
Goffman, E.
    1959   *The presentation of self in everyday life.* Garden City, New York: Doubleday.
Himes, J. S.
    1968   *The study of sociology.* Glenview, Illinois: Scott, Foresman.
Meggers, B. J.
    1968   Environmental limitation of the development of culture. In *Environments of Man,*
           edited by J. B. Bresler. Reading, Massachusetts: Addison-Wesley.
Merton, R. K.
    1957   *Social theory and social structure* (Rev. ed.). London: The Free Press of Glencoe,
           Collier–Macmillan.
Parsons, T.
    1951   *The social system.* New York: The Free Press.
Secord, P. F., and C. W. Backman
    1964   *Social psychology.* New York: McGraw-Hill.
Simon, H. A.
    1966   Thinking by computers. In *Mind and cosmos: Essays in contemporary science and
           philosophy,* edited by R. G. Colodny. Pittsburgh: Univ. of Pittsburgh Press.
Singh, R. N., and K. P. Wilkinson
    1974   On the measurement of environmental impacts of public projects from a sociologi-
           cal perspective. *Water Resources Bulletin, 10*(3).
U.S. Department of Housing and Urban Development
    1974   Environmental review procedures, proposed policies and procedures. *Federal Re-
           gister 39*(198).
U.S. Department of the Interior
    1974   *Preparation of environmental statements: Guidelines for discussion of cultural
           (historic, archeological, architectural) resources.* Washington, D.C.: U.S. Gov-
           ernment Printing Office.
Federal Highway Administration
    Undated   *Social and economic considerations in highway planning and design* (Training
              Course Manual). Washington, D.C.: National Highway Institute.

Willard, B. E.
  1975 *Ecological considerations in transportation systems.* Paper presented at the July
        15–19, 1974 ASCE Transportation Engineering Meeting, Montreal, Canada. *Proceedings of the American Society of Civil Engineers,* Vol. 101 (July 1975).
Wilson, E. K.
  1966 *Sociology: Rules, roles, and relationships.* Homewood, Illinois: The Dorsey
        Press.
Zollschan, G. K., and W. Hirsch (Editors)
  1964 *Explorations in social change.* Boston: Houghton Mifflin.

14

Personal, Interpersonal, and Institutional Impacts

Introduction

As discussed in Chapter 13, a social system is a system of interdependent personal, interpersonal, and institutional components and dynamics. While there is apparently no limit to the diversity of form that can be evidenced by these components and dynamics, the assessment team should consider (a) that the particular form (or social organization) of a particular social system does in fact reflect ongoing relationships among individual people, individual groups, and individual organizations and bureaucracies; and (b) that such relationships are important social realities regardless of the assessment team's evaluation of their intrinsic worth or desirability.

The practical relevance of these considerations to the assessment of project impacts on the social environment can be defined with respect to:

1. The comprehensiveness of overall assessment approach
2. The balancing of local, regional, and national goals and objectives

COMPREHENSIVENESS OF OVERALL ASSESSMENT
APPROACH

A review of computerized models which are currently available for conducting urban studies includes the following observation (Brown, 1974):

When one writes of "socioeconomic" effects [i.e., of projects] in the literature, the economic half dominates the article with a rather brief mention of the need for "recognition" of social problems. Little actual research has been devoted to what impact water resource development has on the social system of a region. The studies that have been completed are explanatory in nature and examine limited aspects of the problem and focus on small areas [p. 12]."

While the observation that social assessment is essentially minimal and myopic is offered with specific reference to the impact of water-resource development on urban centers, it also describes the assessment of impacts of all types of project development on social systems. In general, once an assessment team considers project impacts on aesthetics, public health, and economics (and possibly on community cohesiveness), the team all too often reaches the limit of its sociological imagination and concludes that there is nothing more to consider of humankind as social beings.

There is, admittedly, no absolute measure of comprehensiveness that can be used to gauge the sufficiency or insufficiency of the scope of social assessment. However, a general rule is useful:

If the assessment team cannot name and/or describe real people, real groups, and real institutions that will be affected by the different phases and activities of actual project development, the comprehensiveness of the team's approach to social assessment is open to serious challenge by any social discipline.

Like the concept of ecosystem, the concept of social system must be considered—not in the abstract, but in terms of the actual environment in which the project is to be located.

Of course, depending on the project, it may be quite unreasonable to expect that the assessment team will have intimate knowledge of every individual, group, or institution within a region. As in the assessment of impacts on the physical environment, the assessment team has to be selective in its approach to the assessment of impacts on the social environment. The practical need for selectivity, however, does not justify a hap-hazard, random approach to impact assessment. Rather, the assessment team should be able to justify its selective consideration of individuals, groups, and institutions in terms of the standard methodologies and concepts of relevant social science disciplines.

BALANCING LOCAL, REGIONAL, AND NATIONAL GOALS AND OBJECTIVES

Whether or not any attribute of social organization (e.g., the social cohesiveness of a particular neighborhood) shall be sacrificed by project development for the achievement of some "greater" regional or national goal is a question that must be answered in the overall decision-making process. It is not a question to be answered by the assessment team as it undertakes the analytical and integrative tasks of impact assessment.

If, for example, the assessment team undertakes assessment with the attitude that the interests of individuals must always be secondary to group interests, or that the objectives of local groups must always be secondary to the objectives of the region, or that the concerns of local and regional institutions must always be second to the concerns of the nation, the assessment team is engaged more in the business of rationalizing a proposed project than in appraising the project's impacts on the social environment. Thus, in order to ensure the comprehensive and balanced assessment of project impacts on the social environment, the assessment team must focus on social realities and not on its own conception of what those social realities should be.

Some Examples of Social Impacts

The proximity of proposed projects to local residents is generally considered in the process of assessing impacts on the social environment. Proximity is particularly important with respect to project impacts on air quality, aesthetics, and noise levels in local areas. However, impacts on air quality, aesthetics, and noise levels do not define the total social impacts of projects. The proximity of projects to people can also affect social dynamics.

For example, Fig. 14.1 depicts three alternative locations of a hypothetical project with respect to three individuals. As depicted in the diagram, Individuals B and C are geographically equidistant from Individual A. However, in terms of social reality (reality as perceived by these individuals), the *social distance* (as measured by frequency of visitations, importance of visual contact, etc.) between Individuals A and C is depicted as being much shorter than the social distance between Individuals A and B. It may be, for instance, that the interactions between Individual A and Individual B may be much less frequent than between Individuals A and C, being primarily maintained by telephonic communication rather than by visual communication or actual visit.

Because of the different social distances between the depicted individuals, different project locations can have quite different social impacts. As indicated in Fig. 14.1 alternative location 3 may directly interfere with interactions between Individuals A and C. Such interference may occur as a result of interrupting the visual line-of-sight between them, and/or by making it more difficult and time-consuming for each to visit one another. On the other hand, alternative location 1 may have no effect on the social distance between Individuals A and B. Alternative location 2 may similarly have no effect on the social distance between

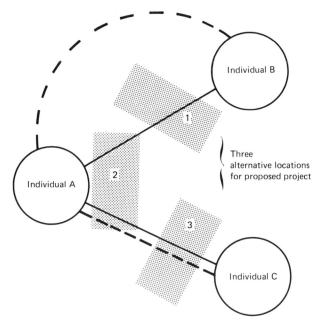

FIGURE 14.1 *Alternative project locations in relation to the geographic and social distance between individuals. Solid line shows geographic distance between individuals; broken line shows social distance between individuals.*

Individuals A and B, and may also minimize (though not remove) project obstruction of visual or personal communication between Individuals A and C.

Of course, there is no reason to restrict consideration of project impacts on social distance to impacts on the social distance between individuals. The degree and frequency of an individual's participation in or identification with groups and organizations in the general project area can also be affected by project development and operation. The important point is that the social distance between any two or more social components (i.e., personal, interpersonal, and institutional components) can be altered by project location.

As in the previous example, one effect of project location is to interfere with the communication and/or identification among interdependent social components. Projects may also, of course, effectively shorten social distance between social components. The shortening of social distance may (a) facilitate the communication between individuals and groups who desire improved access to one another; and also (b) facilitate (or even force) communication between individuals and groups who do not desire improved access to one another.

For example, a project may result in the removal of a natural physical barrier, such as a woodland, between people who, except for that barrier, might enjoy visiting one another—or who might be content with just being able to see someone else and feel, thereby, less isolated. Or, the same project, with the consequent removal of the same natural barrier, might result in actual clashes among groups whose disparate interests and life styles were previously shielded from one another by the natural barrier.

As these examples indicate, *it is important that the assessment team not assume that improved access of people to people is always for the social good.* It depends on who is defining "social good." Individual members of the assessment team may individually believe that the shortening of social distances between people is a basic social goal, but the team must consider in its assessment that social facts hardly ever conform to social postulates. In a particular project area, it is quite possible that people want to maintain social distance from one another and that projects which promote the shortening of that distance may result in social conflict and instability (including both *personal* and *social disorganization).*

In evaluating possible project impacts on social distances between individuals and groups in the project area, the assessment team should consider that the proximity of the proposed project to people has several dimensions. For example, the proximity of the project during its construction phase might be quite different from its proximity during its operational phase. Proximity is not just geographic nearness, but, more importantly, social nearness. Thus, the proximal effects of constructing an underground garage, for example, may include the relatively short-term effect of obstructing home-visitations within a local area. The operational phase of the same project, however (depending on final landscaping, etc.), may result in an increase in the frequency of personal interactions in the local area. The assessment team must, therefore, approach the assessment of project impacts on social distance (and related dynamics) by noting:

1. The mechanisms by which social distance is traversed in the project area prior to the project (e.g., visual signals, visits, informal gatherings, community functions)
2. The relationship between different project activities in different phases of project development and the successful use of these mechanisms (e.g., construction noise and conversations between neighbors, open pits and trenches and visitations of adults and children in the area)
3. The relative sensitivity of some segments of the population to impacts on specific mechanisms (e.g., the use of wobbly planks to facilitate walking across construction areas is fine for healthy children and adults, but what of the physically handicapped or the

aged who have difficulty walking? Would they stay home and not bother to visit their friends?

While projects can directly affect ongoing social distances between social components by directly affecting mechanisms by which those distances are traversed and maintained, projects can also indirectly affect social distances by affecting individually perceived roles. As discussed in Chapter 13, the role of an individual is behavior that others expect of him or her by virtue of his or her status (or position) within society. Failure of an individual to meet role expectations may result not only in psychological stress for the individual (if the person has internalized those roles), but also in changes in the social distance between the individual and other components of society and, therefore, in changes in social dynamics.

As shown in Fig. 14.2, the behavioral responses of an individual to any situation in which he finds himself cannot be defined from the perspective of a simple stimulus–response paradigm. Rather, the individual's internalization of various roles (as defined by the expectations of numerous social components), previous experience, and his or her values, beliefs, and other emotional attributes, all contribute to the person's actual response. Thus, project impacts on the individual's role perception, on his or her experiential knowledge, and/or on values and beliefs have the potential to affect his or her behavior and, thus, the social dynamics in which his or her behavior plays a part.

For example, a particular project (e.g., a major pipeline, such as the Alaskan pipeline) may result in a dramatic change in the distribution of money in a local area (Chapter 15). The income of many individuals, however, can remain essentially unchanged—because they continue in jobs that are unrelated to project development. Yet, because of the general infusion of monies (although they are unequally distributed), the cost of living may increase dramatically and such individuals can experience a severe drop in their real income. For those individuals who perceive their roles within the family as the "breadwinner," any serious drop in real income may be experienced as such a role strain that the individual must reconsider his response to the project-mediated change in his social environment. He may decide, for example, to go to work on the pipeline. Or, he may decide to "stick it out." He may also "get out" entirely, and relocate. The first option may require his being away from his family for long periods; the second may have any number of affects on both his family and his social life, including changes in the family's interaction with other families, group affiliations, and the needs of the family for institutional services. The third option results, of course, in the need for total family readjustment to new surroundings. In short, whatever the option taken, the social dynamics at individual, interpersonal, and institutional levels may be affected.

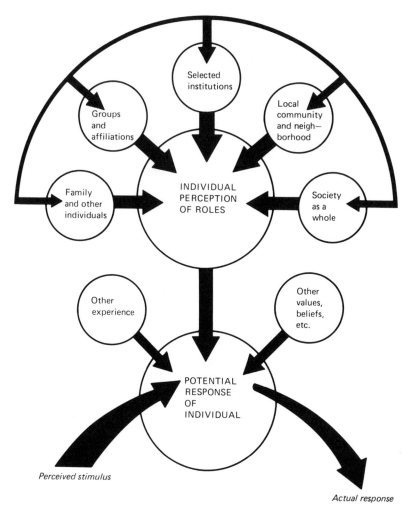

FIGURE 14.2 *The concept of role: the link between the individual and interpersonal and institutional components of society.* [*Adapted from the American Right of Way Association, 1976.*]

Economic influences are not, however, the only influences on role perception that can be affected by projects.

For example, it was pointed out in Chapter 13 that one aspect of a pluralistic society is that the individual is presented with numerous alternatives to the overall socialization process. At least one consequence of increased pluralism is an enhancement of acculturative conditions within a previously cohesive community. Thus, a project (e.g., a subway system, a "new" community) may result in an infusion of different peoples, life styles, and institutions into an ongoing community. These personal, interpersonal, and institutional alternatives may conflict with historic community values and aspirations which have long found their expression in the controlled socialization of the young in a local area by local institutions. The consequence of conflict between historic and new alternative socializations lies not only in the potential for overt clashes between opposing groups and institutions, but also in the minds of individuals, who, faced with alternatives never before available, find their traditional roles challenged. In opting for new roles they can experience role conflict and so must devise approaches for dealing effectively with contradictory expectations of their behavior. In opting to maintain their tradition roles, they may be forced to undergo role stress and must accordingly devise mechanisms for dealing with that stress.

It should be noted that there is no absolute sociological value either of role stress and conflict or of their absence. Nor is there any absolute value of social change, or of the maintenance of the status quo. Finally, there is no absolute sociological value either of personal and social disorganization, or of personal and social organization. The positiveness or the negativeness of each of these conditions is to be determined within the context of a real and ongoing social system—a context that, after all, can be largely determined by what the people who are actually involved think about social change, about their aspirations, and about their frustrations and fears.

While the above examples demonstrate how various phases of various types of projects can directly affect individuals and groups, it is important to emphasize the importance of considering indirect impacts in the assessment of impacts on the social environment.

For example, Fig. 14.3 depicts a number of individuals who are linked together in a particular pattern. The lines drawn between these individuals can represent a variety of dynamic interactions, as, for example, a high frequency of personal communication, or the identification which one person expresses for another and which is, in turn, reciprocated. Such a diagram (i.e., sociogram) may (as in Fig. 14.3) indicate subgroups that cannot be identified merely by noting the relative location of domiciles, or socioeconomic patterns within a community—that is, the social reality of the depicted sociogram is in what particular people do (i.e., communicate and/or identify with one another) and not where they are located or anything else about them.

As indicated in the figure, individuals within Group I are tightly coupled with Individual A1; In Group II, individual members are tightly coupled with Individual D1. Moreover, Individuals B and C, acting through individuals A1 and D1, provide linkage between Groups I and II. If the basis of the depicted sociogram is interpersonal identification (or reference), it is clear that a project-mediated impact on certain individuals may be perceived as an impact on everyone. Thus, the relocation of

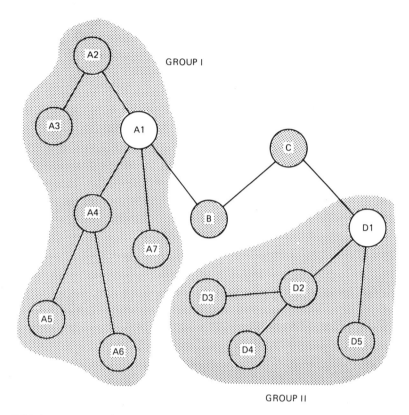

FIGURE 14.3 *Example of a sociogram showing interpersonal and intergroup relation-ships.*

Individual A1, B, C, or D1 can affect individual group structure and overall interactions between Groups I and II. Project-mediated role stress of Individual A4 may be directly transferred to Individuals A5 and A6. And, while it is unlikely (from the figure as depicted) that project impacts on D3 will directly influence A7, there is an obvious pathway whereby the perceptions of D3 can not only be sympathetically transferred to A7, but also can become as real to A7 as to D3.

Direct project impacts on individuals (e.g., through relocation) and groups (e.g., through interference with meetings, group functions, etc.) typically receive most attention by the assessment team. Little if any attention is given to the consequences of individual impacts on *social perceptions*, and thence on *group dynamics*. This is probably due to a general attitude among professionals trained in quantitative disciplines that, in the practical, day-to-day world in which we live, perceptions have only an individual, subjective reality. Thus, the relocation of an individual, whatever its consequences to that individual, has only real consequences to that individual—how others might perceive and react to the relocation of any other individual is beyond the scope of meaningful analysis of objective reality. If such an attitude does influence the assessment of project impacts on the social environment, it is necessary here to reiterate a basic principle of social science—*that social reality is ultimately constructed out of social perceptions.*

According to this principle, the taking of right-of-way through one person's property not only impacts (either positively or negatively) that person's life; it also impacts those who identify or otherwise interact with him. Thus, the individual directly affected may be primarily concerned with the interference of the right-of-way (e.g., for power lines) on his personal use of his land (e.g., for farming). But a group (or groups) may be primarily concerned with what it perceives as an overt demonstration of the power of "city slickers" over "country boys." The impact at the personal level may be psychological and/or economic; however, the same impact, at the level of social groups may be an enhanced sense of social disorganization and may result in the initiation of retaliatory political and/or legal (and even illegal) action. To overlook the importance of social perceptions is to overlook much of the basis of social dynamics.

An Important Consideration

As evidence by the discussions of even the few preceding examples of social impacts, the assessment of impacts on social dynamics requires consideration of numerous subjective elements of social life—including emotive identification (or reference) among individuals and groups. *In assessing social impacts, one cannot, therefore, avoid dealing with human emotions, or with the cognates of human emotions, whether in the form of expressed, or in the form of unexpressed attitudes, values, and general concerns.* Of course, a major problem that faces any assessment team is the identification of those human attitudes, values, and concerns that indicate and/or influence social dynamics.

Some assessment teams have avoided this problem altogether by adopting the view that human attitudes are so notoriously subjective that they should not be considered in any objective assessment of project impacts. As discussed previously, this approach has no place in the conduct of impact assessment under NEPA.

Other assessment teams have faced the problem squarely and attempted to deal with it. They have elicited the attitudes and concerns of local government officials, of individuals and organizations, and of other interested parties in an effort to understand and consider the subjective sequelae of project development. Yet, it would appear that many of these sincere efforts have succumbed to a characteristic frustration of social research—namely, the fact that the most easily elicited and the most articulately expressed attitudes and concerns often do not reflect the controlling attitudes and concerns of the very population considered.

Assessment teams have tended to take various positions with respect to the intransigence of local populations to speak their minds fully and freely, including (but not limited by) the following:

1. If people do not care enough to make sure that their own attitudes and concerns are considered, why should the assessment team worry about them?
2. Local politicians should reflect the attitudes and concerns of their constituents—so why should the assessment team have to consider social issues that have not already been identified in the process of consulting local politicians?
3. Everyone has his own ax to grind. If the assessment team tries to consider everyone, the assessment will never be completed.
4. We put out a notice in the local paper that said we wanted to know what people were really concerned about—and we only got half a dozen responses. The hell with them!
5. If we try to elicit everybody's input we'll just stir up a hornets' nest. They'll have us running around in circles!

The practical consequence of any of these individual or combined positions is predictable—namely, that when human attitudes and concerns about projects are considered in an effort to identify impacts on social dynamics, primary attention tends to be restricted to only a few, highly selected attitudes and concerns. In other words, attitudes and concerns of the public that assessment teams deem important tend to be those which are perceived by the team as normative, desirable, and/or reasonable. This observation is by no means offered as a casual observation; rather, I am of the opinion that any contrived rule of thumb which rationalizes the assessment team's inattention to particular human attitudes and concerns—regardless of their lack of plurality, or regardless of any perceived undesirability or unreasonableness of those concerns—is directly contrary to both the spirit and the law of NEPA.

Persons charged with the responsibility of identifying social impacts of project development have no legislative authority whatsoever for dismissing the attitudes and concerns of people merely because those attitudes and concerns (or the people who hold them) are in the minority. Also, there is no actual legal or sociological justification for giving preeminence to attitudes and concerns of populations in the local project area over the attitudes and concerns of regional and extraregional populations.

While it is admittedly easier to consider a narrow range of attitudes and concerns in the identification and evaluation of project impacts on the social environment, it is more in keeping with the disciplinary knowledge of social structure and dynamics to consider the full range of attitudes and concerns which characterize a human population. The assessment team should, therefore, ensure that:

1. A comprehensive assessment of attitudes and concerns is made in the local project area.
2. Personal, interpersonal, and institutional components and dynamics that influence and/or are influenced by such attitudes and concerns are identified.
3. Direct and indirect impacts of project development on these attitudes and concerns (and/or on related social components and dynamics) are identified.
4. Similar considerations be given to the attitudes and concerns of distantly located regional and extraregional populations, and
5. interrelationships among local, regional, and extraregional dynamics be evaluated in light of project impacts on any one or combination of these dynamics.

As shown in Fig. 14.4, the overall approach can be described by (a) a

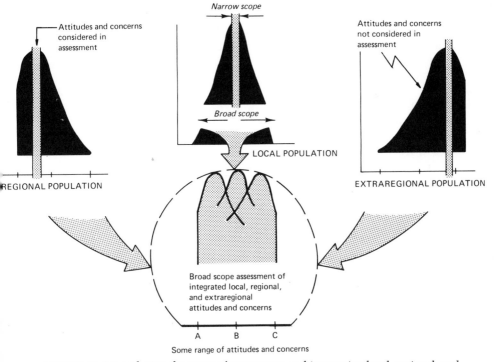

FIGURE 14.4 *Broadening the scope of assessment and integrating local, regional, and extrapersonal attitudes and concerns.*

broadening of the assessment of social impacts at the individual population level; and (b) an integration of local, regional, and extraregional
social components and dynamics.

Social Factors versus Social Dynamics

There are innumerable sources of information on social components
and social issues. Some of these sources have a broad scope (e.g., community development, acculturation); others have a relatively narrow scope
(e.g., recreational opportunities in a local area). An important directory of
national sources of information is *A Directory of Information Resources
in the United States: Social Sciences* (Rev. Ed.), (National Referral
Center, 1973). Important government sources are also listed in Appendix
A.

A common approach to data and information gathering for purposes
of social assessment has been (a) to identify "social factors"; and (b) to
contact those agencies and organizations which can be identified as
having jursdictional interest in those factors. In this approach, the assessment team typically focuses its concern on a list of social factors
previously identified as being most relevant to a particular type of project.
For example, a list of social factors for transportation projects may include (Canter, 1977) both local community and metropolitan factors.
Local community factors include:

- neighborhood boundaries
- cultural patterns
- crime
- fire hazards
- health
- religious institutions
- educational facilities
- recreational facilities
- social services
- public utilities

Metropolitan factors include:

- fire and police protection
- medical services
- educational services
- parks
- historic sites

While such lists are useful for directing the assessment team to
particular local, regional, and even national sources of information, they

can also be counterproductive to an integrated, comprehensive assessment of social impacts. This is due to such practical considerations as: (a) lists of assessment factors generally tend to be quite large; (b) lists do not indicate how individual factors are interrelated in dynamic social processes; (c) assessment teams tend to divide responsibility for individual items in the list; and (d) whatever social issues and concerns may not already be included in a list are usually the issues and concerns that are ignored.

All too often, the end result of this approach to information gathering results in the breakdown of meaningful interaction among assessment team members, and the compilation of voluminous and unintegrated data and information. The fact of the matter is that a static list of social factors (and all the information such a list can generate) cannot be a substitute for systems thinking by the assessment team.

One way in which an assessment team can avoid the pitfalls of basing assessment on a "laundry list" of social factors is to utilize the primary source of social data and information—the public itself. While public involvement in decision-making is a basic goal of NEPA, too often the assessment team approaches the public only after the assessment is completed or well underway. It is necessary, therefore, to emphasize here the importance of the public as a source of information that can guide social impact assessment. As a general rule, *the more direct the involvement between the assessment team and the public, the more direct will be the assessment team's understanding of those social components and dynamics that can be affected by project development.*

Of course, it is highly unlikely that any mechanism for public involvement will result in important contributions of information and data if the assessment team has the attitude that the public is something to be withstood, manipulated, or otherwise exorcised of a supposed innate propensity toward self-interest and obstructionism. Under NEPA, the public is a full partner in the environmental decision-making process. Any attitude to the contrary is counterproductive to the conduct of comprehensive impact assessment and should not be tolerated within the assessment team.

An excellent guide to various mechanisms whereby the assessment team and the public can cooperate meaningfully in the assessment process is the U.S. Forest Service's *Guide to Public Involvement in Decision Making* (U.S. Department of Agriculture, 1974). While this guide is specifically designed for U.S. Forest Service personnel, it is useful to any assessment team involved in social impact assessment. Techniques for achieving meaningful public involvement that are discussed in this guide include:

- press conferences
- open public meetings

- informal group discussion meetings
- workshops
- questionnaires and surveys
- response forms
- advisory committees
- ad hoc committees
- consultation visits with key groups or individuals
- formal public hearings
- letter requests for comments
- use of mass media
- "show me" trips
- utilization of fora provided by other organizations
- service on local and regional committees
- utilization of outside groups to conduct studies

Each of these techniques has its limitations and advantages. The assessment team should consider how each of these techniques (and others) may be used to aid the comprehensive assessment of impacts on ongoing social dynamics. This is not to say that lists of social factors promulgated by various agencies and organizations should be ignored or be given less preference than issues identified through public involvement. Rather, issues and concerns identified through public involvement should be used to supplement, balance, and integrate social factors already identified through agency guidelines. Toward this end, any assessment team is well advised to follow a guideline of the U.S. Forest Service (U.S. Department of Agriculture, 1974): "Discard any notion that actions which will affect environmental quality or the public interest can be judged only by professionals. Although a proposed action may be scientifically (or technically) correct, public concern may well outweigh scientific considerations and justify proposal modification [p. 4]."

In short, in conducting an assessment of social impacts, one must be sure that consideration is being given to real people as well as to abstract guidelines.

The "Social Dummy"

Various methodologies are available for constructing simplified models of complex social components and interactions. As already mentioned, techniques for achieving public involvement, and formal checklists of social factors can generate selective data and information that can be used to model parts and processes of social life. Other methodologies which may be utilized by assessment teams to construct a picture of ongoing social life, and thus a model of project impacts, may include such tools as (Federal Highway Administration, undated):

1. Social feasibility models
2. Social interaction analyses
3. Neighborhood social interaction indices
4. Mobility and stability indices

Whatever tools may be employed, it is important that the generated model of social components and interactions not be so simplified as to be actually lifeless—that is, reduce people to "social dummies." While it may be that no one intentionally sets out to construct "social dummies" as irrelevant images of real people, it is clear that assessment teams too often succeed in doing just that. Simply, real people are too often reduced by computerized algorithms to unidimensional ghosts called "recreators," "consumers," "voters," "taxpayers," and "providers."

A good rule of thumb for any assessment team that is more interested in assessing impacts on real people than on ghosts is that *assessment team members should consider whether or not they themselves might feel grossly misrepresented if the analytical algorithms that they use to describe social components and dynamics in project areas were precisely the algorithms applied for purposes of describing the real context of their own daily lives.*

Concluding Comments

It is my opinion that (a) contemporary impact assessment is largely devoid of even a modicum of sophistication with respect to the evaluation of impacts on the social environment; and (b) that the lack of sophistication is particularly evident in the emphasis of social assessments on individual social components rather than on the dynamic interrrelationships among social components. This is generally not the case with respect to the assessment of project impacts on the physical environment. Ecological concepts, both qualitative and quantitative, have been developed over many years and are increasingly utilized for assessing impacts on the physical environment (Part II). As discussed in this and preceding chapters, the argument is often made that similar integrative concepts have not been developed by social scientists and that the assessment of impact on social dynamics is necessarily superficial. It has also been pointed out that this argument is largely based on a profound ignorance of social science concepts pertaining to role dynamics, small group dynamics, and personal and social disorganization.

Having been so reckless as to offer these opinions about shortcomings of social assessments, I feel obligated to suggest correctives. The following suggestions are offered toward the end of achieving comprehensive and meaningful assessments of social impacts:

1. The assessment of social impacts should not be entrusted solely to individuals who are primarily trained in natural sciences or engineering disciplines, or who (regardless of expertise) see the public as an obstacle to project development.
2. An assessment team having responsibility for the assessment of social impacts should include individuals who have practical as well as academic experience.
3. The total assessment team should have a precise understanding of the relevance of individual social factors to the assessment and evaluation of project impacts.
4. As in the assessment of impacts on the physical environment, the assessment of impacts on the social environment should be based on concepts and methodologies that can be documented in the professional literature of relevant disciplines. In cases where standard concepts and/or methodologies are not utilized, the relevance of substitute techniques to assessment needs should be clearly demonstrated.
5. Whenever analytical models and other paradigms are employed for purposes of data handling and interpretation, all underlying assumptions of the models and paradigms should be described and discussed so that everyone involved in the assessment process (including the public) understands the limitation of the models and paradigms.
6. All identified impacts on individual social components (persons, groups, and institutions) should be examined for their consequences on dynamic social processes (e.g., role strain, personal disorganization).

One final word is necessary—the best analysis of social impacts is totally inadequate if it does not include consideration of the interactions of social and physical components of the total human environment. Yet, as in any systems study, it is quite impossible to study interactions without having a clear understanding of the very components which are the basis of interaction (Barker, 1968, p. 7). In Part II we focused on the physical environment and on the kinds of physical components and processes that characterize that environment. In this section, we are focusing on the kinds of social components and processes that characterize the social environment. While some project impacts on the physical and social environment can be identified by focusing on each environment separately, the comprehensive assessment of project impacts required by NEPA depends on an understanding of interrelationships among components and processes of interacting physical and social environments. Important issues in the assessment of impacts on the total environment will be discussed in Part IV. In the remaining chapters of this section we will focus our attention on some types of social impacts that begin to demonstrate the theme of total environment to be discussed in Part IV.

References

American Right of Way Association
  1976 *Student workbook, right of way and the environment,* Los Angeles: American Right of Way Association.
Barker, R. G.
  1968 *Ecological psychology: Concepts and methods for studying the environment of human behavior.* Stanford, California: Stanford University Press.
Brown, J. W.
  1974 A review of models and methods applicable to Corps of Engineers Urban Studies Projects. In *Proceedings of a Seminar on Analytical Methods in Planning, 26–28 March 1974.* Davis, California: The Hydrologic Center, U.S. Army Corps of Engineers.
Canter, L.
  1977 *Environmental impact assessment.* New York: McGraw-Hill
Federal Highway Administration
  Un-   *Workbook and resource guide, social and economic considerations in highway*
  dated *planning and design training course.* Washington, D.C.: National Highway Institute, Federal Highway Administration.
National Referral Center.
  1973 *A directory of information resources in the United States: Social sciences* (Rev. ed.). Washington, D.C.: U.S. Government Printing Office.
U.S. Department of Agriculture.
  1974 *Guide to public involvement in decision making.* U.S. Department of Agriculture, Forest Service, Washington, D.C.: U.S. Government Printing Office.

15

Economic Impacts

Introduction

In previous chapters it was pointed out that individual physical, chemical, and biological impacts are ultimately impacts on components and dynamics of ecosystems, and that social impacts are ultimately impacts on components and dynamics of social systems. Thus, primary emphasis has been given to the view that comprehensive impact assessment requires a systems-oriented approach. This approach simply reflects the fact that no single environmental factor (whether in the physical or the social environment) exists in isolation. In a similar fashion, it is important to consider individual economic impacts as impacts on an economic system that is inclusive of numerous and interdependent components and processes. If such a systems approach to the assessment of economic impacts is not utilized, the assessment process will likely fail to identify and evaluate many of the indirect consequences of project development.

For example, a particular project may result in both a short- and long-term increase in the number of jobs in a project area. If the increase in jobs is presented only as an individual, isolated impact, it is difficult to imagine how this isolated piece of information can help to improve the decision-making process in project planning and development. It is only when the potential for an increase in the number of jobs is related to other social and physical factors in the total human environment that the impact can be evaluated for purposes of decision-making. For instance, what will be the consequence of an increased number of jobs to local business, to land development in the area, to recreational demand, and to population mobility? Answers to such questions as these help decision-makers to decide the real significance (or insignificance) of a project-mediated increase in the number of jobs. Of course, neither can such questions be asked, nor can answers be given if the assessment team fails to take a systems overview of economic impacts.

Overview

Categories of economic factors for purposes of impact assessment typically include such local and regional factors as (Canter, 1977, pp. 166–167):

- employment/ shopping facilities
- residential property values
- property tax base
- displaced residents
- displaced businesses
- remaining businesses
- new businesses
- multiple uses of local resources

Jain *et al.* (1977, p. 136) have suggested that it is useful to differentiate two basic categories of such economic factors: (*a*) factors pertaining to economic structure (e.g., employment by industry; public versus private sector income; economic base, income–wealth distribution); and (*b*) factors pertaining to economic conditions (e.g., income, employment, and wealth).

While such factors and categories of economic factors are important in any consideration of the economics of a particular area, they do not, by themselves, describe the economic system of that area. In order to describe system impacts it is necessary to understand the dynamic inter-relationships among individual factors.

Galbraith (1973, pp. 1–10) has pointed out that diversity of definitions of "economic system" exist and that they may be conveniently divided into two basic kinds. One kind of definition describes the economic system as an arrangement of people and organizations that functions to provide the goods and services required by the public. In this kind of definition, preeminence is given to the economic system as a means of meeting the needs of the public. Another kind of definition describes the economic system as an arrangement of people and organizations that functions to meet the needs of the public and/or the needs of its own organizations. In this kind of definition, emphasis is given to the self-serving character of major economic powers (e.g., large corporations).

With respect to the broad environmental context of economic systems (and ignoring the relative importance of public and self-serving objectives), Commoner (1976, p. 2) has suggested that the economic system (a) transforms real wealth (i.e., food, manufactured goods, etc.) into earnings, profit, credit, savings, investment, and taxes; (b) governs the distribution of that wealth; and (c) determines what is done with it. For purposes of an overview of economic impact assessment, it is useful to highlight various attributes and functions of economic systems as dealt with in these definitions, including:

1. That an economic system is composed of *interacting individuals and organizations*
2. That an economic system *regulates the transformation* of raw resources into socially required goods, services, and monies
3. That an economic system *determines the distribution* of socially required goods, services, and monies
4. That an economic system *influences social perceptions* of just what goods, services, and monies are required

While it is certainly possible to speak of an economic system of the United States, it is more relevant to the objectives of impact assessment to identify the particular form of the general national economic system that is manifest in the local project—and surrounding regional area. This form will vary with (a) the nature of individual and organizational interactions; (b) the nature of resource transformation into local goods, services, and monies; (c) the local pattern of distribution of goods, services, and monies; and (d) local perceptions of required goods, services, and monies.

While regional variations in these economic attributes and functions can be immense, it is possible to identify certain dimensions that can be evaluated on a site-specific basis. Some of these dimensions are identified in Table 15.1. It should be noted that the dimensions identified in the table are only examples. Different assessment teams may find it more convenient to use different categories and descriptors (e.g., Jain *et al.*, 1977, pp. 297–305; Hopkins, 1973, p. 25; U.S. Department of the Army, 1975, pp. A92–A98) in order to reflect the real economic conditions of a particular locality more precisely.

The importance of assigning real dimensions to general attributes and functions of local economic systems cannot be overstressed. *First,* such dimensions are the basis for identifying types of data which should be collected by the assessment team; and *second,* these dimensions are basic tools for evaluating and relating specific types of data to system dynamics.

TABLE 15.1
Some Examples of Dimensions of the General Attributes and Functions of Economic Systems

Some general attributes and functions of economic systems	Some dimensions of attributes and functions of economic systems
Nature of individual and organizational interactions	• dependency relationships among individuals and organizations • competetive relationships among individuals and organizations • long- and short-term duration of relationships • alternatives to ongoing relationships • mobility of participants • current and projected trends
Resource transformation into goods, services, and monies	• nature of local resources currently utilized • external sources of resources • availability of manpower • availability of untapped resources • perceived desirability of current and potential resource utilization • dependibility and adequacy of external resources • transportation requirements
Distribution of goods, services, and monies	• patterns by sex, age, race, and educational background • population projections • community services • per capita and family income • hiring practices • labor and capital costs • profits
Perceptions of required goods, services and monies	• demographic differences • source of perceptions • reinforcement of perceptions • community values and life styles

For example, one dimension of local individual and organizational interactions identified in Table 15.1 is *dependency relationships.* Dependency relationships may include such relationships as: (*a*) family member dependency on head-of-household; (*b*) the dependency of small markets on the economic stability of neighborhood households; (*c*) the dependency of local manufacturers on regional and/or extraregional markets; (*d*) the dependency of local public schools on local property taxes; and (*e*) the dependency of local and regional populations of young people on local businesses for jobs.

Types of data and information that are important for understanding the nature of such relationships in local areas may include: (*a*) annual family and per capita income; family size; sex, age, and other characteristics of local heads-of-households; (b) number and distribution of house-hold units, current and projected household debt, small-business bank-ruptcy, new business ventures; (c) adequacy and costs of transportation facilities, local tax incentives to business, regional competition; (d) school enrollment; emigration and birth rates, tax base, cost of living; and (e) educational and training facilities, alternative local and regional em-ployment, current and projected labor supply and demand, wage levels.

Once data and information required by the appropriate dimension of economic systems have been collected, the assessment team must seek to understand how all dimensions of the local economic system are interrelated.

While there is no universally accepted, objective formula for integrat-ing individual data and information on economic variables into a precise systems diagram of an economic system, a systems overview of a local economic system (Fig. 15.1) can be achieved if the assessment team adheres to a few basic principles, including:

1. The nature of individual and organizational interrelationships in any local area does influence the type of resource transformations into wealth within that area.
2. The type of resource transformations does influence the patterns of distribution of that wealth.
3. Social perceptions of required wealth are influenced by social interrelationships, the type of resource transformations, and the patterns of wealth distribution.

LOCAL ECONOMIC SYSTEM

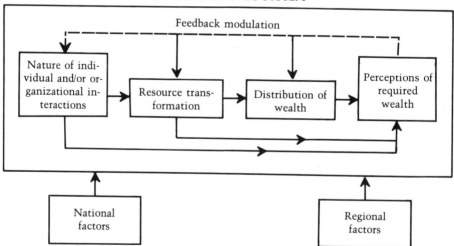

FIGURE 15.1 A systems overview of a local economic system.

4. Social perceptions of required wealth may act as feedback modulators to each of these economic attributes and functions.
5. National and regional economic factors also influence the attributes and functions of local economic systems.

With respect to the practical usefulness of this overview approach to economic impact assessment, it is informative to note that the problem of identifying and describing the basic elements of economic systems on a site-specific basis is not a new problem invented by NEPA; instead, it is a problem that has been faced and dealt with for many years. Experience in the field of economic anthropology is particularly useful for guiding the overall approach to the assessment of impacts on local and regional economic systems. This experience has been summarized by Leclair and Schneider (1968) who state: "There are . . . substantial disputes both in economics and anthropology concerning the proper subject matter of economics, and the nature of economic systems and how they should be studied. But today, all agree on at least this much: that economic activity, properly considered, is a social process of some sort or other. It might be necessary to take technology into account in considering certain aspects of an economic system, but technology is not the economic system itself [p. 3]."

With similar regard for the diverse and dynamic interrelationships among abstract economic factors and real people, it should be emphasized that impacts on jobs, on income, or on any other single economic attribute are not in and by themselves impacts on economic systems. As depicted in the systems overview presented in Fig. 15.1, the assessment team should make an effort to evaluate individual economic impacts in light of ongoing dynamic interrelationships with other economic, social, and environmental factors and conditions.

Benefits and Costs

Impact assessments typically reduce the complexity of economic impacts to a relatively simple model of project costs and project benefits. Therefore, the assessment of economic impacts tends to become an exercise in cost–benefit analysis, in which various project alternatives (including the no-action or no-build alternative) are evaluated for their potential to achieve a maximum of benefits at minimal costs. Since everyone has at least an intuitive appreciation for just what constitutes a benefit and what constitutes a cost, a cost–benefit approach to impact assessment would appear to be quite rational. Yet, like the dichotomous mode of thinking that results in the categories of right and wrong, good and evil, the categorical use of project costs and project benefits can be greatly misleading in a world composed more often of grays than of simple blacks and whites. The basic problem is that cost–benefit analysis typically employs meanings of "cost" and "benefit" that are much more restrictive than are the lay uses of these terms.

For example, for purposes of a cost–benefit comparison of project alternatives (Peskin and Seskin, 1975, pp. 2–3):

1. Benefits are usually defined by how much people would be willing to pay for project outputs.
2. Costs may be defined as the monetary expenditure required (or the monetary value of foregone opportunities) for using resources in one manner rather than in another.

In evaluating benefits, primary attention is given to project effects, such as increases in consumer satisfaction and decreases in the amount or costs of resources required to produce goods and services (i.e., *allocative benefits*); project effects, such as the improvement of the well-being of some at the expense of the well-being of others (i.e., *distributional effects*), are often omitted. Also, costs associated with low probability but disastrous events (e.g., dam failures) are typically omitted (Mark and Stuart-Alexander, 1977). Also, there exist project effects that cannot be easily assigned monetary values (i.e., *intangible costs and benefits*), such as "aesthetic deprivation." While monetary values have sometimes been given to such intangibles, there is much concern over the seeming arbitrariness of such valuations (Peskin and Seskin, 1975, pp. 4–6).

Cost–benefit analyses that focus on allocative benefits of proposed projects, that also ignore distributional effects and that also arbitrarily assign dollar values to the intangibles of daily life have, accordingly, been characterized as having very limited utility for selecting the most efficient of several alternative courses of action (Rowen, 1975, pp. 361–369). An example of the limitations of such narrow-scope cost–benefit analyses is provided in the following scenario by Ehrlich and Ehrlich (1970):

Consider the history of a contemporary housing development. A developer carves up a southern California hillside, builds houses on it, and sells them, reaping the benefits in a very short time. Then society starts to pay the costs. The houses have been built in an area where the native plant community is known as chaparral. Chaparral . . . could not exist as a stable vegetation type unless the area burned over every once in a while. When it does the homes are destroyed, and the buyers and the public start paying hidden costs in the form of increased insurance rates and emergency relief. Of course, there are hidden costs even in the absence of such a catastrophe. The housing development puts a further load on the water supply and probably will be a contributing political factor in the ultimate flooding of distant farmland to make a reservoir . . . . In short, the benefits are easily calculated and quickly reaped by a select few; the costs, on the other hand, are diffuse, spread over time, and difficult to assess [pp. 284–285].

For purposes of impact assessment under NEPA, it is important to stress the fact that cost–benefit analyses of project alternatives are tools for comprehensive and environmentally sensitive decision-making—that is, cost–benefit analyses are means to an end, not the end itself. Thus, the more comprehensive the analysis with respect to allocative and distributional effects, with respect to tangible and intangible benefits and costs, and with respect to short- and long-term impacts, the more useful the analysis becomes for achieving the broad environmental objectives of NEPA.

In an effort to evaluate the comprehensiveness of a cost–benefit analysis, the following questions may be appropriately asked (Howe, 1971, p. 15; Haveman and Weisbrod, 1975, p. 64):

1. Do identified benefits and costs include benefits and costs for which market prices exist and for which these prices correctly reflect social values?
2. Do identified benefits and costs include benefits and costs for which market prices exist, but for which the prices fail to reflect appropriate social values?
3. Do identified benefits and costs include benefits and costs for which no market prices exist, but for which appropriate social values can be approximated in money terms by inferring what consumers would be willing to pay for the product or service if a market existed?
4. Do identified benefits and costs include benefits and costs for which it would be difficult to imagine any kind of market-like process capable of registering a meaningful monetary valuation?
5. Does the analysis specifically identify the "gainers" (present and future) of economic benefits?
6. Does the analysis specifically identify the "losers" (present and future) of economic benefits?
7. Does the anslysis include a rationale for evaluating the desirability or undesirability of the projected redistribution of economic resources in light of society's stated objectives?

Some Guidelines

It has probably most often been the case that responsibility for the assessment of economic impacts has been given to a highly specialized subgroup of the overall assessment team, and that interactions among this and other subgroups having other assessment responsibilities have been largely lacking. Superficially, this would appear to be a rational approach. After all, wildlife biologists are not generally known for their insatiable curiosity about the difference between allocative and distributional aspects of economic systems. Nor are property assessors known as a professional group for their expertise in aquatic or terrestrial ecology. Yet, there are two facts that might well indicate that the assessment of economic impacts solely by individuals having "economic expertise" is in basic error: First is the simple fact that NEPA calls for interdisciplinary assessment; and second, a truly comprehensive evaluation of project-related costs and benefits includes consideration of intangible as well as tangible effects, and distributional as well as allocative effects of project development—effects that cannot be surmised by individual experts working in isolation from other members of the assessment team.

The legal requirement of NEPA for interdisciplinary assessment has been discussed in detail in previous chapters. But the practical requirement for conducting economic assessment in parallel with and integral to the assessment of physical and social impacts is an appropriate topic in this chapter. This requirement is based on the following assumptions that, in light of our previous experience with real projects, seem to be valid assumptions for any impact assessment:

1. Project impacts on the physical and social environment can result in changes in the short- and long-term economic conditions and structure of local and regional areas.
2. Project impacts on the economic conditions and structure of local and regional areas can result in changes in the physical and social environment.

For example, secondary development that occurs as the result of the construction of new highway may, in turn, result in the progressive conversion of local wilderness areas to developed land. If existing wilderness areas have (a) high carrying capacity for game animals; and/or (b) high potential for the production of timber for a lumber industry; and/or (c) an existing or potential value as wilderness camping areas, then subdivision development can result in immediate and future economic constraints. Of course, both immediate and long-term economic gains as a result of subdivision development (i.e., in terms of local tax base, consumer spending) may be much greater than the economic return from alternative uses. However, as discussed in previous sections, a comprehensive economic analysis might consider such issues as:

1. Distribution of benefits and costs of alternatives (e.g., subdivision development as opposed to other uses of wilderness areas)
2. Intangible costs associated with subdivision development (e.g., aesthetics)
3. Long-term costs of subdivision development with respect to future needs for water supply, waste disposal, community services, recreation, etc.

4. Intangible costs associated with social disorganization of existing rural life styles (e.g., acculturation of current values and resultant behavioral patterns)

While these examples focus on some economic factors to be considered as a result of project mediated changes in the physical environment, it is important to consider the converse—that is, how project-mediated changes in economic conditions can cause changes in the physical and social environment.

For example, subdivision development that may result from project development in one area can create consumer demands that will influence economic development in another area. Such demands might include demands for recreational facilities, manufactured goods, and services. The meeting of these demands can result in (a) reduction in wilderness areas (e.g., through the construction of ski runs and boating and bathing areas); (b) increase in pollutional loading of aquatic and terrestrial ecosystems (e.g., through industrial processes, waste deposition, and transportation); and (c) enhanced job opportunities and enhanced social mobility for certain groups of the general population.

While these examples by no means exhaust the possibilities of direct and indirect project impacts on the economic conditions and structure of local and regional areas, they are sufficient to demonstrate an elementary fact about the economic impacts of projects—namely, that in the broadest sense of economics, economic impacts of project development cannot be separated from impacts on the social and physical environment of man. Thus it is essential that the assessment of economic impacts be conducted in parallel with and integral to the assessment of social and physical impacts of project development. Appropriate guidelines that might be followed by assessment teams in pursuance of this objective include:

1. All direct and indirect project impacts on physical and social components and dynamics should be evaluated for their economic consequences.
2. All direct and indirect impacts on the economic conditions and structure of local and regional areas should be evaluated for their consequences on components and dynamics of physical and social environments.
3. Comprehensive economic analyses of project alternatives (including the no-build alternative) should specify and justify the valuation of intangible costs and benefits, and discuss both short- and long-term projections of allocative and distributional effects of project development.
4. All economic effects of project development should be evaluated in light of the goals and objectives of pertinent federal and state legislation.

5. Economic analyses should be inclusive of all phases of project development, from the earliest planning phase throughout operational and maintenance phases.
6. All assumptions and limitations of economic analyses should be clearly identified and discussed with respect to pertinent federal and state legislation, to available data and information, to local and regional social values and objectives, and to project objectives.
7. No one economic criterion should explicitly or implicitly be offered as the single, most important criterion of the desirability of the proposed project.

The difficulties of economic impact assessment that follow from these guidelines are profound. But they are no less profound than the difficulties of environmental assessment that follow from the goals and objectives of NEPA. That is the key issue—whether or not the economic analyses undertaken to aid decision-making in project inception and design are in keeping with the national goals and objectives of NEPA.

In the conduct of economic analyses designed to achieve these national goals and objectives, it is important to note that no one analytical methodology is generally accepted as the preferred methodology—each has its limitations with respect to evaluating project impacts on economic components and dynamics. For example, in a review of rapidly developing techniques of environmental benefit–cost analysis which utilize energy-flow concepts (Lavine and Meyburg, 1976), it is pointed out that: "For any environmental parameter to be considered in energy-flow analysis, it must be quantified in some unit of measure, eventually to be converted into work equivalent units. This requirement tends to leave many ethical–moral, aesthetic, psychological, social–cultural and other information components out of the analyses because appropriate indicators are often not easily quantified" [p. 51].

While practical reasons have been offered for leaving the subjective or normative aspects of many of these issues outside of the scope of assessment (especially as they may influence or be influenced by economic impacts), Pantell (1976) has pointed out that

> There are numerous reasons why it may be desirable to include subjective values in the systems analysis. In the absence of an explicit value system the decision process might be unduly influenced by the pursuasiveness of a demagogue, the pressure of a lobbyist, or the weariness and desperation of the interested parties. It may be easier to detect inconsistencies in decision making when individual preferences are openly discussed and an effort is made to quantify the worth and importance of possible outcomes. It is probably more likely that those variables that are difficult to quantify will receive less emphasis without an explicit attempt to assign a weighting to the variables. Variables such as cost, time, and numbers of people are easier to grasp than factors such as esthetics and privacy, and so in the absence of a weighting scheme the latter may not receive due consideration [p. 131].

There can be no doubt that, as applied to economic analysis, the valuation of subjective issues and/or of distributional (or equity) impacts will generate controversy. Yet, it is by such controversy that the real impacts of any proposed project can be better identified and evaluated. After all, NEPA is not intended as a device to make decisions easier; it is to make decisions better.

References

Canter, L. W.
1977   *Environmental impact assessment.* New York: McGraw-Hill.
Commoner, B.
1976   *The poverty of power.* New York: Alfred A. Knopf.
Ehrlich, P. R., and A. H. Ehrlich.
1970   *Population, resources, environment.* San Francisco: W. H. Freeman.
Galbraith, J. K.
1973   *Economics and the public purpose.* Boston: Houghton Mifflin.
Haveman, R. H., and B. A. Weisbrod.
1975   The concept of benefits in cost–benefit analysis: With emphasis on water pollution control activities. In *Cost benefit analysis and Water pollution policy,* edited by H. M. Peskin and E. P. Seskin. Washington, D.C.: The Urban Institute.
Hopkins, L. D.
1973   *Environmental Impact Statements: A handbook for writers and reviewers.* Chicago: Illinois Institute for Environmental Quality. (NTIS, PB–226 276)
Howe, C. W.
1971   *Benefit–cost analysis for water system planning.* Washington, D.C.: American Geophysical Union.
Jain, R. K., L. V. Urban, and G. S. Stacey.
1977   *Environmental impact analysis: A new dimension in decision making.* New York: Van Nostrand Reinhold.
Lavine, M. J., and A. H. Meyburg.
1976   *Toward environmental benefit cost analysis: Measurement methodology.* National Cooperative Highway Research Program, Transportation Research Board, National Research Council. (Final Report of Project 20–11A). Ithaca, New York: Center for Environmental Quality Management, Cornell Univ.
LeClair, E. E. Jr., and H. K. Schneider (Editors).
1968   *Economic anthropology: Readings in theory and analysis.* New York: Holt, Rinehart and Winston.
Mark, R. K., and D. E. Stuart-Alexander.
1977   Disasters as a necessary part of benefit–cost analyses. *Science, 197* (No. 4309): 1160–1162.
Pantell, R. H.
1976   *Techniques of environmental systems analysis.* New York: John Wiley & Sons.
Peskin, H. M., and Eugene P. Seskin (Editors).
1975   *Cost -benefit analysis and water pollution policy.* Washington, D.C.: The Urban Institute.
Rowen, H.
1975   The role of cost–benefit analysis in policy making. In *Cost-benefit analysis and water pollution policy,* edited by H. M. Peskin and E. P. Seskin. Washington, D.C.: The Urban Institute.

U.S. Department of the Army.
  1975 *Environmental quality: Handbook for environmental impact analysis.* Baltimore,
        Maryland: U.S. Army AG Publications Center. (No. 200–1)

16

Public Health Impacts

Introduction

As with the evaluation of economic impacts, the evaluation of project impacts on public health requires the integration of numerous factors, including physical, chemical, biological, social, and psychological factors. As a general rule, the more comprehensive the assessment of project impacts on the physical and social environment, the more comprehensive will be the assessment of project impacts on public health.

Toward the end of achieving a comprehensive assessment of impacts on public health, it is important for the assessment team to understand that the concept of public health includes dimensions of (a) physical safety; (b) physiological well-being; and (c) psychological (or mental) health.

It is also important to note that project impacts on each of these public health dimensions may occur throughout project development and may be influenced by changing environmental conditions that, even if not related to project activities, can nonetheless potentiate project impacts on public health. For example, a particular project may result in the long-term buildup of potentially stressful chemicals in drinking-water supplies. While projected concentrations may be well within public health standards of drinking water, there may also be a long-term trend in the public's dietary intake of those same chemicals. Certainly the total dietary intake of potentially stressful chemicals is not a direct or even indirect result of any single project. Yet, incremental contributions from a particular project can exacerbate the problem in a local area. This example demonstrates a particularly important point about the assessment of project impacts on public health—namely, that *the assessment team must evaluate the proposed project's impacts on public health in light of the total environmental context in which people must live.*

In keeping with a total environmental perspective of public health, it is essential for the assessment team to rid itself of a rather common attitude—an attitude that has often been pervasive in assessment teams more interested in getting a project built than in doing a comprehensive assessment. This attitude is exemplified by: "The safety risks of this project are less than the risk people take when they get into their cars to drive to the store for a loaf of bread. They don't seem to worry about that risk, so why should we worry about the so-called risks of this project?"

There should be no doubt that *it is not the business of the assessment team to decide which public health risks people should take; rather, it is the business of the assessment team to identify and evaluate the risks which people will likely have to take as a result of project development.*

Systems Approach

A consistent theme throughout previous chapters has been the need for environmental impact assessment under NEPA to integrate numerous and diverse environmental factors. This need is not only based on the legislative requirement of NEPA for interdisciplinary evaluation of project impacts, but also on a realistic appraisal of any environmental process as being interrelated and directly or indirectly coupled with other processes. Accordingly, the approach to assessment has been described in this book as essentially a systems approach. It is necessary, here, to give some emphasis to the notion that any assessment that is likely to realize a systems understanding of overall project impacts is also likely to be precisely the assessment that methodically employs a systems approach to each individual subsystem. Thus, a systems approach to public health not only helps to achieve a comprehensive understanding of project impacts on public health (Lee, 1968, p. 21), but also helps to achieve the integration of public-health impacts with other impacts on the physical and social environments that must be considered by decision-makers. While there is no theoretical necessity for it, practical experience with assessment would suggest that a simplistic checklist approach to the assessment of impacts on public health is a good indication that the overall assessment of project impacts will be poorly integrated and will, accordingly, fail to identify important indirect impacts of project development and operation.

Basic components of a systems approach that are useful for the assessment of impacts on public health are provided by Lee (1968, pp. 17–29). Such an approach includes consideration of:

1. *Environmental factors*, including physical, chemical, biological, and psychosocial factors, and resource (material) and temporal (change in material factors with time) factors
2. *Environmental media*, through which the environmental factors influence human beings (e.g., atmosphere, hydrosphere)
3. *Human activities*, which influence the human experience with different environmental media (e.g., home, work)
4. *Moderating factors*, which qualitatively or quantitatively affect human responses to environmental stimuli (e.g., genetic constitution, susceptibility to various diseases)

These components to a systems approach to public health are summarized in Fig. 16.1.

The need to consider public health impacts of project development from such a total environmental perspective as afforded by the systems overview depicted in Fig. 16.1 is exemplified in a typology of currently important "diseases" by Watt (1973, p. 228):

1. *Pollutional diseases*, such as emphysema, bronchitis, and various kinds of cancers, particularly lung cancer. All of these are in-

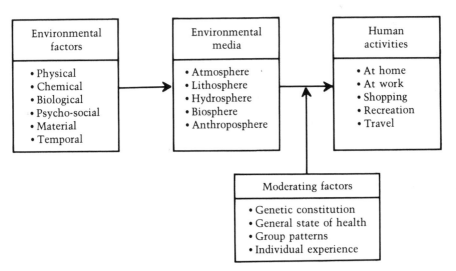

FIGURE 16.1 *A systems overview of public health factors and moderating influences.* [Adapted from Lee, 1968, p. 23.]

creased in incidence by exposure to specific environmental contaminants.

2. *Stress diseases,* such as coronary disease. These are more prominent now than in earlier times and can be related to the social environment of man.

3. *Nutritional diseases,* which are related to social and economic conditions, agricultural practices of man, and food-processing practices

4. *Allergies,* which are clearly, at least in part, the product of environmental factors

5. *Developmental anomalies* (teratologies) that are caused by environmental contaminants that affect the mother while she is carrying the fetus

With respect to any typology of "diseases," it is important to understand that a one cause–one disease concept is inappropriate in any environmental overview of public health. According to MacMahon (1972):

> It has become clear that a particular disease manifestation may have more than one causal antecedent. In addition, we have learned that exposure to a known cause of illness does not always lead to the expression of that illness and that identification of a causal antecedent does not necessarily provide the ability to prevent or control the ailment. We have come to recognize that the one cause–one disease model is too simple [pp. 1–2].

The complexity of multiple-factor etiology of human diseases cannot be overemphasized. This complexity reflects the complexity of interrelationships among physical, chemical, biological, social, and psychological components and processes of the total human environment. Incomplete knowledge and understanding of these interrelationships leads, of course, to disagreement as to the most important factor (or factors) for any particular public health risk. For example, in a recent report of a conference on stress and its role in hypertension (Marx, 1977), it is pointed out that while physicians can usually control hypertension with drugs or diets and that while they have investigated various physiological, environmental, and sociological factors, they still do not understand what causes some 90% of the cases of hypertension.

The complexity of public health issues, and the existence disagreements among experts as to the causes and contributing factors of health risks are properly given emphasis here because of the misuse that assessment teams can make of these two facts. For example, an assessment team may be tempted to ignore a certain public-health risk merely because experts disagree as to the significance of that risk—thus the team's apologia, "If the experts can't agree, how can we, who are not experts, evaluate the risk?" There can, of course, be only one response—

namely, that *neither the complexity of environmental components and dynamics nor our imprecise knowledge of them can justify any failure to consider those components and dynamics in the decision-making process of project development.* With respect to the assessment of public health impacts, this general guideline for impact assessment may be translated into the following rule:

> *Any assessment of project impacts on public health that does not take account of the total environmental context of local and regional populations is inadequate by any measure of contemporary environmental health science.*

Some Examples and Considerations

While projects may contribute directly to the health risks of the public (e.g., industrial contamination of air and water with potential carcinogens), projects may also contribute indirectly (but not therefore negligibly) to the health profiles of local and regional populations. A total environmental perspective of public health impacts is often necessary to identify these indirect impacts.

For example, a particular project may result in enhanced eutrophication of an impoundment (Chapter 9). While the enhancement of eutrophication is usually evaluated in terms of its consequent impact on primary and secondary recreation, the enhancement of eutrophication also has public health implications.

For instance,[1] there is some evidence that submerged aquatic plants which characterize many eutrophic lakes promote swimmer's itch (schistosome dermatitis) by providing habitat to snail species that harbor the causative organism (Mackenthun, 1973, p. 189). Aquatic macrophytes also provide habitat and food for other animal vectors of human diseases, such as malaria, filaria, and fascioliasis (Mitchell, 1974, pp. 54–56).

Dense populations of algae (blue–green species) have also been associated with human health disorders, including contact-type dermatitis, symptoms of "hay fever" (Palmer, 1962, p. 53), headaches, nausea, various gastrointestinal disorders, respiratory disorders, and eye inflammation (Mackenthun, 1973, p. 195).

In addition to these public health hazards that are directly associated with the micro- and macroscopic vegetation of eutrophic lakes and ponds, there are other hazards which may be associated indirectly with the eutrophication of waters. For example, if a lake or pond supports dense

---

[1]The following discussion of public health implications of eutrophic waters is adapted from a previously authored report (Massachusetts Department of Environmental Quality Engineering, 1977).

populations of plants primarily because of human inputs into the water (e.g., septic and industrial runoff and seepage), then the diseases and toxicities typically associated with such inputs may also be associated with the receiving lake or pond. Specific diseases and toxicity hazards will, of course, depend on the nature of the pollutants, concentrations of causative agents, and the potential activity of the agents within a water medium. Among such inputs are bacteria, viruses, heavy metals, pesticides, and various organic chemicals (National Academy of Sciences, and National Academy of Engineering, 1972; Environmental Protection Agency, 1976). Under certain conditions some chemicals can concentrate in fish that may be consumed by humans.

As this example indicates, any realistic appraisal of the public health implications of project-mediated eutrophication requires consideration of a variety of factors, including:

1. The specific ways in which the project enhances eutrophication and which may have direct impact of public health (e.g., the contamination of surface waters with septic wastes, industrial toxins)
2. The uses of water resources undergoing an enhanced eutrophication (e.g., primary or secondary recreation, public water supplies)
3. Habitat made available (through eutrophication) to disease-bearing organisms
4. Regionally identifiable disease vectors (e.g., contact with disease agents through recreational use of water, ingestion of water food, etc.)

Of course, the most important consideration of all is the simple consideration of whether or not eutrophication does have implications for public health. Unless those responsible for the assessment of public health impacts take a total environmental perspective of public health, it is likely that such issues as eutrophication will be left to the consideration of those principally interested in fish, water quality, and aquatic ecology. In other words, the proof of an assessment team's total environmental perspective of public health is in the continual exchange of information between those having primary responsibility for assessing public health impacts and those having primary responsibility for assessing other types of impacts, including impacts on the physical and social environment. The simplest rule to follow is:

Each project impact, whether on the physical or social environment, must be evaluated for its direct and indirect influence on public health and well-being.

This cannot be done if public health is considered as a separate issue that can best be handled by people who are isolated and insulated from the overall assessment process.

In considering the public health ramifications of all project impacts on the physical and social environment, it is important for the assessment team to understand that "positive impacts" on one component of the environment can nevertheless be "adverse impacts" on public health. Thus, the creation of new wetlands in borrow areas, for example, can enhance regional habitat for waterfowl. However, the same wetlands can result in health and safety hazards. They may serve as habitat to disease-bearing animals and insects; they can attract waterfowl that, once in the local area, may contribute large amounts of fecal material to public water supplies; they can also become "attractive nuisances" to local children.

All too often, assessment teams fail to consider the full range of such potential effects of so-called "positive impacts" of project development. As the above examples suggest, a myopic concern for wildlife can lead to serious public health risks. While there is no objective technique for balancing physical and/or social "pluses" with public health "minuses," it is incumbent upon the assessment team to assess all project impacts in light of all their consequences to the total human environment.

It is highly likely, of course, that the assessment team will not include personnel who are professionally trained in public health sciences. As pointed out in previous chapters, it is quite unreasonable to expect that every assessment team will include individuals trained in each of the disciplines that must be considered in a comprehensive impact assessment. Thus, the previously stated principles with respect to the assessment of public health impacts present very real practical difficulties to the assessment team. For example, just how can the assessment team identify the public health implications of all project impacts? Just how can the assessment team identify those "positive" impacts on the physical and social environment that can result in "negative" impacts on public health? As in so much of assessment under NEPA, the answer to these questions does not lie necessarily in the disciplinary expertise of assessment team members; rather, more often, the answer lies in organizational skills that can facilitate the team's access and utilization of knowledge and expertise that are available from sources external to the assessment team. Organization aspects of the assessment team that are particularly important for the comprehensive assessment of public health impacts are discussed in the concluding section of this chapter.

Some Guidelines

For purposes of the following discussion, it will be assumed that the responsibility for assessing public health impacts is delegated to a specific individual or subgroup of the total assessment team. General guidelines for conducting the assessment of impacts on public health, therefore, include ( ) procedural guidelines for integrating the efforts of those responsible for assessing impacts on public health with those responsible for assessing other types of impacts; and (b) substantive guidelines for actually assessing impacts on public health.

PROCEDURAL GUIDELINES

1. In the earliest phases of project development (i.e., predesign phases), those responsible for assessing public health impacts should establish liaison with federal, state, and local agencies and organizations having jurisdictional interest in public health issues and environmental concerns. Such agencies may include (a) U.S. Environmental Protection Agency (regional office); (b) state and local public health agencies and commissions; (c) regional and municipal centers for research; and (d) regional and local medical centers. Once established, it is important that liaison continue throughout all phases of project development.

2. Management of the assessment project should ensure that those responsible for assessing impacts on public health are immediately and fully aware of all impacts on the physical and social environment that are identified in the course of the assessment process, as well as all proposed mitigation and enhancement measures. This may be accomplished by (a) in-house memos; (b) written reports; and (c) oral progress reports. Mechanisms should also be established whereby personnel responsible for assessing impacts on public health may initiate requests for information from other members of the assessment team.

3. Those having responsibility for assessing impacts on public health issues should continually update other members of the total assessment team on public health and safety issues related to project development. Information transmitted to the total assessment team should include such information as (a) source(s) of risk; and (b) environmental factors (physical and social) that can influence the risk. This approach will help to ensure that all members of the assessment team, regardless of their individual areas of responsibility, will be cognizant of the influence that physical and social impacts can have on public health. They will, therefore, be in a better position to collect data and information that will be needed for a comprehensive assessment of project impacts on public health.

SUBSTANTIVE GUIDELINES

1. All phases of project development (from predesign through operational and maintenance phases) should be considered for their influence on public health (including the psychological as well as the physiological well-being of the public).

2. Each individual activity which is to be undertaken in each phase of project development should be evaluated for its direct and/or indirect, long- and/or short-term, and cumulative impacts on public health.

3. Federal, state, regional, and local agencies, commissions, and organizations should be fully utilized for establishing baseline overviews of the current health and well-being of local, regional, and extraregional communities, and for identifying relevant physical and social factors that might be affected by project development.

4. All environmental standards for public health (e.g., water quality standards, air quality standards, noise standards) should be utilized by the assessment team in evaluating potential impacts on public health. Environmental parameters (including physical and social parameters) that are likely to be affected by project development and for which no standard has been promulgated should be evaluated in liaison with recognized experts and in light of the developing scientific literature.

In following these general guidelines, assessment personnel will find their task greatly eased if they organize their analytical and integrative efforts so as to answer the following questions. It will be noted that the form of these questions is the basic form of the historic reportorial questions designed to enhance precision and conciseness in writing:

1. Specifically, what are the physical and social consequences of each action in each phase of project development (e.g., an increase in turbidity of water during construction, air pollution during the operational phase, introduction of toxic materials such as pesticides during the maintenance phase)?

2. Where will these consequences and their subsequent (i.e., indirect) effects be manifested (e.g., in the hydrosphere, atmosphere, lithosphere)?

3. How will these consequences be manifested (e.g., trophic magnification of toxic materials in higher-order consumers, atmospheric precipitation and dryfall of particulates, architectural amplification of ambient noise)?

4. What quantitative and qualitative factors can influence the manifestation of these consequences (climatic factors, population density, topography)?

5. Who is likely to be subjected to these consequences of project development (e.g., present and/or future populations, local and/or regional populations, lower socioeconomic groups)?

6. How can the consequences of project development "enter into" each population (e.g., absorption through skin, respiration, sensorial inputs, ingestion)?

7. What social, physiological, and social–psychological factors can

influence the effects of these inputs into each population (e.g., age, occupation, recreational and other behavioral patterns, stress)?

8. Considering these factors, what effects are these environmental factors likely to have in each of the populations considered?

In order to answer these questions, personnel who are responsible for assessing public health impacts will have to consult both in-house members of the total assessment team and external experts in government, industry, and research and educational institutions. In either case a simple fact is paramount—one cannot get information without giving information. For example, one cannot expect to find out from, say, a regional planning commission who it is that is likely to be subjected to an environmental change if one does not tell the commission first what "the change" is. Thus, those who conduct assessments of public health are probably best advised to consider themselves coordinators of information—that is, they have to coordinate specific information about project impacts on physical and social components and dynamics with specific information about the health and well-being and the behavior of human populations. Unless professionally trained in public health, the assessment team definitely should not "play expert" with other peoples' lives. Nor should the assessment team, regardless of its training in public health, "play God" by determining which risks other people *should take* and which risks, therefore, lie beyond their need to consider. As "coordinators of information," it is simply the responsibility of the assessment team to present to decision-makers, as concisely and as comprehensively as possible, the risks that people *will take* because of project development.

A Final Note

Public health considerations of project development are often restricted by a historic perspective of health as the sole disciplinary domain of the natural sciences. According to this perspective, threats to human health can be fully described in terms of microbes, chemicals, and those physiological aberrations that are the common lot of mankind. Certainly this perspective permits no nonsense about mind, emotions, or the intricacies of social organization as being important or even relevant factors of human health. Thus we find that impact assessments tend to deal with the categories of biological and ecological impacts, social impacts, economic impacts, aesthetic impacts, and public health impacts as separate and independent categories of human existence.

It is necessary to emphasize that *any perspective of public health that does not accommodate social and behavioral disciplines as well as natural science disciplines is long outdated, and has no place in contemporary impact assessment.* This premise has been succinctly stated in the form of two concepts identified in a report on research planning in environmental health science (U.S. Department of Health, Education and Welfare, 1970, p. 201):

1. People's reactions to the people around them, and to the social groups of which they are members, may have a major influence on any disease process.
2. The effect of the environment on illness cannot be fully understood unless people are considered within the context of the social groups of which they are a part and unless their perceptions of their environment are considered also.

In view of these concepts of contemporary environmental health science, it is necessary that those having responsibility for assessing public health impacts learn to deal with a real world in which physical and social phenomena actually interrelate. Such a world is certainly more confusing than a neatly compartmentalized world—but it is also the world in which real people live.

References

Environmental Protection Agency.
  1976   *Quality criteria for water.* Washington, D.C.: U.S. Government Printing Office (EPA–440/9–76–023).
Lee, D. H. K.
  1968   The nature of environmental health sciences. In *Environmental problems: Pesticides, thermal pollution, and environmental synergisms,* edited by B. R. Wilson. J. B. Lippincott, Philadelphia
Mackenthun, K. M.
  1973   *Toward a cleaner aquatic environment.* Washington, D.C.: U.S. Environmental Protection Agency, Office of Air and Water Programs, U.S. Government Printing Office. (Stock No. 5501–00573)
MacMahon, B.
  1972   Introduction: Concepts of multiple factors. In *Multiple factors in the causation of environmentally induced disease,* edited by D. H. K. Lee and P. Kotin. New York: Academic Press.
Marx, J. L.
  1977   Stress: Role in hypertension debated. *Science, 198* (4320): 905–907.
Massachusetts Department of Environmental Quality Engineering
  1977   *Draft environmental impact report, control of aquatic vegetation in the Commonwealth of Massachusetts.* Boston: Massachusetts Department of Environmental Quality Engineering.
Mitchell, D. S.
  1974   The effects of excessive aquatic plant populations. *Aquatic vegetation and its use and control,* edited by D. S. Mitchell. Paris: UNESCO.

National Academy of Sciences, and National Academy of Engineering
  1972   *Water quality criteria, 1972.* U.S. Government Printing Office. Washington, D.C.:
         (Stock No. 5501–00520)
Palmer, C. M.
  1962   *Algae in water supplies.* Washington, D.C.: U.S. Department of Health, Education
         and Welfare, Public Health Service, Division of Water Supply and Pollution Con-
         trol. U.S. Government Printing Office. (PHS Publication No. 657)
U.S. Department of Health, Education and Welfare (HEW)
  1970   *Man's health and the environment: Some research needs.* Washington, D.C.: U.S.
         Government Printing Office.
Watt, K. E. F.
  1973   *Principles of environmental science.* New York: McGraw-Hill.

17

Criteria of Adequacy of Assessment: Social Environment

Introduction

Criteria for evaluating the adequacy of assessments of impacts on the physical environment have been previously noted. A number of these criteria are as equally applicable to assessments of impacts on the social environment as they are to assessments of impacts on the physical environment. Rather than repeat the relevant criteria in this chapter, the reader is advised to consult Chapter 12.

In this chapter, we will focus on some criteria that are particularly important to assessments of project impacts on social components and dynamics. Again, it is necessary to emphasize that we are dealing with the assessment process and not the assessment report. Thus, many of the following criteria can be seen as "quality control" mechanisms and devices that are available for use by in-house members of the assessment team. Of course, insofar as actions that are undertaken by assessment team members in pursuance of assessment objectives become part of the public record, then to that same degree the following criteria may also be utilized by the public in order to gauge the adequacy of the overall assessment effort.

For purposes of the following discussions, relevant criteria may be described as criteria (*a*) pertaining to the organization of the assessment team for purposes of assessing impacts on the social environment; (b) pertaining to analytical efforts; and (c) pertaining to integrative efforts.

ORGANIZATIONAL CRITERIA

Important organizational criteria for evaluating the adequacy of social impact assessment include criteria pertaining to the skills, experience, and attitudes of personnel, and to the management of the overall assessment effort. While there is no consensual method for determining just who may be best suited for undertaking assessments of social impacts, or just how the assessment process is best conducted, there are some general questions which should be considered—both by the manager of the assessment project and by the public:

1. Is the disciplinary training of personnel representative of a cross-section of contemporary social, psychological, and behavioral disciplines?
2. Do personnel have actual field experience with social science techniques and methodologies?
3. Do in-house mechanisms exist to ensure the communication of needs and findings between personnel responsible for assessing impacts on the social environment, and all other assessment and project personnel?

With respect to the disciplinary training and field experience of personnel involved in the assessment of social impacts, it may be useful to highlight two important issues.

First, impact assessment requires skills (e.g., communication skills, analytical skills, integrative skills). Regrettably, the fact that an individual may have a substantial academic background in a particular discipline does not necessarily ensure that the individual also has the requisite skills for applying disciplinary knowledge to real problems and issues. Surely an academic background in social science disciplines that is composed primarily of introductory survey courses cannot be expected to have imparted skills as much as information. And it might as well be openly said that many of the new undergraduate and even graduate programs in "environmental affairs" that attempt to cut across disciplinary barriers often seem to produce more people who are expert in pontificating environmental concerns than people who can translate those concerns into specific analyses and technical recommendations. A well-managed assessment team for social impact assessment will, therefore, ensure a balance of theoretical and practical training in those personnel who may be specifically selected for their academic backgrounds in the social and behavioral sciences.

Second, it should be clearly understood that the most important reason for having personnel who are trained in the theory and application of social and behavioral sciences *definitely is not* that such personnel have "total capability" (a favorite marketing term among assessment firms) with respect to assessing social impacts; but rather, that such personnel are likely to possess the attitudes, skills, and knowledge that will facilitate maximal professional use of sources of information which are external to the assessment team (e.g., government agencies, educational and research organizations, community and neighborhood groups). Nothing can be more contrary to the broad intent of NEPA than an assessment team that becomes so impressed with its own expertise (real or imaginary) that it sees no need to consult with external organizations, groups, and individuals. Management practices that fail to ensure intimate liaison between the assessment team and external sources of social expertise and experience are poor management practices, regardless of the technical sophistication of the in-house team.

Unless one believes that the public lives in sterile test tubes, it is difficult to imagine how a comprehensive assessment of project impacts on the social environment of the public can be conducted in the absence of specific information about project impacts on the physical environment of the public. Yet, this is often done. It is done because (a) in cases where portions of the overall assessment are contracted out to private firms or other assessment organizations, it is difficult to coordinate in-house and external operations; and (b) even in cases where portions of the assessment are not contracted out, both management and technical personnel often fail to appreciate the interdependence of many social and physical impacts of project development. With respect to the broad objectives of NEPA, neither circumstance can excuse any isolation of personnel charged with assessing social impacts from personnel charged with assessing impacts on the physical environment.

The responsibility for ensuring meaningful interaction between personnel involved in assessing social impacts and personnel involved in assessing physical impacts is clearly the responsibility of the managers of assessment projects. Management responsibility for staffing is also central to any comprehensive assessment of project impacts on the social environment. While the role of management in assessment has been previously discussed (Chapter 5) and will be examined further in Chapter 23, it is important to stress here that the managers of assessment projects cannot absolutely ensure the comprehensive assessment of social impacts, but they can surely ensure that comprehensive assessments will not be achieved. Too often, poor assessments are attributed to technical and scientific shortcomings of assessment personnel, when, in fact, poor assessments may be the result of project managers who misuse the technical and scientific manpower at their disposal.

In order to provide an organizational framework for the comprehensive assessment of project impacts on the social environment, project managers should carefully consider the following general principles, in addition to those already discussed:

1. The scope of work for the assessment should be decided upon in consultation with technical assessment personnel trained in social and physical science disciplines, as well as with legal staff.
2. Allocations of manpower, budget, and other resources should reflect (a) a balanced assessment of social and physical impacts; and (b) major commitment to assessing interactions and interdependencies among social and physical components and dynamics.
3. Progress reports, review meetings, and other appropriate in-house mechanisms should be utilized for updating assessors of social impacts with information on physical impacts.
4. Personnel having responsibility for assessing social impacts should clearly understand their responsibilities to (a) coordinate their assessment with local, state, regional, and national organizations having jurisdictional or other interests in social components and dynamics; and (b) document relevant social and behavioral concepts and methodologies in appropriate disciplinary literature.
5. No social impact should be ignored or otherwise considered superficially merely because (a) it cannot be quantified objectively; or (b) it cannot be evaluated without consideration of political or other socially sensitive issues.

ANALYTICAL CRITERIA

As discussed in Chapter 6, it is essential to conduct analytical tasks efficiently in order to conserve adequate time and budget for such important integrative tasks as the evaluation of the significance of project impacts and the design of appropriate mitigation and enhancement measures.

The efficiency with which analytical tasks (Chapter 6) are performed is influenced by several factors, including (a) the familiarity of assessment personnel with promulgated environmental standards and guidelines, and with disciplinary concepts, principles, and methodologies that are relevant to social analysis and assessment; (b) the team's access to data and informational sources (at local, regional, and national levels); and (c) the quality of interdisciplinary effort within the total assessment team.

It must be noted that familiarity with promulgated environmental standards and guidelines is not sufficient to guarantee an adequate assessment of social impacts; standards and guidelines with respect to most social components and dynamics simply do not exist. Thus, the sole reliance of the assessment team on promulgated standards and guidelines and on informational sources selected for their relevance to environmental standards and guidelines can greatly increase the inefficiency of the analytical phase of impact assessment.

For example, personnel responsible for assessing social impacts may focus early in the assessment process on potential impacts of project-induced noise on local residents and community functions; or, on potential impacts of project-induced changes in air quality on public health and agricultural productivity; or, on potential impacts of project development on historic or natural landmarks. In each of these cases there are standards and guidelines that can be utilized for identifying and evaluating impacts having social ramifications. Yet, for restricting analytical efforts to these issues, the assessment team may very well discover, upon the completion of its assessment, that people in local and regional areas are primarily concerned with impacts on such things as:

- community cohesiveness and identity
- conservation of local terrestrial and aquatic resources
- individual privacy and life style
- long-term population growth, and potential for social disorganization
- individual social mobility

In such a circumstance there is no justification whatsoever under NEPA for an assessment team to excuse its failure to consider any of these issues simply because there are no relevant promulgated standards. Reviewing agencies and the public will rightly demand that they nevertheless be considered.

The necessity for redoing assessments because of major conceptual failings can be avoided if project management requires assessment personnel to be as familiar with the concepts of social and behavioral disciplines as they are with legislative requirements for social assessment. As a general rule, *not one bit of social data and information should be collected until these social concepts and phenomena have been identified and integrated with legislative requirements.*

With respect to the quality of interdisciplinary interaction as a factor which influences the efficiency of analytical tasks, it is time we recognize the elementary fact *that there cannot be interdisciplinary assessment in the absence of disciplinary knowledge and understanding.*

Assessment teams that purport to be interdisciplinary, but which are composed of individuals who are not well grounded in one or more disciplines, may be excellent mechanisms for providing job security for the untrained, but they are certainly poor mechanisms for identifying, collecting, and analyzing the different kinds of detailed data and information required for comprehensive impact assessment.

As has been pointed out throughout this book, it is quite unreasonable to expect assessment teams to be made up of experts from all the disciplines that are relevant to the analysis of social and physical impacts. Yet, it is also unreasonable to entrust interdisciplinary assessment to individuals who do not have the faintest notion of what "disciplinary" knowledge is. Certainly teams composed of individuals who are basically ignorant of social and behavioral disciplines are typically those teams that consistently fail to recognize the need for collecting and evaluating relevant social data and information.

While no discipline can be identified as being more important to comprehensive assessment than any other, project managers are well advised to ensure (a) a balance of social science and natural science disciplines among assessment team members; and (b) a degree of training in these disciplines which is greater than that afforded by introductory or survey courses.

INTEGRATIVE CRITERIA

Integrative tasks of assessment (Chapter 6) include the evaluation of the significance of impacts and the identification and evaluation of appropriate mitigation and enhancement measures.

While there is no objective measure of "significance," various criteria have been utilized for evaluating the relative significance of social impacts, including:

1. The number of people who will likely experience the impact
2. The short- and long-term relevance of impacts to public health and economic stability
3. Political or other local institutional and organizational reactions and sensitivities
4. The relationship between potential impacts and the goals and objectives of existing federal and state legislation

In general, social impacts that affect relatively few people, that have little effect on public health or on economic processes, that arouse little political or other interest or concern, and that do not directly contradict legislated goals and objectives are considered to be "insignificant." This is certainly an eminently practical approach to the assessment of social impacts. It may also be suggested that such an approach is more insensitive to project impacts on people than it is to project impacts on the public exercise of political and juridical power and influence on project development.

It must be emphasized that NEPA was not purposely designed as one more tool for implementing majority over minority interests—it is a tool for making better-informed decisions than have been previously made.

Thus, any assessment approach that merely perpetuates historic patterns of decision-making is, at the best, irrelevant; and, at the worst, contrary to the objectives of NEPA.

Some additional criteria for evaluating the significance of social impacts, and which reflect the broad goals and objectives of NEPA, might reasonably include:

- The inability of individuals and groups to avoid project impacts
- The irreversibility of impacts, regardless of their magnitudes
- The long-term commitment of future generations to a social environment that such generations have no voice in determining and that will be difficult to alter should they choose to do so
- The diversity of social environments, and the ability of populations to utilize that diversity in their own interest.

There can be no doubt that many assessment teams will consider such criteria to be ludicrous—ludicrous because they include consideration of philosophical issues which are more relevant to "social engineering" than to "project engineering." However, as discussed at the very beginning of this section, *each proposed project that falls within the purview of NEPA should be seen as an exercise, however small, in social engineering.* From this perspective, it is quite impossible to avoid consideration of project impacts on fundamental social issues, including the importance of the individual, as well as the social collectivity of which the individual is a part. To argue that some project impacts are insignificant for purposes of decision-making merely because they will affect only one person or a few families or one neighborhood is to argue that the individual is irrelevant to governmental decision-making. *If this is, in fact, the case, decision-makers should say so—not members of the assessment team.*

Some Concluding Remarks

Criteria discussed in this and other chapters (Chapters 12 and 23) have been selected on the basis of my appreciation of how the very real difficulties of impact assessment can overwhelm even sincere efforts to conduct comprehensive and in-depth assessments.

With respect to the assessment of impacts on the social environment, primary attention is given to criteria that reflect on the management of the assessment process (directly and/or indirectly), and on the attitudes and sensitivities of assessment personnel.

This emphasis should not be interpreted, however, as an indication that management skills, or that the attitudes of those who conduct assessments, are somehow less important for the adequate assessment of physical impacts than they are for the adequate assessment of social impacts. It is only to highlight the fact that, when dealing with the social environment, the assessment team must be particularly well organized and otherwise prepared to deal as seriously with intangible perceptions and emotions as with tangible goods and services—as seriously with the subjective values and attitudes of people as with objective, quantifiable economic indicators. Regrettably, there is some reason to believe that many assessment teams are not so organized or prepared.

For example, in an effort to persuade this author of the "absolute foolishness of the public," a state highway official related the following story. It seems that a proposed new highway project called for the abandonment of a portion of an existing road in a small town. In the course of public hearings on the proposal, it became obvious that there was much public dissatisfaction with the planned abandonment. To the utter disbelief of the state highway official and members of his assessment team, this dissatisfaction was finally coalesced into a formidable opposition by a teen-aged girl who took her turn at the podium. Her argument against the proposed abandonment was straightforward—the abandoned roadway would become an ideal "lovers' lane"; and she could guarantee that more than a few pregnancies would result.

"Now they want to blame the highway department for pregnancy!" the state official shouted.

Well, of course, the girl did not say that highway projects caused pregnancies. She merely noted that a particular highway project could cause a situation that might result in pregnancies. In fact, she was probably looking at the project more realistically than was the highway engineer. He was looking at the project as a means of moving goods and people efficiently and safely from one point to another. She was looking at a change in the physical environment that, regardless of its stated purposes, had broader implications for the social environment of local people.

This story has been retold many times for many different types of projects in all parts of the country. The particulars change, but the supposed moral is the same—namely, that the impacts of any particular type of project are totally defined by the stated objectives and specific engineering design of the project; that any impacts that are suggested but which cannot be solely and directly related to the project's stated objectives or engineering details are figments of overworked imaginations; and that people who nevertheless persist in blaming individual projects for all their problems should be absolutely ignored for purposes of serious assessment.

With respect to social impact assessment, the first postulate is nonsense; the second is more nonsense; and the third is still more nonsense.

Projects do not cause impacts on the abstract social environment, but on site-specific social components and dynamics. Thus, the social impacts of projects are not defined by the type of project alone, *but by the project in the context of a particular social setting of personal, interpersonal, and institutional factors.* Also, a project may indeed be only one of several sources of the same type of social impact in a local or regional area—*but NEPA does not exempt decision-makers from considering the incremental contributions from a proposed project, or their cumulative consequences.* Finally, *NEPA does not authorize any assessment team to ignore any member of the public for any reason whatsoever.*

IV

Assessing Impacts: The Total Human Environment

18

The Total Human Environment

Introduction

The total environment is the totality of interrelationships between the social environment (including personal, interpersonal, and institutional components and dynamics) and the physical environment (including abiotic, biotic, and ecological components and dynamics). While the analytical tasks of impact assessment require team members to focus on those disciplines and subdisciplines that are most relevant to the study of individual social and physical components and dynamics, integrative tasks of assessment require interdisciplinary consideration of interrelationships among different components and dynamic systems. Thus, as has been emphasized throughout preceding chapters, impact assessment under NEPA ultimately requires consideration of the total environment.

Toward the goal of achieving comprehensive assessments that can influence decision-making, it is generally agreed that there are two kinds of practical tasks which should be given emphasis:

1. The estimation of the actual risks associated with the development of proposed projects
2. The judgment of just which risks are acceptable

Assessment teams typically have important responsibility for identifying the risks of project development. Assessment teams can also influence decision-makers (including the public), who must finally decide which of the identified risks are acceptable. Because of the extreme complexity of the total environment, assessment teams are particularly helpful to decision-makers if they can provide decision-makers with:

1. Highly integrated scenarios of all risks associated with alternative actions
2. Alternative criteria for evaluating the significance of individual and collective risks with respect to the total environment
3. Estimates of the potential for mitigating and/or enhancing those risks likely to be identified as significant
4. Estimates of the limitations of the assessment with respect to the various social and physical components considered

In order to deal with these issues effectively, assessment teams must understand that the concept of environmental risk is not as simple a notion as it might appear to be. This is particularly true when the assessment team adopts a total environmental perspective. For, in considering the effects of project development on dynamic interactions among social and physical components of the environment, assessment teams will come to realize that subjective as well as objective factors are (and must be) involved in identifying and evaluating environmental risks.

The Concept of Risk

The question of whether or not a project will result in significant environmental risks (i.e., exposures to danger) presumes, we might imagine, that we have clearly established criteria by which risks can be (a) objectively defined; and (b) objectively measured one against the other. But this is by no means the actual situation. Should we, for example, consider something as prohibitively dangerous if that something can be expected to kill 3000 people in 10 years? Or, rather, if it did it in 1 year? Maybe if it did it in 1 month? Obviously, no action is completely without danger, without risk to the actor or to bystanders. Equally obvious is the fact that throughout human history, man, in his consideration of risks, has given priority sometimes to himself and his immediate family, and sometimes to other humans and to the future of the species as a whole. In some instances he has considered his friends to be more worthy than his sons; and, in others, his pets more worthy by far than any friend. Danger, then, depends on human values.

But what of science? Can science set criteria and priorities for evaluating and selecting risks? Certainly to a nation that has accomplished the prodigious feat of placing men on the moon it would seem reasonable that science should give us some definite sense of direction as we deal with environmental risks. But no. Science cannot do this. It can state some of the possibilities (and fewer of the probabilities) we might face should we undertake one or another alternative action. Science can do nothing more.

There are a number of reasons why science is limited when it comes to defining and evaluating the risks involved in the manipulation of our environment. Two of the more general reasons are:

1. We have not developed the methodological and analytical skills for dealing with those problems and questions that go beyond rather immediate and narrow intents and concerns and that require us to consider a very large number of variables. All of that physics which got us to the moon, for example, can be studied in terms of a relative handful of parameters. But ecology and sociology have a seemingly infinite number of variables. Also, unlike a rocket trip to the moon, we have no objective measure of the final success of the journey that is social life.

2. We have not been generally concerned with how one can go about evaluating quantitative, objective information in light of qualitative, subjective experience.

In short, the problem of assigning and evaluating and acting upon risks has been and continues to be relatively simple only as long as the range of our concerns (Fig. 18.1) goes only very little beyond our immediate lives and interests and the time of our immediate experience. But more and more, we are beginning to recognize the necessity for expanding our horizons beyond our immediate selves. And we are finding that priority setting for the taking of risks is a major problem indeed.

Those who believe that environmental decision-making is essentially the task of the scientist, or who believe that decisions to be made as to what environmental risks should or should not be taken should follow from objective scientific principles and knowledge, would do well to consider the following two pronouncements on science and its limitations. The first is by Max Born (1968), a Nobel laureate physicist; the second is by Felix Cohen (1960), one of the most important legalists of this century.

> I am convinced that ideas such as absolute certainty, absolute precision, final truth, and so on are phantoms which should be excluded from science. . . . This relaxation of the rules of thinking seems to me the greatest blessing which modern science has given us. For the belief that there is only one truth and that oneself is in possession of it, seems to me the deepest root of all that is evil in the world. [pp. 182–183].

> It is clear that those who pretend to derive moral judgments from the facts of science have somehow slipped a doctrine of ethics into their science. If they are not clearly aware of the doctrine so much the worse is the doctrine likely to be. But false ethics is still ethics, and those who, in the name of some science of human

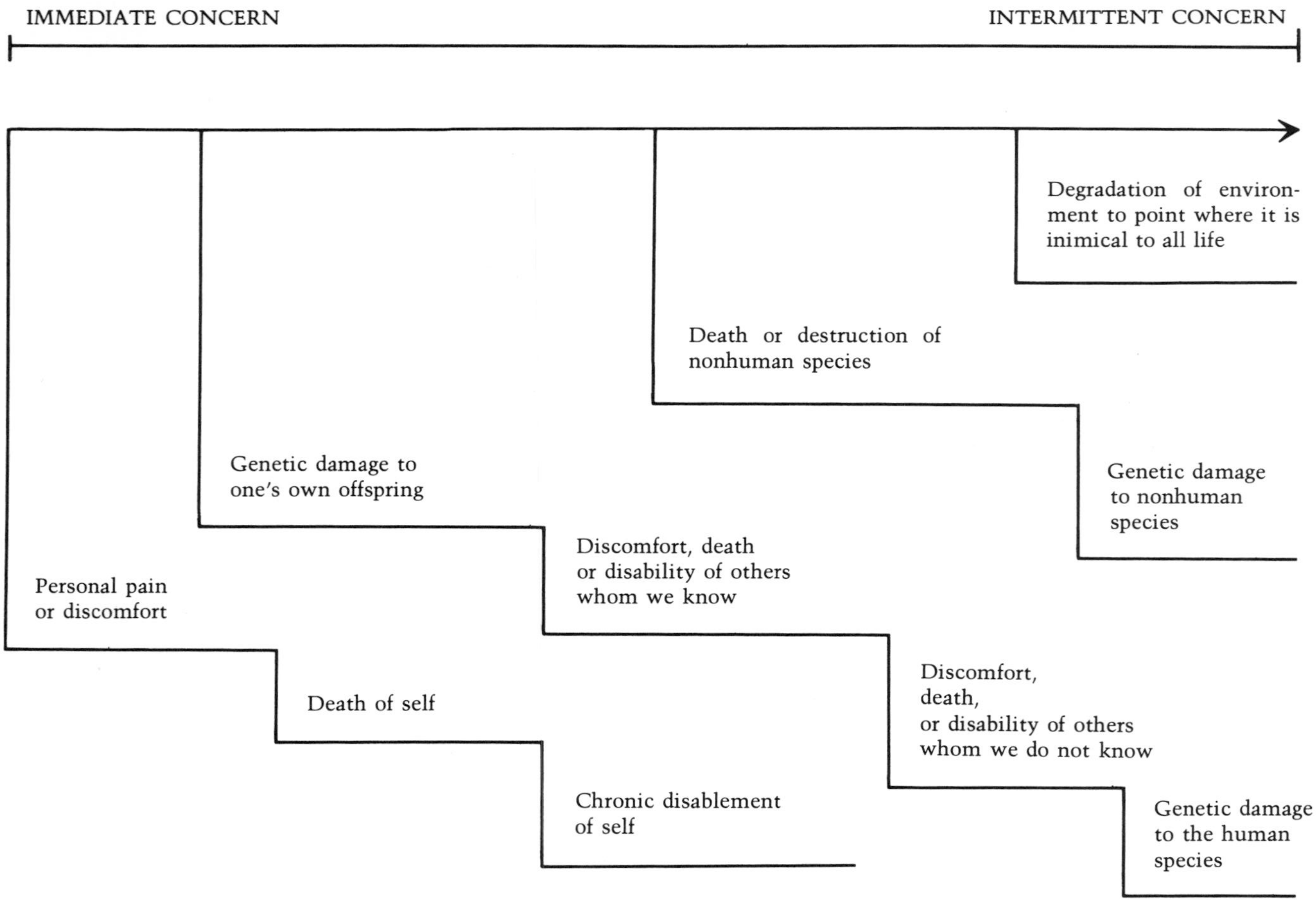

IMMEDIATE CONCERN

INTERMITTENT CONCERN

Personal pain
or discomfort

Genetic damage to
one's own offspring

Death of self

Discomfort, death
or disability of others
whom we know

Chronic disablement
of self

Death or destruction of
nonhuman species

Discomfort,
death,
or disability of others
whom we do not know

Degradation of environ-
ment to point where it is
inimical to all life

Genetic damage
to nonhuman
species

Genetic damage
to the human
species

FIGURE 18.1 *Some usual risk priorities.*

conduct, offer advice on how we ought to treat law breakers, or how we ought to control or fail to control the exchange of goods and labor, are in fact dealing in ethics, whether they know it or not [p. 24].

What these men are saying is, in short, that science never gives an absolute answer and that science never gives a priority for human action. In light of these considerations, it is evident that the prioritization of human effort with respect to the environment must be influenced by the social, political, and moral judgments of the lay public as well as by the recommendations of the scientist.

If social, political, and moral judgments of environmental import are actually to be made by the people, then people have to examine and clarify their concept of environmental quality, and do so from both the short and the long-range points of view.

Any such reexamination and clarification of human attitudes and values with respect to the quality of the environment over time, and the evaluation of risks to be either taken or not taken at any point of time, have to be ranked as among the more complex tasks undertaken by man. A good example of the complexity involved, and of how the short- and the long-range concerns of man can conflict with one another and confuse his prioritization of risks can be easily found. We have, for example, only to look to the contemporary use of the chemical compounds known as synergists in insecticidal preparations.

A *synergist* is any chemical compound that multiplies the physiological effect of another chemical. When used with commonly employed insecticides, synergists typically multiply the toxicity of those insecticides by a factor of from 2 to 15. This characteristic of synergists is of great import when one considers the continuing evolution of insects, and is, therefore, important to any long-range consideration of man's total environment. However, the fact that some synergists have also been shown to have important effects on specific mammalian tissue is also of central and even more immediate importance to the question of human well-being and safety.

The concept of Darwinian evolution is based on the premise that there are individual differences in any population of organisms. While nothing was known of genetic mechanisms in Darwin's time, today we understand and accept this premise as based on the fact that in any population of a given species there are differences (however minute) in genetic makeup. The environment acts in such a way as to select out from such a population those individuals having specific characteristics that are most suitable for ensuring survival within that environment. Over time, the genetic characteristics that give individual organisms an enhanced opportunity for survival will become more and more predominant within the surviving population. Relatively fast evolution is, therefore, most likely to occur whenever (a) the environment presents a population with stringent rules for survival; and (b) the environment is constantly testing that population for survival over long periods of time. Relatively slow evolution is most likely when the environment only weakly selects out survivors.

In terms of the characteristics of insecticidal compounds, relatively fast evolution within a species or group of species is most likely to occur when the insecticide, serving as a selector within the environment, is both extremely toxic to the species and persistent within the environment. A good example of such a strong selector is DDT.

The use of a synergist in an insecticidal formulation permits the use of lesser amounts of the active toxic agent that would otherwise be required and also the use of compounds that are relatively nonpersistent within the environment. Thus a synergist, in combination with an insecticide, allows man to combat insect pests without unduly increasing the rate at which pests develop resistance to the toxic agents (i.e., through evolution). But there is another consequence of the use of synergists that is necessary to consider in any total environmental perspective.

It has been found, for example, that certain synergists affect the normal functioning of liver tissue in some mammals. The effect is such that the rate at which liver tissue metabolizes or breaks down certain chemical compounds, including pesticides which have found their way into the organism, is depressed. If compounds are not broken down in the body (i.e., detoxified), those compounds are stored in body tissues and liquids. Thus the use of some synergists can directly result in the buildup of potentially dangerous chemicals within the organism.

How do we balance the evolutionary, long-range "pros" against the short-range toxicological "cons" and arrive at a decision with regard to the acceptable risk of a particular synergist? Any answer to this question would presume that we have more information and more unified goals than we actually do. For example, some synergists are also natural dietary constituents. Setting aside their commercial uses in insecticidal preparations for a moment, we do not know even the ordinary levels of human dietary intake of many of these compounds, nor do we understand either the changes these compounds undergo in the natural environment, or the consequences of a long-term buildup of these compounds in human systems.

People often tend to act in behalf of the greatest immediate safety whenever they are faced with a problem of choice that is characterized, as in the above example, by theoretical "pros" and "cons," and a dearth of hard, factual knowledge. But sometimes the popular concept of "immediate safety" is curiously narrow. The Delaney clause to the Federal Food, Drug and Cosmetic Act, for instance, states that any food which contains any additive shown to be capable of inducing cancer in experimental animals must be removed from interstate commerce. People are obviously concerned about cancer. But were this clause rigidly enforced (e.g., as it has been attempted with respect to saccharin), gone from our markets this very day would be various meats, all dairy products, eggs, fowl and fish!

This example of conflict between short- and long-range concerns about man-made and natural synergists highlights a type of problem that is indeed recognized in contemporary impact assessment—namely, balancing short- and long-term consequences to public health that can occur as a result of releasing chemicals into the environment throughout project development and operation.

But projects do not have to release chemicals or other materials into the physical environment in order to cause impacts on the social environment. As discussed in previous chapters, projects may cause a variety of direct impacts on personal, interpersonal, and institutional components and dynamics. Unfortunately, there is often little attention given to conflicting short- and long-range consequences of such impacts.

The sociological and anthropological literature is filled with examples of how decisions that have been made for the purpose of achieving immediate goals have also resulted in future consequences that can only be described as calamitous. However, it is probably more instructive to assessment teams and project managers to demonstrate how concerns for future risks have, in fact, sometimes outweighed concerns for immediate risks.

For example, at the beginning of the twentieth century, a cholera epidemic swept across Malaysia. Public health officials from the United States were summoned to help and soon uncovered a bizarre sequence of events that had already resulted in the deaths of hundreds, and that threatened thousands more.

It seems that a fisherman from one of the islands had observed a bubbling-up in the ocean and cautiously paddled nearer to the disturbed waters in order to investigate. As his canoe approached the area, he noticed that the bubbles formed the sign of the cross. Believing that he was witness to a miracle, the fisherman hurried back to his village to share his wonderful news. In a short time, a large number of canoes converged on the area, and hundreds of people saw the miracle for themselves. Not only did they look at it; they collected it in bottles and distributed the bottles throughout the islands.

What the public health officials soon discovered was that a sewer line between the islands had burst and that what the islanders were distributing among themselves as holy water was, in fact, diluted human fecal material.

The question before these officials, then, was whether or not to recommend that martial law be immediately declared in the islands, and that the bottles containing both the religious faith and the risk of death for thousands of people be collected by whatever force might prove necessary.

In the end, the officials decided against martial law. They recommended, instead, that the government undertake the task of educating the islanders to the danger hidden in the bottled waters. In time, this effort succeeded and the epidemic was broken. Of course, in this same period of time, many died who would not have died had sterner action been taken. Yet, in a world so often characterized by the wanton disregard of one culture's values by another, this decision must be reckoned a truly remarkable decision. For it was a decision expressly made out of consideration for the future of the ongoing culture. The implementation of martial law for the purpose of saving as many lives in as short a time as possible would have been a direct assault on the religious integrity of a living and complex island culture. In all probability it would have severely disrupted the system of beliefs and values of the islanders, even while upholding someone else's morality concerning the value of individual human life.

In the progress of any impact assessment, it is tempting, of course, to assume that the risks that may result from ordinary projects are hardly likely to require the consideration of such extraordinary concerns as discussed in this example. It should therefore be pointed out *that the ordinariness of a particular project is based on perceptions of the engineering and design features of the project, and not on an understanding of the project's consequences to the total human environment.* The simple fact is that the ordinariness of project engineering and design is no measure whatsoever of the total risks that can be incurred by a particular project in a particular location. For, until the advent of NEPA, there was no legal requirement for identifying and evaluating such risks. Thus, perhaps the most salient characteristic of so-called ordinary projects is that we are largely ignorant of the total environmental risks that they produce.

Risks and Opportunities

The concern over risks that can be associated with project development should not preoccupy assessment teams to such an extent that the assessment team overlooks the opportunities for human betterment that can also be associated with project development. While individual team members will, of course, have their own opinions with respect to ulterior motives for undertaking the design and construction of a particular project, it is necessary for the assessment team to view the assessment process itself as the primary means for identifying both the risks and the opportunities that must be evaluated and balanced in the decision-making process. As indicated in Fig. 18.2, the objective of this approach is to present decision-makers (including the public) with the total environmental costs and benefits that can be expected as a result of project development.

As in the identification of environmental risks, the identification of opportunities for human betterment is limited by current analytic skills and by the methodological difficulties of dealing with subjective experience and human values. For example, a particular project may require the relocation of an individual family presently domiciled in a proposed right-of-way. The Uniform Relocation Assistance and Real Property Acquisition Policies Act of 1970 is a basic tool for ensuring that the relocated family will have safe and sanitary housing. In many instances, the provisions of this Act can result in an overall improvement in domicile. The Act is therefore widely acclaimed as an important opportunity for human betterment which would not occur in the absence of a proposed federal project.

However, in certain circumstances, this Act can result in serious risks for some persons. For instance, in one particular project it proved impossible to find substitute safe and sanitary housing for an indigent head of household and his six children who were living in a run-down shack within a proposed right-of-way. The alternative was to construct a new domicile for this particular family. Yet, upon receiving the keys to his new, safe and sanitary house, the head of household had to run the risk of having the house taken from him by previous creditors. In this instance, a law designed to promote the well-being of a family seems instead to have promoted the well-being of local businessmen and to have exposed the family to even greater stress than it had had to face before.

As has been stated throughout this book, general principles, concepts, and doctrines are less useful for purposes of impact assessment than is an understanding and appreciation of site-specific facts and conditions. The assessment team just cannot deal with risks and opportunities as abstract possibilities; rather, the assessment team must examine how risks and opportunities are specifically related to actual project development in an actual environmental setting.

Dimensions of Environmental Risks and Opportunities

In the effort to identify and evaluate the real risks and opportunities to be associated with project development, it is useful for the assessment team to employ certain dimensions of environmental risks and opportunities. Careful consideration of such dimensions helps to ensure a

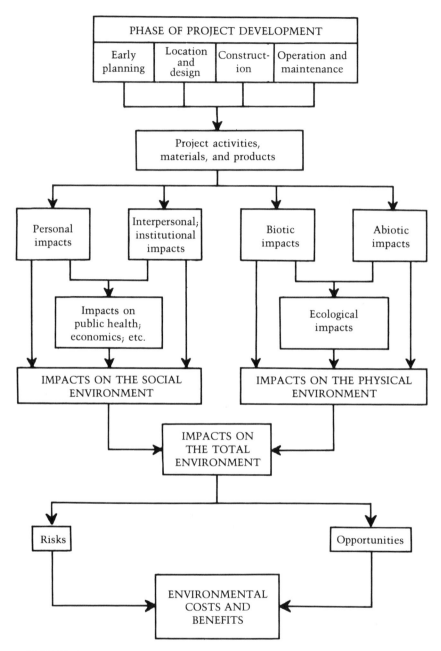

FIGURE 18.2 *Impact assessment as an effort to determine environmental costs and benefits.*

balanced evaluation of the effects of project development on the total human environment. These dimensions are:

1. Who will be on the receiving end of project risks and opportunities?
2. How long will identified risks and opportunities be manifested?
3. What environmental factors and circumstances that are not related to project development may nonetheless mitigate and/or enhance identified risks and opportunities?
4. What capacity does an individual have (in both present and future generations) to choose from among the identified risks and opportunities?
5. What criteria of "significance" have been or may be used to evaluate project induced risks and opportunities?
6. How can it be documented that the public knows and understands the risks and opportunities which will result from project development?

With respect to these dimensions, it is necessary to emphasize certain premises that must underlie the efforts of an assessment team that is seriously concerned with an objective evaluation of total environmental risks and benefits of project development. These premises are predicated, if not in the letter of the law of NEPA, then certainly in its spirit:

1. NEPA gives precedence to no one component of the physical or social environment, but rather, to our understanding of all project impacts, regardless of majority or minority concern for those impacts.
2. NEPA is as concerned with long-term environmental costs and benefits as it is with short-term environmental costs and benefits.
3. Jurisdictional limits of agencies proposing projects do not define the limits of the assessment to be made in order to understand the total environmental impacts of proposed projects.
4. NEPA is meant as a device for expanding the public's power to influence environmental decision-making, and is not meant as a device for limiting or otherwise constraining that power.
5. There is no one technical, scientific, or other measure of significance for project impacts that must take priority over any other measure; project development must serve a plurality of concerns and needs.
6. Even the most sophisticated, balanced, and sensitive assessment of project impacts is an absolute waste of time if the public does

not understand the pertinence of the assessment's findings to individual and social life and well-being.

Some Approaches

The complexity of the total environment, and such methodological and conceptual limitations as have been discussed, require careful consideration of the role of the interdisciplinary team and the role of the public in comprehensive impact assessment. Some aspects of the roles of interdisciplinary teams and the public will be examined in greater detail in Chapters 19 and 20. In this section, we will highlight some general approaches that can be specifically utilized by the assessment team in order to ensure a comprehensive evaluation of project-related risks and opportunities. These approaches may be characterized as (a) the "So what?" approach; (b) the "What if?" approach; and (c) the "Says who?" approach.

THE "SO WHAT?" APPROACH

In this approach, each and every individual impact of project development is subjected to the simple though scornful question, "So what?"

For example, one effect of excavation may be the increase in suspended particles released into streams. Thus, the one direct impact of excavation is an increased turbidity in receiving water. But, so what? In order to answer this question, it is necessary for the assessment team to identify consequences (i.e., secondary impacts) of enhanced turbidity. For instance, the enhanced turbidity may lead to occlusion of sunlight. But, so what? Just what are the possible consequences of occluding sunlight with respect to plants, water temperature, etc.?

The end result of a persistent posing of the question "So what?" is, of course, a chain, or network, of interrelated impacts. As the chain or network expands, it is highly likely that physical impacts of project development will also be seen to have ramifications for social components and processes in the total environment. Thus, turbidity may influence the well-being of aquatic plants and fish, but turbidity can also influence the recreational (e.g., fishing), and aesthetic use of streams and rivers.

In pursuing the "so what?" approach to assessment, the assessment team can quickly come to identify possible cause–effect relationships in both the physical and social environments, and thus the types of data and information that will be required to evaluate the probability and significance of individual impacts. A second consequence of this approach is that the assessment team must consider numerous secondary impacts for every identified direct impact of project development. This is in keeping with NEPA's concern for secondary or indirect impacts of project activity. Sometimes, the secondary or indirect impacts of project development can prove to be more significant than the preceding, direct impacts.

However, the advantages of a "so what?" approach to impact assessment can be effectively short-circuited if the team is not truly interdisciplinary. For example, it is unlikely that fisheries biologists will be able to identify very many of the social "so whats" to direct and indirect impacts on fish. Likewise, it is unlikely that sociologists will be able to identify very many of the physical "so whats" to direct and indirect impacts on social, psychological, and economic parameters of human life. Even if such impacts are identified in the absence of an interdisciplinary team, the chances that each impact will receive serious consideration are maximized by the interdisciplinary nature of assessment and are minimized when interdisciplinary effort is absent.

THE "WHAT IF?" APPROACH

The result of a well-conducted "so what?" approach to impact assessment is the identification of interrelated direct and indirect impacts on the total human environment. However, in pursuing a series of cause–effect relationships, it is quite a simple matter to overlook the real complexity of the real environment. The simple fact is that impacts that are serially linked according to identified cause–effect schemes are typically oversimplified. Thus, in a chain of impacts such as the following

```
                                          ┌→ D1
                          ┌→ C1 ──────────┴→ D2
A ──────→ B ─────────────┤
                          └→ C2 ──────────┬→ D3
                                          └→ D4
```

it might happen that the assessment team may fail to consider the following possibility:

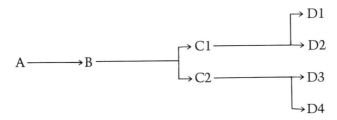

In order to achieve a comprehensive assessment of project impacts, it is therefore useful for the assessment team to ask "what if?" with respect to all identified cause–effect chains. For example, clearcutting may be a project activity that can reduce the carrying capacity of an area to some level that is nonetheless sufficient for a particular population to maintain itself. But "what if" the use of certain pesticides during operational phases of project development also results in an enhanced mortality of the same population in a period of 1–2 years after clearcutting? The cumulative impact of two or more different activities may, in fact, be quite different from the impact of either activity alone.

The "what if?" approach to impact assessment not only facilitates the identification of the total impacts of total project development, but also serves to remind the assessment team that project impacts can actually be influenced by factors unrelated to project development. Thus, a project-mediated loss in carrying capacity has to be evaluated in light of other stresses on carrying capacity that might occur as a result of ongoing regional development that can be expected to occur regardless of project development. For example, local effects of laying a pipeline through an area may be potentiated by ongoing local development which is totally independent of a pipeline designed primarily to transport resources from one distant region to another distant region.

As in the "so what?" approach, the "what if?" approach can be effectively short-circuited. This typically happens whenever assessment team members and other contributors to assessments are allowed to substitute opinions in lieu of knowledge and actual experience. As a general rule, whenever the question "what if?" is answered with immediate and uniform derision by the assessment team, the project manager has good reason to believe that the assessment team is lacking either in knowledge or good sense, and probably in both.

THE "SAYS WHO?" APPROACH

Altogether too many impact assessments have been conducted by teams that have persisted in the belief that impact assessment is essentially an exercise in applying conventional wisdom to meet new legislative requirements. Thus we find that assessment teams often utilize, explicitly or implicitly, certain axioms for purposes of identifying and evaluating impacts on the total environment, including:

1. Expert opinion is to be given more credence than nonexpert opinion.
2. The existence of expert disagreement releases decision-makers from all responsibility for considering issues involved in the disagreement.

From a total environmental perspective, a perspective that includes subjective as well as objective considerations, and a perspective which is inclusive of social goals and objectives as well as of technical and scientific understanding, there is only one appropriate response: "Says who?"

It is necessary, of course, to utilize experts in the analysis of highly technical and scientific issues. However, it has often happened that experts have been utilized beyond the limits of their expertise. For example, a professional ecologist may perform a series of field and laboratory experiments in order to measure the productivity of a wetland. With respect to conclusions on the impacts of a proposed project on that productivity, or on the relationship of wetland productivity to other natural resources, the question "says who?" is rightly answered, "Says a professional ecologist who by training and experience is best equipped to carry out and interpret the relevant scientific experiments." Yet, should the issue not be the productivity of the wetland or the wetland's influence on other local resources, but rather, the social implications of protecting wetland areas from development; then the opinions of even the expert ecologist are no more (or less) important than those of local citizens, businessmen, and private developers. NEPA, after all, is not a mechanism whereby biologists and social scientists shall remake the human environment to fit their private visions.

The "says who?" approach is not only important for ensuring that relevant expertise will be distinguished from irrelevant expertise, but also for ensuring that decision-makers will receive a balanced presentation of scientific and technical disagreements with respect to project impacts. The assessment team should understand that there is likely a greater probability for scientific and technical disagreement about environmental phenomena than there is for consensus. The "says who?" approach, therefore, serves to document these disagreements, as well as the technical and scientific basis for them.

The purpose for documenting scientific and technical disagreements is not, of course, to provide decision-makers with a basis for ignoring contested issues in project development. As an extreme example of the fallacy of this approach, we might consider the responsibility of a project engineer when a number of experts disagree with one another about the potential for a proposed dam to burst. Would such disagreement be reasonable justification for the project engineer to go ahead with the proposed dam design as though the entire question of safety had not been raised?

As in this extreme example, the purpose for documenting technical and scientific disagreements in impact assessment is to provide decision-makers with a full understanding of all possible risks that may be associated with project development. That there may not be a consensus as to the possibilities and probabilities of risks is a simple fact of life that decision-makers always have to consider. Good documentation as to the range and sources of disagreement is a basic tool whereby decision-makers can seek to evaluate the significance of disagreement and identify possible alternative actions for which there may be stronger consensus with respect to environmental consequences. There can be no excuse whatsoever for an assessment team's failure to identify issues about which there is disagreement, or to provide relevant documentation.

A Special Issue

In recent years, many universities and colleges have developed programs for the purpose of training students in the techniques of impact assessment and/or in interdisciplinary evaluation of contemporary environmental problems and issues. In many instances, students can earn undergraduate and advanced degrees in "Environmental Affairs," "Environmental Science," or in similarly designated academic areas. At the same time, job descriptions in private and public agencies and organizations have also been altered to note specific and general responsibilities for environmental assessment and evaluation, as in the job titles "environmentalist," "environmental engineer," "environmental coordinator," etc.

Certainly one must applaud the efforts of universities and colleges to enhance the interdisciplinary training of students and also the apparent desire of these academic institutions that such interdisciplinary training be applied to the practical solution of environmental problems. Similarly, one must applaud the efforts of governmental agencies and private corporations to expand their roles and responsibilities in environmental matters. Yet, with respect to the practical problem of conducting assessments of project impacts on the total human environment, one might also suggest that such academic and bureaucratic "retooling" can actually be counterproductive.

Such a suggestion is not based on any educational or management theory or principle, but rather on various observations of actual assessment personnel in various private and governmental agencies and organizations, including the observations that:

1. By far the majority of personnel having academic degrees in an "environmental" area are primarily trained and experienced in the natural sciences.
2. Those bureaucratic positions that are clearly identified as having environmental responsibility tend to be filled by persons having academic degrees in an environmental area.
3. Bureaucratic positions that are not clearly identified as having

environmental responsibility tend to be filled by persons who do not see themselves qualified to deal with environmental issues.

Clearly, *the end result of these factors is that environmental impact assessment becomes largely the bureaucratic preserve of those who are primarily trained in the natural sciences and that this bureaucratic structure tends to be upheld and strengthened by the attitudes of personnel trained and experienced in other disciplines.*

While such a situation is eminently reasonable if we assume that the word "environment" means essentially "bugs 'n bunnies," it is clearly unreasonable for a total environmental perspective!

As discussed throughout this and other chapters, any mechanism that mitigates against a comprehensive evaluation of interconnected social and physical impacts is contrary to the goals and objectives of environmental impact assessment under NEPA. To the extent that academic degrees and bureaucratic compartmentalization actually serve to equate the human environment with "bugs 'n bunnies," then to that same extent are such degrees and bureaucratic structures detrimental to the conduct and utilization of impact assessment under NEPA.

References

Born, M.
  1968   *My life and my views*. Charles Scribner's Sons, New York.
Cohen, F.
  1960   *The legal conscience*. New Haven, Connecticut: Yale Univ. Press.

19

The Role of the Interdisciplinary Team

Introduction

Throughout many of the preceding chapters, various suggestions and guidelines have been presented for conducting assessments in conformity with the broad goals and objectives of NEPA. Some of these suggestions and guidelines have been discussed as being relevant to specific analytical or integrative tasks (Chapter 6) that are undertaken during the assessment of impacts on either the physical or the social environments. Some have been discussed as being generally relevant to a broad range of project-management and attitudinal issues that greatly influence both the efficiency and the quality of the overall assessment process (e.g., Chapter 18). This chapter will focus narrowly on the role that the interdisciplinary team plays in the assessment of project impacts on the total human environment. In discussing this role it will be necessary, of course, to reiterate some of the basic guidelines already discussed in other chapters.

Before proceeding with this discussion, however, it is necessary to emphasize a fundamental characteristic of the role of the interdisciplinary team in the assessment of impacts on the total human environment, that is, *that the role of the interdisciplinary (ID) team in the assessment of impacts on the total human environment is not a simple summation*

of the team's functions and responsibilities with respect to the assess-
ment of individual impacts on the physical and social environments.

This characteristic of the ID team's function with respect to the total environment derives from the fact that the total human environment, which includes physical, aesthetic, historic, cultural, social, and economic dimensions (CEQ, 1976, p. 49), is not as compartmentalized as those sciences and disciplines that we have invented for studying its dimensions.

The total human environment is a dynamic system that, even when viewed in its local manifestations, integrates, adjusts to, and influences objective and subjective elements. As a system, it exhibits a qualitative and quantitative complexity which we are only just beginning to appreciate. If we could be sure that our present disciplines were complete in their individual knowledge and that the collectivity of these same disciplines actually spanned the full scope of the total human environment, then certainly we might hope that a summation of our current disciplinary perspectives would yield a comprehensive overview of the total human environment. But this is hardly the case.

The knowledge base of individual disciplines continues to increase exponentially—which is another way of saying that we are currently discovering the extent of our ignorance more rapidly than we have in the past. The number of new disciplines and subdisciplines also continues to increase—which is another way of saying that we are also discovering just how conceptually and methodologically limited any temporal disciplinary knowledge of our environment can be. From the practical standpoint of impact assessment there is, therefore, an apparent dilemma.

Either interdisciplinary assessments of project impacts are to be carried out simply by adding up the individual and separate impact reports of individual and separate experts, or by inventing a new and integrative knowledge on a project-by-project basis. The first approach may be limited by the limited knowledge and concerns of individual disciplines, but it is eminently practical from the standpoint of finite budgets of time and money. The second approach may be ideal with respect to the desire to make ultimately wise decisions, but it is also highly impractical from the standpoint of finite budgets of time and money. Having taken the first approach, and being criticized for it, a project manager of an ID team may well respond to academic critics: "Look! We deal with questions that can be reasonably answered. You want us to deal with questions that can't be answered at all . . . not unless you want to pay for basic research. Even then you'd be lucky to get any meaningful answers."

As has been emphasized throughout this book, impact assessment under NEPA is definitely not intended as job security for academic researchers. Impact assessment is meant as a practical tool for improving decision-making. As a practical tool, it must be usable within practical constraints of time and money. Thus, we cannot expect the ID team to focus on "impossible questions." Still, is the only alternative to begin assessment by concentrating on questions that can be "reasonably answered"? The premise of this chapter is quite different. It is, simply, that the major role of the interdisciplinary team regarding the total environment is to *begin and conduct assessment in such a way as to deal with questions that, from a total environmental perspective, may reasonably be asked.*

Team Components versus Team Dynamics

The interdisciplinary team has rightly been called "the critical link in the assessment process," a link made up of "resource experts *who interact with one another and the public* to provide technical planning information to project decision makers [Soil Conservation Service, 1977, emphasis added]." Yet, for purposes of identifying the role that interdisciplinary teams play in assessing impacts on the total environment, it is useful to point out that in actual practice an *interdisciplinary team* is something quite different from what often meets bureaucratic definitions of an *assessment team.*

For example, an expert on water quality may be a member of an assessment team. But, unless he or she interacts professionally with other members of the assessment team who have different expertise, and interacts in such a way that the team as a whole considers issues which otherwise would not have been considered, that water quality expert is not part of an interdisciplinary team. This is a crucial point. The definitive aspect of an interdisciplinary team is not the types of different expertise that can be brought to bear on a problem, but rather, the actual results of the team's interactions. *If the results of an assessment team's study are conceptually different from the collective results likely to have been achieved by individual experts working independently,* then one might reasonably conclude that the assessment team was, in fact, an interdisciplinary team.

It is my opinion that many assessment teams that have had their genesis in NEPA fail to meet this operational criterion for interdisciplinary teams. The emphasis has been largely on team components which give the appearance of interdisciplinary study (e.g., the diversity of expertise among team members, advanced degrees in academic disciplines), *rather than on actual, day-to-day interactions and interdependencies among team members.* It would therefore appear that many middle- and upper-echelon executives and managers who are responsible for staffing and directing the efforts of such teams have yet to learn the lesson learned by the universities—that where management gives precedence to individual credentials rather than to group performance, a collection of diverse experts tends to evolve into a weak confederation of separate and contending departments.

The need for the interdisciplinary (as opposed to multidisciplinary) assessment of project impacts on the physical and social environments has already been summarized in Chapters 12 and 17. Also, basic approaches that might be taken in the interdisciplinary assessment of impacts on the total human environment have been discussed in Chapter 18. Rather than repeat specific points made in these chapters, it is more useful here to identify the more generic functions which truly interdisciplinary teams can perform in the assessment of impacts on the total environment. These generic functions are based on the principle that environmental assessments should give rise to sound project planning and implementation. They include:

1. Maximizing the informational value of assessments for project decision-makers and the public
2. Minimizing the risks of unnecessary delay and project expenditures that can result from an incomplete or otherwise incompetent assessment of project impacts
3. Facilitating the evaluation of project alternatives and the consequent design and implementation of ameliorative measures

Each of these functions will be discussed in the following sections.

Maximizing Informational Value of Assessments

It is important to realize, when discussing the broad goals and objectives of NEPA, that there is a practical difference between *data* and *information*. Data are the factual results of assessment studies; information is the interpretation of these data. For example, that the turbidity of receiving waters in a project area is 75 turbid units is a water-quality datum. However, the significance of this datum with respect to fish reproduction, primary productivity, and/or human uses of the water is information. This difference between data and information is not offered as a mere semantic difference; it derives from the underlying rationale of impact assessment. Decision-makers need information; they do not require (nor can they typically use) raw data.

Many assessment teams apparently believe that they are providing information to decision-makers and the public when they provide them with ream after ream of data. This is typically the case when the assessment is conducted by a *multidisciplinary team* whose members exhibit the tunnel vision with respect to their own interests that is so characteristic of the independent specialist.

The *interdisciplinary team* is the basic mechanism whereby data on project impacts on physical and social components and dynamics of the environment are transformed into information about total environmental impacts. This transformation of physical and social data into information that can be utilized by decision-makers and the public takes place in a number of ways, including:

1. Identifying key environmental components and dynamics (physical and social) that should be evaluated in order to define baseline conditions of the total environment and potential project influences on those conditions.
2. Scheduling of individual studies of individual components and dynamic systems so as to maximize (a) the exchange of findings among team members, the public, and other informational resources during the entire assessment process; and, thus, (b) the potential for identifying interconnections between the physical and social environments.
3. Requiring each participant—expert to justify the need for collecting specific kinds and amounts of data in terms of existing regulatory guidelines, legislative requirements, and also in terms of those disciplinary concepts that have a high potential for generating conclusions relevant to project development and its effects on the total human environment.
4. Allocating budgetary and manpower resources so as to achieve a balanced consideration of the social and physical environment.
5. Monitoring of the progress of individual disciplinary studies so as to ensure the relevance of generated data to the total assessment effort.
6. Identifying all social and physical factors and conditions which can trigger individual impacts, or otherwise influence impact chains and networks.
7. Translating all findings of individual studies into concise statements that can be understood by people having diverse training and experience.

Minimizing Unnecessary Delays and Expenditures

A truly interdisciplinary team is a microcosm of the overall assessment review process (Chapter 23) which includes both governmental and private organizations and the public. Just as agency and public comments received during the review process should be utilized to improve the relevancy and quality of assessment, so should day-to-day interactions among members of the interdisciplinary team significantly expand the relevance and quality of assessment.

The key point is that the *interaction among team members should actually result in the team's serious consideration of at least those issues that are of some concern to persons who are not directly involved in the assessment process as assessment team members.* This is not likely to happen if the team functions in a typically multidisciplinary manner; individual specialists are largely left unimpeded by the real interests and concerns of others. Perhaps the *best operational measure of the true interdisciplinary nature of an assessment team is simply that the review process does not result in conceptual, methodological, or informational "surprises," that is, key issues that have not already been identified and evaluated by the assessment team.*

Whenever the review process does identify key issues and information that have not been considered by the assessment team, we must accept the fact that the assessment team has failed in its specific tasks and that whatever quality control mechanisms were designed to prevent failure have actually proven ineffective. In such a situation, consideration should be given not only to redoing the assessment, and/or redoing project planning, but also to the revamping of management and operational structures that have caused or permitted the failure.

Toward the goal of minimizing the unnecessary delays and costs of redoing assessments and project planning and design, it is essential that the interdisciplinary team actively seek to:

1. Maintain close liaison and coordination with all relevant federal, state, and local government agencies during the assessment process
2. Maintain close liaison and coordination with all public and other groups and organizations, as well as with individuals who can provide information and guidance during the assessment process
3. Ensure that the tasks of information-giving and information-getting (which are requisite to good liaison) are assigned to team personnel, who, regardless of their technical specialties, are effective communicators and who are also sensitive to the real need for external inputs into the assessment team
4. Include as members of the assessment team those personnel who have practical experience and extensive knowledge of all phases of project development and project-related activities
5. Ensure that all members of the assessment team are equally aware of all developments pertaining to the overall assessment of impacts on the total human environment

With respect to Item 5, it can be suggested that the distribution of information among interdisciplinary team members on the basis of "need to know" is both irrelevant and counterproductive to the assessment of impacts on the total environment. It presumes that someone (e.g., the project manager) already knows just what everyone else has to know. This presumption is inconsistent with the objectives of interdisciplinary assessment and the current state-of-the-art of impact assessment and should not be tolerated in any form.

Facilitating the Evaluation of Project Alternatives

It has been pointed out in previous chapters on physical and social impacts that there is no one consensual and objective measure of the significance of impacts. This problem, as well as some of the available means for dealing with it, will be discussed again in Chapters 20 and 21. However, it is useful here to point out that an interdisciplinary team can contribute a basic tool to decision-makers who must, in the end, decide (a) what impacts will be considered significant; and (b) which will be considered insignificant for purposes of project planning. This basic evaluative tool is the "connectedness" among individual impacts. By "connectedness" is meant the nature and extent of dynamic interactions among individual physical and social impacts.

An individual impact, when considered in isolation from other impacts, may well be judged to be insignificant. However, the same impact may trigger or otherwise influence other physical and social impacts that might be judged to be highly significant. Thus the connectedness of impacts is a criterion that decision-makers can utilize in order to decide which impacts should reasonably be avoided or otherwise ameliorated and controlled by project location and design, or by operational and management practices. Therefore, in considering project alternatives, this same criterion can be an important tool for identifying those alternative actions that are most prudent and feasible with regard to environmental and project-objectives and goals.

Again, the piecemeal approach to assessment which typifies most multidisciplinary efforts is not likely to result in the definition of the nature or extent of connectedness among different types of impacts. In fact, the *improbability* of identifying the connectedness of physical and social impacts, and thus of providing decision-makers with an important evaluative tool, is usually directly proportional to the degree that individual members of an assessment team can work independently of one another.

Some basic guidelines for helping decision-makers evaluate project alternatives are:

1. Identify all relationships among physical and social impacts *for all alternative locations and designs*
2. Identify all relationships among physical and social impacts *for all alternative construction, operational, and maintenance activities*
3. Relate possible mitigation and/or enhancement measures *to existing and projected interrelationships* among social and physical components and dynamics
4. Compare and contrast all mitigation and/or enhancement measures in terms of *probability for success, economic costs, duration of effectiveness, and possibility for undesirable effects*
5. Ensure that this information is available to decision-makers as early as possible in project development.

Some Additional Considerations

These rules and guidelines pertain directly to assessment teams that function in a truly interdisciplinary manner. However, there are various ways in which an interdisciplinary team may be structured into or related to an overall assessment team.

For example, the interdisciplinary team may constitute the entire in-house assessment team. Or, the interdisciplinary team may be a relatively small in-house group that directs the activities of other members of the total assessment team. In the first case, members of the interdisciplinary team must become individually involved in specific analytical tasks (usually at the level of a single discipline), as well as in interdisciplinary deliberations. In the second case, interdisciplinary team members can usually concentrate on the coordination and assimilation of diverse information and data generated by others, that is, the interdisciplinary team functions in an executive capacity with respect to the overall assessment team.

While the guidelines discussed in this chapter are directly applicable to those situations in which the interdisciplinary team constitutes the entire assessment team, it may appear that they are not relevant to an assessment team in those instances where the team is subdivided into an executive interdisciplinary group and an operational group. This is not the case. The efficiencies of project management that may result from separating an executive, interdisciplinary component from other operational components of the assessment team do not reduce the importance of active and meaningful interaction among all team members.

This is not to say that operational personnel should be included in all the deliberations of the executive, interdisciplinary team. But to totally exclude operational assessment personnel from interdisciplinary interactions in the progress of an assessment is to introduce an element of compartmentalization into the assessment process. The wise project manager will want to assure himself that any bureaucratic efficiencies that may result from this approach far outweigh the losses in assessment quality that could conceivably result by totally excluding operational expertise and experience from interdisciplinary deliberations of potential project impacts.

Whether or not assessment teams are substructured into interdisciplinary and operational components, it is important to emphasize one final point about the contribution of the interdisciplinary team to the identification and evaluation of project impacts on the total human environment.

The quality of objectivity of assessment is certainly one of the most important criteria that has been used to evaluate the adequacy of assessments conducted under NEPA (Mallory, 1976, pp. 43–48). *The criterion of objectivity pertains to the interdisciplinary team's approach to assessment, and not (as many assessment personnel have imagined) to the types of environmental issues and concerns which the team must consider.*

The legal requirement for objective assessment is not a license for ignoring project impacts that cannot be quantitatively analyzed, measured, or evaluated; rather, it is a mandate for the assessment team to conduct assessments objectively, that is, for the purpose of determining the consequences of project development and not for the purpose of justifying project development. As discussed previously (Chapters 12 and 17), it is vital that interdisciplinary teams include personnel who are equipped by training and experience to identify the subjective aspects of the total human environment and to integrate a consideration of these aspects with the interdisciplinary evaluation of other physical and social impacts of project development.

References

Council on Environmental Quality
　　1976 *Environmental Impact Statements: An analysis of six years' experience by seventy federal agencies.* Washington, D.C.: Executive Office of the President, Council on Environmental Quality.
Mallory, R. S.
　　1976 *The legally required contents of a NEPA Environmental Impact Statement.* Stanford, California: Stanford Environmental Law Society, Stanford Law School, Stanford Univ.
Soil Conservation Service
　　1977 Guide for Environemental Assessment. *Federal Register, 42*(152, Part IV).

20

The Role of the Public

Introduction

The role of the public in the assessment of impacts on the total human environment is determined by a variety of legal and practical issues which influence the overall assessment process. With respect to NEPA itself, "the public" is addressed in four separate sections, as follows (italics added):

- **Section 101 (a)**
"The Congress . . . declares that it is the continuing policy of the Federal Government, *in cooperation with State and local governments, and other concerned public and private organizations* . . . to use all practical means and measures . . . to create and maintain conditions under which man and nature can exist in productive harmony."

- **Section 101 (c)**
"The Congress recognizes that *each person should enjoy a healthy environment, and that each person has a responsibility to contribute* to the preservation and enhancement of the environment."

- **Section 102 (C)**

"Copies of . . . [the EIS] . . . shall be made available to the President, the Council on Environmental Quality, *and to the public* as provided by Section 552 of Title 5, United States Code."

- **Section 102 (G)**

"The Congress authorizes and directs that, to the fullest extent possible, all agencies of the Federal Government *shall . . . make available to States, Counties, municipalities, institutions, and individuals advice and information* useful in restoring, maintaining, and enhancing the quality of the environment."

Other legal and informal mechanisms by which the public is given specific rights *to have access to environmental information, to contribute information, and to challenge decisions* which affect environmental quality have been identified in Chapters 1 and 3. It is reasonable to assume that those mechanisms that greatly increase the public's access to the courts have resulted in agencies giving increased attention to facilitating citizen involvement and participation in the overall assessment process. The effort clearly has two objectives: to reduce the potential for litigation initiated by concerned public groups and individual citizens; and to facilitate the day-to-day analytical and integrative tasks of the assessment process. Toward the goal of achieving these objectives, agencies have increasingly subscribed to the wisdom of the following dicta:

1. A public which is left out of the assessment process is likely to be the very public that can find its way into the courts.
2. To leave the public out of the assessment process is equivalent to disregarding key data and informational sources.

Perhaps some of the most positive statements with respect to citizen involvement in the assessment process can be found in the Federal Aviation Administration's instructions for processing airport development projects which affect the environment and in FAA's policies and procedures for considering environmental impacts under NEPA:

> The involvement of the community at large is a *necessary element in the decision-making process.* Communities, organizations, and others affected by airport development proposals submitted to the FAA shall be provided an *effective opportunity to comment at all appropriate stages in the decision-making process;* and, in all cases, they shall be provided an opportunity to review and comment on draft statements and to receive and inspect available final statements [Federal Aviation Administration, 1975, p. 15, italics added].

> Citizen involvement should be initiated at the *earliest practical time* and *continued throughout the development of the proposed project* in order to obtain

meaningful input . . . [and] . . . *may be appropriate in defining the scope of work of an environmental impact assessment* report developed by an applicant for aid or a consultant, or of a draft EIS being developed by FAA [Federal Aviation Administration, 1976, p. 34226, italics added].

These regulations and policies of FAA, as well as similar regulations and policies of other agencies such as the Federal Highway Administration and the U.S. Forest Service, clearly reflect the original intent of the Council on Environmental Quality that the assessment process be a key mechanism for direct public involvement in environmental decision-making; and also, that such involvement be recognized not as a mere ploy whereby the public may be placated, *but as a positive and constructive attempt to improve decision-making.* The CEQ defined this approach in 1971 as follows:

> The ability of citizens and citizen groups to make their views known and to participate in government decision-making on the environment is *critically important.* Often individuals and groups can contribute data and information *beyond the expertise of the agency involved.* . . . The new openness to citizen involvement is bound *to check, stimulate, and test future Federal agency activities.* Citizen concern cannot substitute for assumptions of environmental responsibility by government and industry. Nor can it provide the mechanism to resolve the many policy issues involved. What it can provide, however, is a *highly potent quality control and "feedback"* [CEQ, 1971, p. 163, italics added].

Of course, policy statements about the importance and value of public involvement in environmental decision-making do not necessarily reflect or influence what actually happens in an actual assessment project. With respect to the thousands of assessments which have been completed up to the present time, *it is not unreasonable to suggest that public involvement has largely remained an unfulfilled ideal. Even 9 years after NEPA became law, assessment teams typically view the public more as an adversary than as a partner in the assessment process.* This perspective precludes any serious consideration of the constructive roles that the public may, in fact, play in the assessment of impacts on the total human environment. Such an attitude cannot be tolerated in any assessment team.

Instead of concentrating on the risks and difficulties of including the public in the assessment process in a meaningful way assessment teams must concentrate on the benefits that can be derived from an enhanced and much expanded communication among team members and the public. This is the key to public participation—*communication* between the assessment team and the public. Not just the team's giving of information to the public; and not just the public's giving of information back to the assessment team. Communication is more than a simple transmission of information from one group to another. It is an active and constructive exchange of information, meaning, and opinions. A communicator is not simply a tape recorder whose sole function is to make prerecorded announcements. A communicator is a dynamic integrator of information who functions so as to *identify and disseminate available information that is useful for meeting human needs, and also to identify human needs that may require the generation and assimilation of new or additional information.*

The importance of communication as a means of achieving constructive public involvement in the assessment of impacts on the total human environment is increasingly recognized by government agencies and has been translated into three practical guidelines (Federal Highway Administration, 1976, p. 65) which all assessment teams would be well advised to follow:

1. Communicate with the public *as early as possible.*
2. Communicate with *as many people as possible.*
3. Communicate through *as many different means as possible.*

Some Potential Roles of the Public

In considering these three guidelines for communicating with the public, the assessment team must understand that the objective of communication is to facilitate that constructive participation of the public in the assessment process which legislation, executive orders, the courts, and federal agencies have deemed critical to the quality and adequacy of assessment. Thus, the timing, scope, and means of communication with the public are to be determined by the practical needs of impact assessment under NEPA, and not by agency rules and procedures that have been adopted for fulfilling agency responsibilities prior to NEPA.

Within the broad scope of NEPA's goals and objectives, the public can perform each of the following functions. Thus, *communication between the assessment team and the public must be undertaken in order to facilitate the public's performance of these functions.*

1. Provide data and information that is essential for the assessment of impacts on the total human environment.
2. Help identify local citizens and groups having special expertise that might be utilized by the assessment team.
3. Identify local and regional environmental issues that should be addressed in the assessment process.
4. Provide historical perspective to current environmental conditions and trends in the local and regional area of proposed project development.
5. Help generate field data.
6. Help provide "field-truth" for key data generated during assessment.

7. Provide criteria for evaluating the significance of identified project impacts.
8. Identify project alternatives.
9. Suggest and help organize other forums and mechanisms for public participation in the assessment process.
10. Monitor the relevancy and adequacy of ongoing assessment efforts.
11. Review interim assessment reports and findings for public readability and relevance to local issues and concerns.
12. Help analyze and evaluate direct, indirect, and cumulative impacts of project development.
13. Help define the scope of work and schedule for overall assessment.
14. Provide liaison between assessment team members and key organizations and other public groups and individuals.
15. Review, comment, and make recommendations with respect to the qualifications and planned approach of consultants to be utilized by the assessment team.
16. Identify and evaluate potential mitigation and/or enhancement measures which might be incorporated in project design and/or management.

It is clearly unreasonable for the assessment team to presume that any one public group or organization can or would desire to undertake all of these 16 functions. Different groups and individuals in a local area may undertake one, several, or none of these functions, depending on individual interests, availability of time, knowledge, and experience. *It is, therefore, the responsibility of an assessment team that is committed to the enhancement of public participation in environmental decision-making to help diverse public groups and individuals identify assessment functions that each group and individual may profitably undertake.* This responsibility must be given the highest priority by the assessment team. In fact, actions undertaken in order to fulfill this responsibility should be among the very first actions initiated by the assessment team.

Techniques versus Objectives

Some of the basic techniques for involving the public in environmental decision-making were identified in Chapter 14. These techniques, and others, are discussed in a variety of publications, including those by the U.S. Department of Agriculture (1974), the Federal Highway Administration (1975), Walton (1971), Fielding (1972), and Schofer *et al.* (1975). It is evident that the number of publications on citizen-participation techniques is rapidly increasing, especially publications focusing on the needs of individual agencies (e.g., Federal Aviation Administration, 1978). As this type of information becomes increasingly available, it is reasonable to expect the development of agency guidelines for selecting and implementing specific techniques that are most appropriate for different types of public groups and different phases of project development.

An example of one such guide that is currently available to assessment teams concerned with proposed highway projects is included in Fig. 20.1. It will be noted that this figure offers specific recommendations on the type of technique that is most appropriate for (a) general type of public audience; (b) type of function to be performed; and (c) the timing of the function with respect to the different phases of project development. Such guides are extremely useful and should be seriously considered by the assessment team. However, it is necessary to point out that our current experience with public participation in environmental decision-making is extremely limited. Thus, *the assessment team would be wise to consider that such guides as are currently available are themselves limited—and that they therefore should not be used to restrict the team's efforts to experiment with and devise other and different approaches.*

Not only should assessment teams experiment with various techniques for achieving citizen participation in the assessment process, they should also experiment with different means of employing the same technique. For example, informal public meetings may be conducted in a variety of formats and using a variety of assessment personnel. Sometimes, personnel having a technical or engineering background may be effective group-leaders. Sometimes, personnel having very different backgrounds may be even more effective.

Assessment teams which have actually utilized techniques for involving the public in assessment functions have often tended to put too little effort in analyzing how the technique might be effectively implemented in a specific situation. Thus, the objectives and goals of citizen participation have often been sacrificed, if only because the team was more concerned with selecting the "right" technique than with applying that technique skillfully.

The following role-playing scenario is a good example of how a sound technique (i.e., informal public meetings) can be utilized in such a poor manner as to be actually counterproductive to the objectives and goals of the assessment team. This scenario, developed for the American Right of Way Association, Inc., has been enacted and evaluated by several hundred right-of-way specialists throughout the country. These professionals have consistently identified some issues which strongly influence the successful use of informal public meetings as a means of enhancing citizen involvement in environmental decision-making. These issues, which the

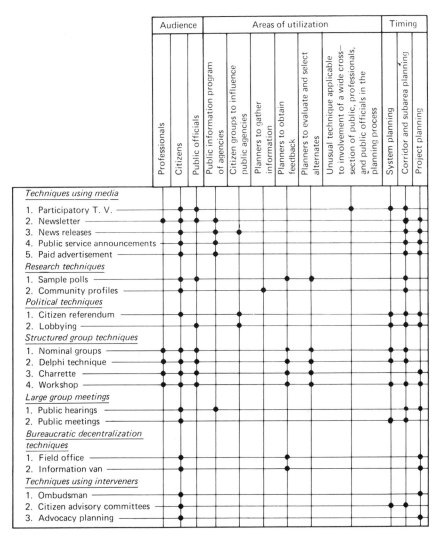

FIGURE 20.1 *Some applications of citizen involvement techniques.* [Adapted from Technical Council Committee, 1975.]

reader should keep in mind while reading through the following scenario, include:

1. Public perceptions of a group-leader with respect to his or her capability, credibility, and project responsibility
2. The adequacy of a group-leader's preparation for the public meeting
3. The group-leader's apparent sincerity (or insincerity) with respect to eliciting and using public inputs into the assessment process

4. Basic communication skills, and the apparent desire (or lack of desire) to communicate with the public

5. The group-leader's recognition (if any) of follow-up actions that must be undertaken in consequence of previous and ongoing citizen participation.

ROLE-PLAYING SCENARIO[1]

The following scenario comprises a dialogue between a landowner and an environmental group-leader. The dialogue takes place during an unofficial public hearing which the state has called for the purpose of increasing public participation in the EIS process. The hearing is being conducted in the form of small groups of local citizens (7–10 per group), each of which also includes members of the interdisciplinary team with EIS responsibility.

At the time of this hearing, the draft EIS has not been written. The public has, therefore, been told that their concerns, ideas, and information are to be considered inputs into the interdisciplinary team's efforts. This hearing is the third unofficial hearing held on the proposed new highway.

We are to imagine that the group convened about an hour ago, and several issues have already been discussed. The group-leader is now attempting to review these issues so that they might be more closely examined.

SCRIPT

Group-Leader: Let's see if we can list the ones we've already talked about. Then we can add to the list if we want, or go back and look at them individually.

Landowner: Mr. Group-leader! I've got a list. I'll read mine if you want.

Group-Leader: Go right ahead!

Landowner: Thank you. Number one: We're concerned that too much attention is being given to supposed positive effects of this highway on the large urban areas of the state, and too little to the negative impacts here in the local area of the proposed highway. Number two: We're concerned that you fellows are always talking about experts studying this and experts studying that, and that these so-called experts probably don't know beans about what we think is important. And Number three: We're concerned that you fellows are more interested in getting good PR with

---

[1] Adapted from American Right of Way Association (1976).

all these meetings and interviews you've been holding, than in getting our suggestions and acting on them. Now, Mr. Group-Leader, can we go back and (as you say) examine these concerns more closely? Or do you want some more?

Group-Leader: As far as I'm concerned, I think these will produce quite a bit of conversation. Personally, I'd like to start with Number three.

Landowner: That's okay with us. Go ahead.

Group-Leader: Well, first of all, as you all know, there is no legal requirement that we hold these meetings. They're not required by law. Second, I think you can appreciate the cost in money and man-hours that we've put into these meetings over the past several months. I think these two points are important to show that we're serious about using the inputs we can get from you people.

Landowner: Excuse me, Mr. Group-Leader. It's all very noble that you're doing something you don't have to do, and spending a heck of a lot of public funds doing it! But that's not the question. We want some real, concrete examples of how our inputs are being used.

Group-Leader: I can promise you that when you examine our draft EIS you will find that all of the issues you people have identified will be addressed.

Landowner: Can't you give us an example now?

Group-Leader: Well, take the issue of noise. Some of you were quite concerned about the noise from heavy vehicles on the steeper grades. We've got people looking at various designs that can reduce that noise.

Landowner: When we first discussed that issue (I think it was probably at the first of these meetings, several months ago) we were a little more specific. If my memory is right, we wanted to know how you were handling noise with respect to several different things—people living in the area, public meetings, outings, and other social gatherings, and also wildlife. Now, you have people working on each of these? Is that correct?

Group-Leader: Yes. But you see, this is a difficult area. In some cases there are specific standards that we can go by. But in other cases, like in wildlife, there aren't. But we are trying to pull it together.

Landowner: You're using experts in noise, then?

Group-Leader: Yes. Measurements have been made of the existing areas, and we're making projections of these levels to the time when the highway will be operational so that we can see the difference.

Landowner: No one's been around asking me what I thought of one noise level or the other. Maybe I'm an exception?

Group-Leader: No. I don't think the use of interviews has been considered as part of the noise studies.

Landowner: Why? Aren't my ears expert enough?

Group-Leader: Well, I can't answer that. I don't know why. I do know it's a very technical subject.

Landowner: Look. Let's leave my ears out of it for a minute. A lot of us here are hunters and have been out in those woods around the project area for a long time. Don't you think we might know something about how different noises affect different kinds of animals?

Group-Leader: I would imagine that you do. In fact, we're also gathering information on the kinds of animals in the area. Once we know that, we can examine the effects of noise on them.

Landowner: Cripes! We already know what animals are out there. Want a list now?

Group-Leader: I didn't mean to say that you didn't know this. In fact, I know our people can use that list of yours. Get it to me and I'll pass it on to the right people.

Landowner: Well, you see, that's just the point here. Three months ago we talked about noise—and you just brought this up as an example of how you are using our inputs. But now, 3 months later, and only by accident, you suggest we send a list in of the animals in the area that might be affected. Now I don't mean to pick on you personally. But, if you fellows were really serious about this public involvement stuff, why haven't you come around earlier for that list? Well, that's a rhetorical question. And I don't expect even you know why it hasn't been done. But let me jump on this "right people" thing. It's related. I suppose you mean the experts?

Group-Leader: Specifically, I meant the members of the EIS team who have responsibility for coordinating that kind of information. In turn, they would feed it to the wildlife experts we're using.

Landowner: Well, okay. That's fine. And I don't doubt that the experts know what they're talking about. But how do they know what we're most concerned about? I hunt deer. Not rabbits. How can we be sure they're studying things that affect us? You see, if they're not, why should we spend the effort to give them information?

Group-Leader: I want to emphasize here, because I think it pretty important relative to what we're talking about, that you—each and every one of you—will have an opportunity to read and evaluate and criticize the draft report. Now, if we haven't focused on the issues that you're concerned about, or if you think we haven't done it well, then you point to specific instances and demand that they be addressed.

Landowner: Frankly, we think that by the time that draft report is written, it's already too late.

Group-Leader: It's not. By no means.

Landowner: I'd like to believe you. I say, I'd like to. But now look at the facts. All the speeches and little talks and even the newspaper articles that have dealt with this new highway all have one thing in common. They all emphasize the necessity of this highway in terms of the economy of this region of the state. Now, the effect of this is that every time we raise an issue that affects us directly and which leads us to think they can damn well put this highway somewhere else, we come out looking like narrow-minded, selfish SOBs. Personally, I stand to make some money from the deal. They have to take a pretty good strip from me that I'm not using anyway. Maybe in the end that will decide me on it. But meanwhile, youngster, I know a snow job when I see it.

Some Key Considerations

In selecting and implementing any technique for involving the public in the assessment process, the assessment team must assure itself that adequate consideration has been given to such questions as:

1. What specific assessment objectives (including objectives of both analytical and integrative tasks) can be achieved by the proposed technique?
2. What are key criteria (e.g., physical setting, timing, nature of public groups) for the successful utilization of the proposed technique?
3. What follow-up actions and related budgetary, manpower, and informational resources will be required if the proposed technique is implemented?
4. How can the implementation of the proposed technique be monitored in order to ensure the timely correction of any counterproductive conditions and tendencies?
5. How can local conditions (e.g., attitudes, previous experience with public participation measures) influence the successful utilization of the proposed technique?
6. What criteria can be used to ensure the most appropriate assignment of team personnel with respect to the successful utilization of the proposed technique?

As evidenced by the role-playing scenario, the appropriateness of a particular type of citizen participation program, and the particular skills required for the successful implementation of that program, are largely determined by local conditions, including local attitudes and concerns, as well as by the previous experience of citizens with public participation programs. Thus, any recommendations for utilizing a particular technique for involving public groups and individuals in the assessment process must be made in light of the team's knowledge of these local conditions. As a general rule, the potential for initiating counterproductive programs of public involvement is greatly increased whenever assessment teams fail to define site-specific factors. In short, *any failure of a public participation program must be attributed to the shortcomings of the assessment team, and not to the public.* Any other approach will invite an infinite number of excuses as to why the public should be excluded from environmental decision-making.

It is also important for the assessment team to understand that the use of any technique for enhancing public participation is essentially an exercise in manipulating the public. While such "manipulation" is a function of the assessment team that is necessitated by the practical analytical and integrative tasks of assessment, "manipulation" also has understandably ugly connotations to the public. The assessment team may, therefore, well consider the usefulness of seeking the advice of community groups and local citizens with respect to the appropriateness of various techniques that might subsequently be utilized for enhancing citizen participation. As with all issues pertaining to the public's involvement in environmental decision-making, the only rational approach for the assessment team is to be direct and honest in its communication with the public.

Finally, the attitudes of members of the assessment team about public involvement cannot be overemphasized. Team attitudes are key factors in implementing a successful public participation program. The idea that the assessment team is doing the public a favor by dealing with the public at all, or the idea that the assessment team is more expert, more knowledgeable, and more interested in the total environmental ramifications of project development than is the public itself, has no place in impact assessment under NEPA. Any public which rightly perceives these attitudes among assessment team members is fully justified in assuming that the public is being used, rather than served. In such a circumstance, who would reasonably expect the public to do anything else but to turn to the courts?

References

American Right of Way Asociation, Inc. (ARWA)
    1976   *Student workbook, right of way and the environment.* Los Angeles: American Right of Way Association.
Council on Environmental Quality (CEQ)
    1971   *The second annual report of the Council on Environmental Quality.* Washington, D.C.: U.S. Government Printing Office.

Federal Aviation Administration
  1975 *Instructions for processing airport development actions affecting the environment.* (Order 5050.2A). Washington, D.C.: U.S. Department of Transportation, Federal Aviation Administration.
  1976 Policies and procedures for considering environmental impacts. *Federal Register,* *41*(157, Part II).
  1978 Request for proposal: Development of workshops and manual for community participation in aviation Environmental Actions. (RFP No. LGR-8-3680).
Federal Highway Administration
  1975 *Social and economic effects of highways.* Washington, D.C.: U.S. Department of Transportation, Federal Highway Administration.
  1976 *An introduction to community involvement in the transportation planning process.* Washington, D.C.: U.S. Department of Transportation, Federal Highway Administration.
Fielding, G. J.
  1972 Structuring citizen involvement in freeway planning. *Highway Research Record,* Number 380. Washington, D.C.: Highway Research Board
Schofer, J. L., et al.
  1975 *The judgmental impact matrix approach: A framework for evaluating the social and environmental impact of transportation alternatives.* Washington, D.C.: Technical Analysis Division, National Bureau of Standards.
Technical Council Committee 6-Y-7
  1975 Methods for citizen involvement. *Traffic Engineering,* August 1975, p. 12.
U.S. Department of Agriculture
  1974 *Guide to public involvement in decision making.* Washington, D.C.: U.S. Department of Agriculture, Forest Service.
Walton, L. E., Jr.
  1971 *What Virginia learned about public hearings.* Charlottesville, Virginia: Virginia Highway Research Council.

21

The Assessment Report

Introduction

There can be little doubt that, in the first few years after NEPA, the primary objective of the assessment process was perceived to be the production of a document called the *environmental impact statement* (EIS). The EISs were produced in all colors, sizes and thicknesses. Editorial formats also varied, depending of the understanding and/or misunderstanding of assessment teams with respect to their own reading of Section 102 (2) (C) of NEPA. Many of the early EISs were blatantly superficial public relations documents which sought to justify the continued development of projects already far advanced at the time of NEPA's enactment. As the courts began to define the legal inadequacies of such documents, assessment teams sought to regain credibility by producing massive compendia of undigested data and engineering details. While early CEQ guidelines sought to introduce some rational order into the "war-of-paper" that was rapidly developing between EIS writers and EIS reviewers (including the public and the courts), it was obvious that trial and error was to be a highly instructive method for determining the operational adequacy of EISs. Simply, EIS, were sufficient if they kept the project out of court and on schedule. They were clearly inadequate if they resulted in litigation or otherwise led to project delay.

A perception of the assessment process as a bureaucratic "shuffling of papers" that is largely irrelevant to the national commitment to environmental quality still persists. For example:

> The tragedy of NEPA is that it turned energy, attention, and effort away from a redefinition of agency authority and spent it on proliferating paper. It truncated discussion of environmental protection in terms of authorizing statutes which define the existence and mission of executive agencies, and it directed attention to the preparation and filing of reports [Fairfax, 1978].

Insofar as assessment teams, project managers, and decision-makers (including the public) actually view the EIS as the primary means for achieving compliance with NEPA, then indeed the entire assessment effort is nothing more than a public session of "bureaucratic origami."

However, the point is that *the assessment report is not now—nor was it ever—the primary means for achieving compliance with the total environmental objectives of NEPA.* In fact, in a competently designed and executed assessment, the EIS can be of relatively minor importance. This is not to say that the EIS does not perform key functions in the assessment process; but rather, that whatever function the EIS performs must be secondary to the basic goal of the assessment process—which is the improvement of decisions, not the writing or the filing of reports.

Form and Function

For purposes of this discussion, the term "assessment report" will be used synonymously with (a) draft and final EISs that are produced in compliance with Section 102 (2) (C) of NEPA and relevant CEQ guidelines; and (b) environmental reports that are typically defined by agency guidelines and regulations pertaining to such documents as "negative declarations," or documents other than the formal draft and final EIS. Regardless of the type of assessment report, *it is necessary for the assessment team to distinguish between the form of the report and the function of the report.* If this is not done, assessment teams are likely to spend their major effort putting things onto paper and minimum effort putting environmental considerations into actual project development.

Some key criteria that have been consistently used for evaluating the adequacy of assessment reports have been discussed in Chapter 2. It is important to point out that such criteria are based on CEQ and federal agency guidelines and regulations, and on court decisions (e.g., CEQ, 1973; Anderson, 1973; Mallory, 1976). *These criteria are therefore subject to change, and assessment teams must be continually aware of legislative, regulatory, and juridical developments that can influence the measures of adequacy that will be applied to any assessment report.*

Numerous publications are available to assessment teams for this purpose, including the *Federal Register* and the *102 Monitor,* which are available from the U.S. Government Printing Office. Computerized abstracts of a large number of publications on the preparation and evaluation of EISs are also available through the National Technical Information Service (1978). Each of these publications identifies and discusses various guidelines and requirements for the editorial format and substantive content of environmental reports. General and specific suggestions for writing EISs are also available in such texts as those by Canter (1977) and Jain *et al.* (1977). In fact, the literature pertaining to the writing of EISs is so extensive that assessment teams could well devote so much effort to designing the editorial format of their reports that little time would be left for anything else.

While there has long been evidence that the EIS was never intended as the primary means of achieving NEPA goals and objectives, or as a paper substitute for real actions undertaken to achieve those goals, it is only recently that this key point has been given strong executive emphasis. For example, Executive Order 11991 states:

> [The CEQ shall] issue regulations to Federal agencies. . . . Such regulations . . . will be designed to make the environmental impact statement process more useful to decisionmakers and the public; and to reduce paperwork, and the accumulation of extraneous background data, in order to emphasize the need to focus on real environmental issues and alternatives. They will require impact statements to be concise, clear, and to the point, and supported by evidence that agencies have made the necessary environmental analysis [The White House, 1977].

Members of the CEQ also emphasize the functional utility of the EIS:

> [There is universal agreement that] statements should be shorter, less technical, easier to read. They should deemphasize background material—the so-called "dandelion counts"—and emphasize the alternative choices for action facing decisionmakers and the public. . . . Moreover, as we try to improve EIS procedures, we shall keep in mind that such statements are only a means to an end. . . . Better actions, not better statements. That I think is a useful summary not only of CEQ's role in NEPA reform, but of the purpose of your [i.e., environmental professionals] work [Speth, 1977].

Also

> EISs tend to be extremely bulky, to be burdened with masses of detail that obscure rather than facilitate analyses. A fat volume or series of volumes is apt to serve nobody's interest. The decision-maker never reads it, which hardly accomplishes NEPA's goal. The business man or woman complains that it takes too long to prepare. The environmentalist asserts that agencies use the bulk to hide the critical issues. EISs should be concise, to the point, and concentrate on the real alternative choices that confront the decision-maker and the public. . . . It was and is the goal of the Council to change this emphasis on paperwork to an emphasis on

substance—better environmental decision-making which is the purpose of the Act [i.e., NEPA] [Warren, 1977].

Of course, it should be pointed out that CEQ has declared that EISs are only means to an end before. For example:

> Properly conceived and written, the EIS is an extremely useful management tool. But too many statements have been deadly, voluminous, and obscure, and lacked the necessary analysis and synthesis. They have often been inordinately long, with too much space devoted to unnecessary description rather than to analysis of impacts and alternatives. CEQ guidelines state that the "agencies should make every effort to convey the required information succinctly in a form easily understood, both by members of the public and by public decisionmakers, giving attention to the substance of the information conveyed rather than to the particular form or length of the statement" [CEQ, 1975, pp. 632–633].

In an effort to achieve the functional goals and objectives of NEPA with respect to environmental decision-making and to minimize bureaucratic preoccupation with the mere paper products of assessments, the CEQ undertook a series of hearings on NEPA reform in 1977. Regulations which emanate from such hearings are, of course, continually subject to revisions that may be made in the future. Yet, for the reasonably foreseeable future, it is likely that the concerns expressed during the 1977 hearings will strongly influence the writing of environmental reports under NEPA. These concerns, translated into proposed mechanisms for dealing with them, are summarized in the following excerpt from the congressional testimony of the chairman of the Council on Environmental Quality [Warren, 1977].

Excerpt from Congressional Testimony of Charles Warren, Chairman, Council on Environmental Quality

> There are several areas in which we [i.e., the CEQ] expect to act, which may be summarized under the general headings of (1) reducing length, (2) reducing delay, and (3) facilitating better decisions. Let me take them one by one.
>
> 1. Reducing length. The multivolume EIS should be reduced in size.
> a. First we expect to greatly reduce the lengthy discussion of the existing setting—the bulky "dandelion counts" that serve neither the decision maker nor the public. We must make clear that such lengthy catalogues do not add up to an adequate impact statement. Concise analysis is what is useful.
> b. We hope to eliminate repetition. There is an unfortunate tendency slavishly to adhere to the five subsections of NEPA rather than organizing the discussion to serve the decisionmaker. As a result the same fact can be repeated five times in five separate discussions. Once is enough.
> c. Universal page limits appear too inflexible to prescribe. The same law, we

must remember, covers a bank building in Woodstock, Vermont, and a Trans-Alaska pipeline. However, a mechanism should exist for setting page limits in individual cases.

2. Reducing delay

a. We should get everybody involved in the NEPA process early. If a project manager ignores the environment in his planning and adds it on as an afterthought, NEPA will be an add-on and will cause delay. If, however, the environmental planning takes place at the same time as other planning, NEPA should cause no delay, and the environment will become a meaningful part of decision-making.

b. Often several Federal agencies are involved, and much time is taken deciding who is to take the lead in preparing the EIS. We intend to prescribe a prompt and fair means of lead agency designation.

c. Scoping. We need a means to settle on the scope of an EIS before it is prepared. We intend to propose a scoping process to involve all those interested in a project. This should lead to early decisions about what areas need emphasis.

d. Replacing an adversarial emphasis with a cooperative emphasis. We hope agencies will cooperate in the preparation of an EIS rather than holding their fire and letting go with adverse comments on somebody else's completed product.

e. Elimination of duplication. This has two aspects:

(i) It is possible to eliminate duplication between Federal agencies by providing that one adopt the environmental documents prepared by another.

(ii) It is possible to eliminate duplication between Federal and State and local requirements by providing for joint statements.

f. Time limits. While universal time limits, like page limits, appear needlessly inflexible, we expect that a scoping meeting could serve as the appropriate occasion for setting time limits specific to the particular action.

g. Tiering. We expect to encourage the process of tiering, whereby general questions are dealt with in a program EIS covering the entire program and need not be repeated in site specific EISs, which may then be considerably briefer documents which concentrate on the issues peculiar to the specific site.

3. Ultimately, of course, it is not better documents but better decisions that count. The Council intends to take several steps to facilitate better decisions.

a. First, we intend to emphasize all of NEPA—not just Section 102 (2) (C), the subsection that deals with EISs. There has been so much attention to EISs in recent years that NEPA seems to have acquired the reputation of the law that establishes environmental impact statements, rather than the law that establishes a National Environmental Policy with the EISs only the most conspicuous of several action-forcing devices to insure implementation of that policy. We hope to restore that original intent—so that the procedures of Section 102 (2) (C) are means to achieve the requirements of Sections 101 and 102 (1). We expect to emphasize the actual decision and its relation to the purpose of the Act.

b. We want to insure that the decision is based on impartial analysis. We expect to place restrictions on preparation of EISs by applicants who have a conflict of interest in their preparation.

c. We intend to emphasize follow up. If the NEPA process results in some worthy mitigative measures, we should like to insure that those measures are in fact incorporated and carried out.

d. Finally, Congress has provided for the referral to the Council of actions found by the Administrator of EPA to be environmentally unsatisfactory pursuant to Section 309 of the Clean Air Act. Since enactment of the provision in 1970, the previous Councils never adopted procedures governing such referrals. We have now done so. Our interim guidance on the subject was sent to the heads of all Agencies in August, and it is our intention to make formal that guidance in our regulations.

This is not all we have in mind. Replacing the 1973 Guidelines with 1977 Regulations has been a long and arduous task into which we have placed great effort. We trust the result will better implement the Act and better serve the public.

Some Additional Considerations

Contemporary concern for past preoccupation with a mere paper compliance with NEPA should not be interpreted as de facto evidence of the irrelevance of the EIS or other environmental documents to national environmental goals and objectives. The point is, these documents have not been as useful as they might have been—not that they should be thrown out of the assessment process altogether.

Some of the key functions of these documents (and functions which are likely to be given increased attention in the future) include:

1. They are basic public and legal evidence that project planning has been undertaken in light of national environmental goals and objectives.
2. They are an important means for eliciting information and comments from reviewing agencies, organizations, and individuals, and for otherwise involving the public in environmental analysis, evaluation, and decision-making.
3. They provide a basis for public review, consideration, and deliberation of long-term trends and tendencies in the environmental decision-making and planning of governmental agencies.
4. They provide a written record of planning considerations and priorities for proposed projects which can be compared and contrasted with the actual in-place design and operational details of completed projects.
5. They are an important means for enhancing the environmental knowledge and understanding of the general public, and for enhancing agency knowledge and understanding of public needs, concerns, and aspirations.

While it is tempting to assume that the broad environmental goals and objectives of NEPA can be achieved through an unpoliced, cooperative interaction among governmental agencies, social institutions, and public groups and individuals, it is essential to remind ourselves that *NEPA was enacted precisely because of the perceived failure of government agencies to consider adequately the total environmental consequences of their proposed programs and projects.* It can, therefore, be suggested that whatever shortcomings and misappropriations of effort and concern may in fact be associated with environmental documents, such documents are nonetheless essential for providing public access to ongoing governmental decision-making processes which affect the total human environment—either through the courts or through other democratic processes.

A simple rule of thumb of practical politics is: If the bureaucrats have to put "it" on paper, then "it" and the paper can usually be used to get to the bureaucrats. A converse of this rule is that the potential for an independent and self-serving bureaucracy is increased whenever bureaucrats do not have to commit themselves in writing. Thus, the functions of assessment reports identified above are important not only because of their relevance to the quality and efficiency of the environmental assessment of a particular project, but also for their general relevance to such fundamental issues as (a) citizens' rights to the legal and political redress of grievances and injury; (b) executive and other governmental complicance with public laws; and (c) public accountability of public servants.

In light of these considerations, it is important for assessment teams and government agencies to ensure that any streamlining of assessment reports that is expressly undertaken for the convenience of project decision-makers not be accomplished to the serious detriment of the rights and interests of those other key decision-makers, the public. This is a critical point, especially in view of the fact that, for proper environmental planning to take place at all, *project decision-makers should be well informed of the results of assessment as they are obtained and prior to the writing of even a draft EIS.* The public should not presume that project decision-makers having environmental responsibility are only to be called into action just at the time that the draft EIS becomes available. It is more realistic to presume (a) that even prior to the submission of a draft EIS, many design and location decisions have already been made and have already been influenced by the assessment process; and (b) that these decisions are the very basis for considering the alternative actions identified in the draft EISs as reasonable and serious alternatives (Peterson, 1976). If this is not the case, and if project decision-makers have their first introduction to the environmental consequences of their projects only in the draft EIS, then it is fair to question how any amount of streamlining of the draft EIS can reasonably enhance serious environmental planning of projects and project alternatives.

Of course, the streamlining of environmental documents so as to preserve their utility for meeting public interests as well as the needs of project decision-makers is an editorial and writing task that calls for uncommon skills. It is a serious mistake to presume that environmental expertise is the primary and requisite expertise for writing EISs clearly and concisely. In fact, it requires only a cursory reading of completed EISs to reach the conclusion that probably the easiest way to enhance the practical utility of assessment reports is to take writing chores away from illiterate engineers, scientists, and the other technical personnel to whom they are often entrusted, and entrust them instead to people who know how to communicate on paper, and who also enjoy doing it. All the guidelines for streamlining environmental documents are likely to accomplish precisely nothing if we persist in the idiotic notion that everyone can write effectively.

The Question of "Significance"

If impact assessment reports are to focus the attention of decision-makers on "real environmental issues and alternatives" (as required by Executive Order 11991), and if decision-makers are to use environmental findings in the earliest planning phases of project development (even before the writing of the EIS), it is reasonable to assume that the question of just what constitutes a significant impact has been raised and clearly answered in the progress of interdisciplinary deliberation of impacts. As discussed in previous chapters on the physical, the social, and the total human environments, there is no single, objective, or consensual measure of the significance of impacts. Thus, the assumption that the question of significance has both been raised and clearly answered infers that a variety of criteria of significance have in fact been identified and utilized for making decisions in the earliest planning phases of project development.

There can be no more critical question in the entire environmental assessment process that this issue—*just which criteria were utilized for evaluating the significance of just what impacts?* Certainly decision-makers will ignore so-called "insignificant" impacts for purposes of early project planning. Consequently, *if the public is actually to evaluate the reasonableness of proposed project alternatives, the public must understand how the significance of some impacts and the insignificance of others were weighed and balanced in the determination of real alternatives.* In order to ensure meaningful public review of environmental decision-making, assessment teams would be well advised to adopt the maxim: That it is far more important for the public to be given concise information on how impacts were actually evaluated, prioritized, or "weighted" one against the other than it is for the public to be given raw, undigested technical and social data.

The draft EIS is a key mechanism for providing this information to the public. Other mechanisms may also be utilized, including interim assessment reports, newsletters, and other special-purpose documents which may be developed during the progress of the actual assessment effort—especially those documents that are directly related to a public participation program.

The public disclosure of just how potential impacts have been evaluated in early project planning is a task that cannot be lightly undertaken. For, in order to perform this service to public decision-makers, the assessment team and project decision-makers must:

1. Clearly identify individual evaluation criteria (including such criteria as the probability, duration, and magnitude of possible impacts, their connectedness, potential for influencing public health, economic, and other social parameters, and potential for amelioration and reversibility).
2. Clearly identify how these criteria can be individually and collectively applied to each identified impact (e.g., are the criteria assigned equal or different "weight," and can the weighting of any criterion be varied, depending on the type of impact considered?).

There can be no doubt that, in performing these tasks, *assessment teams and project decision-makers must make judgments—they cannot expect to rely totally on clear legislative directives, objective standards, or magic formulas when they say: "This impact is important, and we're going to consider it seriously in project planning," or "this impact is unimportant, and we're going to ignore it in project planning."*

Decision-makers who are primarily interested in hiding their personal responsibility for early project design in the anonymity of bureaucracy are understandably cautious about making public judgments that cannot be easily and objectively verified. Such judgments tend to "make waves" which can disrupt the business-as-usual routine of any bureaucracy. However, as understandable as their caution may be, it has no place in decision-making under NEPA. The simple facts are that:

- There are many decisions that are made prior to the writing of environmental reports, such as negative declarations and the draft EIS.
- These decisions actually result in further consideration being given to some alternative actions and designs (e.g., the proposed alternative actions in a draft EIS) and no consideration being given to other alternative actions and designs.
- Environmental considerations that influence these decisions typically involve human judgment of the significance of possible impacts.
- The public appraisal of—and contribution to— environmental decision-making must be predicated on the public's awareness and review of these judgments.

In light of these considerations, the important goal of environmental reports is not to make life miserable for decision-makers by publicizing the subjective judgments they must make; but rather *to test the adequacy of those judgments with respect to national goals and objectives with respect to the total human environment.*

Decision-makers who are unwilling to seek the advice of the public and other agencies in achieving these goals and objectives, or who are emotionally or intellectually incapable of thinking that their own understanding of the total human environment might be incomplete, shortsighted, or even prejudiced, have no business making decisions that affect the total human environment. In effect, NEPA has declared that such decision-makers are "dinosaurs," and that they should be consigned once and for all to the past.

References

Anderson, F. R.
1973 *NEPA in the courts: A legal analysis of the National Environmental Policy Act.* Resources for the Future, Baltimore, Maryland: Johns Hopkins Univ. Press.
Canter, L. W.
1977 *Environmental Impact Assessment.* New York: McGraw-Hill.
Council on Environmental Quality (CEQ)
1973 Preparation of Environmental Impact Statements, guidelines. *Federal Register, 38* (147, Part II). Washington, D. C.: U. S. Government Printing Office.
1975 *Environmental quality: The sixth annual report of the Council on Environmental Quality.* Washington, D. C.: U. S. Government Printing Office.
Fairfax, S. K.
1978 A disaster in the environmental movement. *Science, 199*: 743–748.
Jain, R. K., L. V. Urban, and G. S. Stacey.
1977 *Environmental impact analysis: A new dimension in decision making.* New York: Van Nostrand Reinhold.
Mallory, R. S.
1976 *The legally required contents of a NEPA Environmental Impact Statement.* Stanford, California: Stanford Environmental Law Society, Stanford Law School, Stanford Univ.
National Technical Information Service (NTIS).
1978 *Current published searches from the NTIS bibliographic data file* (NTIS Search No. PS-76/0898/7PMB, Preparation and Evaluation of Environmental Impact Statements, 1970-Nov. 1976). Springfield, Virginia: National Technical Information Service, U. S. Department of Commerce.
Peterson, R. W.
1976 Memorandum on "Kleppe V. Sierra Club" and "Flint Ridge Development Co. V. Scenic Rivers Association of Oklahoma." *Federal Register, 42* (231). Washington, D. C.: U.S. Government Printing Office.
Speth, G.
1977 Remarks before the National Association of Environmental Professionals, Dirksen Senate Office Building, Washington, D. C., Thursday, June 30, 1977. *102 Monitor,* Vol. 7, No. 6. Washington, D. C.: U. S. Government Printing Office.

Warren, C.
    1977   Statement of Charles Warren, Chairman, Council on Environmental Quality, before the Subcommittee on Fisheries and Wildlife Conservation and the Environment of the House Committee on Merchant Marine and Fisheries. *102 Monitor. 7* (9). Washington, D. C.: U. S. Government Printing Office.
The White House.
    1977   Executive Order 11991, Relating to protection and enhancement of environmental quality, May 24, 1977.

22

The Review Process

Introduction

Assessment teams have typically viewed the review process under NEPA as a gauntlet they must run, rather than as a mechanism that can facilitate the comprehensive evaluation of environmental impacts of proposed projects and programs. Unfortunately, this perception of the review process is largely based on fact.

Agencies that review the EISs of other agencies often appear to have used NEPA as an opportunity for taking potshots at the lead agency, rather than as an opportunity for offering constructive advice and criticisms. Also, agencies have sometimes given the responsibility for reviewing EISs to individuals who, by training and experience, are precisely the least qualified to pass judgment on any type of professional study. Sometimes the responsibility has been given to individuals who are so highly trained in one aspect of the environment that they cannot imagine that impacts on anything outside of their own narrow areas of expertise are worth considering. The result has been the proliferation of written review comments that are largely irrelevant to the serious assessment of project impacts on the total human environment. In fact, if typical review comments have done anything to affect assessment, it is more likely to have been to increase the waste of time and money spent in pursuit of irrelevant issues for irrelevant reasons than to increase either the quality of assessment or the quality of subsequent decisions.

Review of EISs by the public has likewise reinforced assessment teams' perceptions of the review process as a masochistic exercise. However, unlike the failures of agency review, the failures of public review of EISs are typically exacerbated, if not generated, by the failure of assessment teams to accept the public as a necessary and desirable component of the decision-making process (Chapter 20).

As the national experience with NEPA increases, and as guidelines for agency review of EISs and for public participation in environmental decision-making are tested, tried, and improved, it is likely that assessment teams will change current perceptions of the review process. It is highly desirable that the first of these changes will be with respect to the scope of the review process under NEPA.

Scope of Review Process

The review process under NEPA is typically initiated at the time a draft EIS is submitted to the public and to interested federal, state, and local governmental agencies for the purpose of soliciting their comments and recommendations. Such comments and recommendations, which may be in written or oral form, must then be considered by the assessment team. Upon appropriate consideration of these comments and recommendations, the draft EIS (dEIS) is revised and resubmitted in the form of a final EIS (fEIS). There are several important consequences to the formal review process, which is conducted in such a manner, including:

1. The initiation of public and agency review only after the preparation of a dEIS typically means that the public and agencies first become aware of the proposed action and its consequences only after the major assessment effort has been completed.
2. Because the dEIS represents a substantial investment of time and effort by the lead agency, the lead agency tends to take a defensive posture with respect to comments and criticisms that are generated through the review process.
3. The public, if not some agencies, often perceive the formal dEIS as a fait accompli and are often sufficiently motivated by this perception to assume an adversary role in the review process, regardless of the merits of the dEIS.
4. Since there is always the possibility that the review process will result in the input of data and information that have not been previously examined by the assessment team, there is always the possibility that the review process will result in a major reevalua-

tion of project impacts, and a consequent increase in time and budgetary requirements for assessment.

There can be no doubt that any review process should be a quality-control mechanism, or that an effective quality-control mechanism should result in a "redoing" of an assessment when an assessment is deficient, regardless of the costs in time and money. Yet, it can be suggested that a *review process can be even a more effective quality-control mechanism if it is utilized prior to the development of the dEIS.* Such an approach to the review process under NEPA would also serve to reduce the tendency for those defensive and offensive posturings of agencies and public groups that are actually counterproductive to the goals and objectives of impact assessment.

In the following sections we will focus on how the process of assessment review can be utilized by assessment teams in order both to enhance the quality of assessment and to decrease the potential for counterproductive adversarial confrontation. For purposes of this discussion it is useful to subdivide the overall review process into: (a) in-house team review; (b) agency review; (c) public review; and (d) court review.

In-House Team Review

It is my opinion that assessment teams often devote most of their attention to the performance of analytical and integrative tasks of assessment and too little attention to the quality control of analytical and integrative efforts. Of course, quality-control functions are typically associated with project managers, and not with each member of a team. Yet, the interdisciplinary assessment mandated by NEPA requires the integration of such diverse and specialized considerations that a single individual can exercise hardly more than rudimentary quality control. Realistically, quality control in the progress of impact assessment is often manifested in a concern for schedules, costs, and procedural compliance with assessment guidelines and report outlines.

It should therefore be emphasized that:

1. The interdisciplinary team is ideally suited for conducting a continual and adversarial review of its own progress.
2. Such a review, conducted in-house, can be the most effective quality-control mechanism in the entire assessment process.

Of course, there can hardly be constructive, adversarial review within the in-house team if team members are constrained by a bureaucratic structure which gives automatic and unquestioned deference to one expert or another, depending on job title, seniority, or other measures of rank. As pointed out in Chapter 19, a truly interdisciplinary team that is committed to a serious consideration of project impacts on the total human environment must be free to question the relevance of any "conventional wisdom." Thus, in an interdisciplinary deliberation of impacts on the total human environment, each individual must be able to defend the relevance of his work to the goals and objectives of NEPA against the criticism of all other members of the team. *If each individual is not required to defend his work successfully within the forum provided by the assessment team, then the probability that he will fail to defend his work within the public forum provided by NEPA is enhanced.*

Project managers might indeed recoil from the suggestion that assessment personnel must be "let loose to tear each other apart." It would appear to many managers that such an approach is in direct opposition to the managerial sense of team spirit and cooperative effort. This is nonsense. If assessment team members assume adversarial roles with one another for the purpose of improving the quality of the team's assessment, and do so only at the risk of destroying the team's capacity to function as a team, then they are clearly not capable of acting as professionals and should be relieved of all assessment responsibility.

Agency Review

There is nothing in NEPA that prevents a governmental agency that is considering a project from consulting with other agencies, including federal, state, and local agencies, *prior to the preparation of a dEIS.* In fact, some legislation (e.g., The Fish and Wildlife Coordination Act) actually requires interagency consultation throughout project development and regardless of the project's status with respect to NEPA.

The importance of early liaison between governmental agencies cannot be overemphasized. Not only does early liaison (i.e., liaison before the preparation of a dEIS) enhance the potential for coordinated federal projects affecting the environment (a key consideration in NEPA), but it also enhances the potential for identifying alternative actions which may be undertaken to further (a) the broad national environmental goals of NEPA; and (b) the goals and objectives of other federal, state, and local environmental legislation.

The lack of early liaison between an agency that is considering project development and other agencies that have jurisdiction and/or interest in areas relating to project development is likely to lead to one or more of the following situations:

1. The range of alternative actions will be unduly restricted.
2. Key environmental data and information that may already be

available in one agency will be overlooked and/or duplicated by another agency.

3. The assessment of cumulative impacts will be incomplete due to the failure to learn of projects and programs that are under consideration by other agencies.

4. Assessments completed by other agencies for similar projects or for projects in the same or similar local areas will be overlooked.

5. The potential for cooperative actions among different agencies with respect to collecting base-line data, monitoring impacts, and designing and implementing ameliorative measures will be minimized.

In light of these considerations, it should be emphasized that all federal agencies must act so as to accomplish the goals and objectives of NEPA. Assessment teams should, therefore, assume that all federal agencies will contribute as best they can to the sound environmental design and operation of projects undertaken by any other agency. Of course, before they can do this, the assessment team must request their contributions. The simple fact is that to limit such a request to the dEIS—which, after all, is a document prepared only after some alternative actions have been effectively ruled out, and only after basic planning and design have been accomplished, and only after a major assessment effort has been completed—is to appear to request their consent rather than their advice!

Public Review

The role of the public in impact assessment has been discussed in Chapter 20. In this chapter it is only necessary to suggest some general considerations and guidelines for maximizing the constructive participation of the public in the review process. These considerations and guidelines are predicated by a general concern over both the underuse and misuse of the public as a partner in the environmental decision-making of governmental agencies.

1. As in the case of agency review, public review of assessment efforts should begin in the earliest phase of project planning and continue throughout project development, regardless of the specific provisions in NEPA for public hearings. In short, NEPA's stipulations with respect to public hearings do not preclude other public review of assessment efforts. Also, the probability that formal public hearings will yield constructive contributions to the assessment effort is enhanced by a public participation program that extends from the earliest through the latest phases of project development.

2. There is nothing in NEPA that precludes the use of public groups and individuals as members of (or as adjuncts to) the actual assessment team. Either as members of or as adjuncts to assessment teams, the public can play a key role in the quality control of the various analytical and integrative tasks of assessment. Especially important are (a) the public's efforts to ensure the relevance of the assessment to actual conditions and concerns in local areas; and (b) the readability of informal and formal environmental reports prepared by the assessment team.

3. There is nothing in NEPA that precludes the use of public groups and individuals in a liaison capacity between assessment teams and governmental and other agencies and organizations. This approach has the important advantage of integrating individual in-house, agency, and public review mechanisms into a comprehensive review program. Also, public organizations and individuals often have more direct experience with some governmental agencies than do members of an assessment team.

Court Review

If assessment teams tend to view the agency and public review process as being definitely masochistic, they must certainly perceive court review as something bordering on the sadistic. In many assessment teams, "the judge" has become an appellation of "blatant ignorance and rampant presumptuousness." The attitudes of assessment teams toward court review of assessments no doubt reflect a general and historic national concern over the relative powers of executive, legislative, and judicial branches of government. Yet, it is also apparent that assessment teams often suffer from an overpopulation of "legalistic tinkers" who are only too willing to offer their general complaints about the courts as substitutes for professional contributions to the environmental assessment of specific projects. After 9 years of NEPA, *it is time to dispense once and for all with the notion that the assessment team is the primary authority on what is environmentally desirable and what is environmentally undesirable.* The courts, the public, governmental agencies, and the assessment team must all be involved.

Because of this joint responsibility for environmental quality, it is necessary for assessment teams to view the court review process as one means (along with in-house, agency, and public review mechanisms) of ensuring the compliance of assessments with the environmental goals and objectives of NEPA. Toward this end, assessment teams might well consider these guidelines and suggestions:

1. Most governmental agencies have an office of legal liaison for

NEPA-related matters (see Appendix B). The assessment team should be aware of the criteria of legal adequacy each office is likely to use should its respective agency review the team's dEIS, or otherwise participate in the assessment effort. These criteria can be assumed to reflect court findings with respect to numerous assessments which may be relevant to the team's ongoing assessment. It is important that the assessment team also become aware of the environmental issues and concerns that underlie the legal criteria of adequacy in each agency. The object of this liaison, after all, is not merely to meet the legal criteria of adequacy used by agencies and the courts, but to incorporate the consideration of underlying environmental concerns into the actual assessment of project impacts.

2.   All legal considerations (including agency regulations and recent court decisions) that influence the assessment team's efforts should be clearly defined with respect to (a) actions that are legally required; and (b) actions that are not legally prohibited. Too often assessment teams seek only to fulfill legal requirements. But the mere meeting of minimal legal requirements does not necessarily mean that assessments cannot be improved if additional (though not legally required) actions are taken. The point is that assessment teams may conform to agency regulations and still do more than is required by those regulations. As pointed out in Chapter 21, actual compliance with NEPA goes beyond procedural compliance to NEPA.

3.   Frequently, courtroom deliberations identify important issues in a particular case that should be evaluated in other assessments, regardless of the ultimate decision in the tested case. From this perspective, it can be suggested that assessment teams might profitably spend more time examining issues raised in the progress of a legal case—and how such issues might influence an ongoing assessment—and much less time on the final resolution of that case.

4.   As our judicial experience with NEPA-related cases grows, it can be expected that there will be an increasing number of monographs which discuss developing legal trends and issues in impact assessment. Some of the monographs that can already be profitably read by the layman have been cited in Chapters 3 and 21. Assessment teams should consult all such monographs in order to update their own understanding of NEPA and their own obligations under NEPA. The objective is not that assessment teams should become part-time lawyers, but that such legal monographs can play the same role for assessment teams as do the comments and recommendations of agencies and the public which are solicited during the review process—they can inform the assessment team on issues and concerns which the assessment team may otherwise unknowingly overlook.

Concluding Remarks

Previous sections have discussed the positive contributions that the review process can and should contribute to the assessment of impacts on the total human environment. This section will concentrate on an aspect of the review process that has been largely ignored—namely, that the review process can cause impacts on the total human environment as surely as can the construction and operational phases of project development.

Whether or not impacts that result from the review process are properly considered within the framework of the overall assessment of project development is an open question. A rationale for including a consideration of such impacts within this framework might be outlined as follows:

1. Impact assessment is triggered by a federally proposed project, but once triggered the assessment process becomes an integral component of subsequent project development.
2. The review process is an integral component of the overall impact assessment process and, therefore, of project development.
3. Any impacts that occur or are likely to occur as a result of the review process may, therefore, be considered to be impacts of project development.
4. A consideration of such impacts should, therefore, be integrated with the decision-making process of overall project development.

Impacts that can result from the review process might include such impacts as:

1. Increasing alienation among the public at large due to the public's perception of its powerlessness to contribute meaningfully in key phases of environment decision-making by governmental agencies
2. Increasing alienation among the public in local project areas due to the public's perception that some local or regional interest groups and individuals have more influence in environmental decision-making than other groups and individuals
3. Decreasing interagency cooperation during early project planning due to a greater reliance on formal EISs and a diminished reliance on other informal mechanisms of information transfer
4. Increased expenditure of public funds due to multiagency duplication of analytical and integrative tasks of assessment
5. Increased expenditure of public funds due to litigation that might have been avoided by means of timely review of assessment results

The suggestions and guidelines that were discussed in earlier sections of this chapter and which were presented as basic mechanisms for improving the quality-control potential of the review process are also important means whereby such impacts as these may be avoided. In this respect, the review process is a quality-control mechanism which has an intrinsic alarm, for

- If assessment teams perceive the review process as a positive input into assessment and utilize it accordingly, their assessment of impacts on the total human environment will more likely conform to the total environmental objectives of NEPA.
- But, if assessment teams perceive the review process as a negative input into assessment and utilize it accordingly, the review process itself will generate impacts which will clearly signal non-conformance to the total environmental objectives of NEPA.

23

Criteria of Adequacy of Assessment: The Total Human Environment

Introduction

It is tempting to argue that there is only one meaningful criterion by which we can measure the adequacy of the assessment of impacts on the total human environment—namely, the quality of decisions and projects that result from the assessment process. The temptation to take this position is in keeping with the strong emphasis which the executive branch of government, the courts, and the Congress have given to the actual achievement of the goals and objectives of NEPA, as opposed to mere procedural compliance with NEPA. Yet, to measure the adequacy of impact assessment in the quality of subsequent decisions and completed projects is to confuse the goals and objectives of NEPA with the very mechanisms created by NEPA for achieving those goals and objectives. From a practical point of view, must we really wait until decisions are made and projects are designed and constructed before we pass judgment on the adequacy of assessment? If so, NEPA becomes primarily a device for assigning praise or blame (as the case may be) to decision-makers in an irretrievable past, rather than a tool for helping decision-makers enhance the human environment of the present and the future.

The premise of this and preceding chapters is that the assessment process itself can and should be effectively monitored to ensure a high probability for (a) the early identification and evaluation of project impacts on the total human environment; and (b) the integration of total environmental considerations with decision-making processes throughout all phases of project development.

It is definitely not the premise of this chapter that the adequacy of a total environmental assessment will be assured by mere procedural conformity to NEPA-inspired guidelines and regulations—or that the adequacy of impact assessment will be assured by some benign and pervasive spirit of public and governmental cooperation which can hardly be improved upon by cynically contrived checks and balances. To the contrary, a more realistic assumption is that the most likely direction for any impact assessment to take is toward irrelevancy. Thus it is the concern of this chapter to summarize key issues that can and should be considered in order to ensure that assessments do not become irrelevant to environmental decision-making, and that they become, in fact, the very basis for that decision-making.

Comments on Early Identification and Evaluation of Impacts

Throughout all phases of impact assessment it is quite simple to demonstrate whether the assessment team is concerning itself with impacts on the total human environment, or with piecemeal impacts on the social and/or the physical environments. It is only necessary to ask key assessment personnel the following questions:

1. What are the social consequences of project-induced impacts on the physical environment?
2. What are the physical consequences of project-induced impacts on the social environment?

From a total environmental perspective, it is absolutely necessary to assume that there will be indirect, social consequences to most if not all direct impacts on the physical environment and that there will be indirect, physical consequences to most if not all direct impacts on the social environment. Of course, it can happen that such indirect impacts may be quite improbable and/or immeasurably small should they occur. However, depending on site-specific conditions, they can be highly probable and highly significant. The key point is that assessment teams must begin the task of assessment with a heightened sensitivity to dynamic interrelationships between the physical and the social environment; and they must organize the overall assessment effort so as to permit the early identification and evaluation of project impacts on these interrelationships.

The early identification of impacts on interrelationships among physical and social components and dynamics is highly unlikely unless *relevant* data and information are collected early. Note that the emphasis is on the word "relevant." It is one thing to collect environmental data and information. It is quite another thing to collect data and information that are relevant to understanding (a) specific dynamic interrelationships between the physical and social environments; and (b) specific means whereby projects can affect those interrelationships.

The difference can easily be discerned in the following hypothetical example—an example that is particularly relevant to the assessment of potential impacts of national policy with respect to technologically undeveloped societies, but which can also be extrapolated to domestic projects and domestic populations.

Let us imagine that a proposed federal action is a program which will directly or indirectly result in an American "presence" in the midst of a society characterized by (a) lack of any technological development beyond primitive agriculture (in the inland highlands of a tropical island) and subsistence fishing (along the seashore); and (b) a system of foodstuff exchange between inland and seashore communities. Let us also imagine that some concern has been expressed over the effects that an American presence may have on this society, whether that presence is manifested in direct American involvement in the society's affairs (e.g., the giving of agricultural advice and technology), or in indirect American involvement (e.g., through the individual actions of American personnel that do not directly reflect American policy). Because of this concern, it has been decided that an assessment should be undertaken and that decisions concerning the design and implementation of the proposed program are to await the results of this assessment.

A piecemeal approach to such an assessment would require the professional inputs of a variety of experts. Specific areas of interest— especially for the purpose of defining "base-line conditions" (i.e., conditions prior to any American presence)—might well include, but not be limited to:

- Marine fisheries
- Tropical agriculture, medicine, etc.
- Rain-forest ecology
- Cultural anthropology
- Political science
- Social and cross-cultural psychology
- Economics

From a practical viewpoint (i.e., from the perspective of finite time and finite budgets) there is an infinite amount of data that can be generated by experts who focus on each of these areas and which therefore must be assigned priorities. Hopefully, when all the data have been generated according to some plan of priority, it will be possible to (a) identify project impacts on the components and processes within each area of interest; and (b) relate these impacts to the day-to-day lives of people who actually live in the studied environment and to the proposed American presence. But will the data and information that are to be collected in the study of individual environmental compartments be sufficient for the task of integrating those compartments into a comprehensive overview of just how the human environment works under site-specific conditions? After all, there is no logical or experiential rule which says that the more data and information we have on marine fisheries, the more relevant the collective data and information on marine fisheries become to the human condition—or, that the more detailed our scientific understanding of rain-forest ecology, the easier it becomes to understand how and why a human population uses that ecology in a particular way.

As a general rule, *whenever an assessment team relies upon the sheer volume of compartmentalized data as the basic means for relating project impacts to the actual human condition, that assessment team is likely to fail.* It is not a question of the amount of data, or even of the diversity of data. *The key issue in assessing impacts on the total human environment is the selectivity of data—that is, what are the specific data required for identifying just how social and physical environmental components and process are interrelated?*

It is therefore necessary for the assessment team in this example to allow even their earliest data collection activities to be guided by the following considerations:

1. The social significance of marine and rain-forest ecology is manifested, at least in part, in specific human relationships and human uses of specific raw materials.
2. Human relationships are influenced by various social checks and balances, such as laws, customs, rituals, and religion.
3. Social checks and balances can, in turn, be influenced by a variety of physical and social factors.

In light of these considerations, the society under discussion may be generally described by means of the systems diagram shown in Fig. 23.1.

A more specific description of how physical and social components and dynamics are interrelated in this particular society may be derived from additional information such as:

1. The society has a matrilineal kinship system [a child (Ego) owes

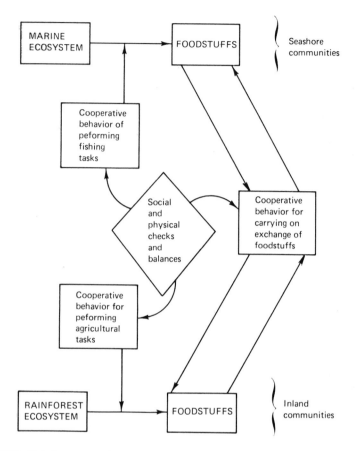

FIGURE 23.1 *Overview of a total human environment, including interrelationships among physical and social components and dynamics.*

his primary allegiance to his mother's side of his family, and Ego's mother's brother becomes an especially important male in Ego's life].

2. The society is patrilocular (upon marriage, the female moves to her husband's village).

3. Agricultural tasks are largely performed by the nuclear family (biological parents and offspring), and

4. the kinship system (see 1 above) and rule-of-domicile (see 2 above) serve to subdivide the nuclear family into two components:

   (a) Ego's biological father (who identifies himself with his own mother's side of the family)

   (b) Ego and Ego's mother (who identify themselves with Ego's mother's family).

In such a society, the rules of kinship play an important role in the exchange of goods between inland and seashore communities and can also serve as a key mechanisms for enforcing cooperative behavior within each agricultural and fishing groups. The possible influence of a kinship system in such a society is depicted in Fig. 23.2. This figure, based on symbols introduced by Odum (1971), attempts to trace the flow of sunlight energy through marine and rain-forest ecosystems. Humans perform the work required (i.e., through *workgates*) for eventual transformation of sunlight energy into desired products (e.g., processed fish, a crop of yams).

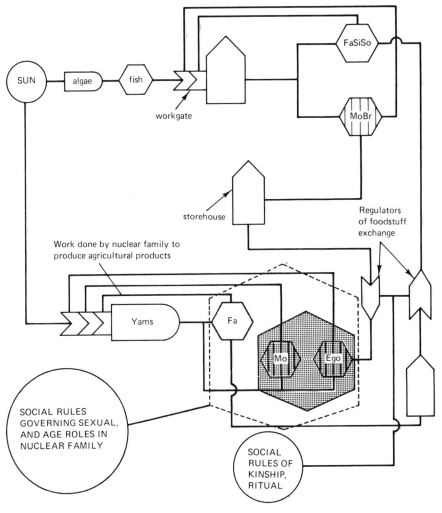

FIGURE 23.2 *Example of possible energy flow and important kinship terms and other social mechanisms for integrating the social and physical environments.*

Social rules (which govern sexual and age-related roles) and kinship rules serve as regulators for the human behavior required for the continued flow of energy and products through workgates. Again, it should be noted that the example depicted in Fig. 23.2 is largely hypothetical, even though it is based on classic anthropological studies.

It should be emphasized here that no model of a total environmental system can be complete. Also, it is largely impossible to use most available modeling techniques when dealing with subjective components and processes. *The key point is that the assessment team should attempt to formulate some comprehensive overview of interrelated physical and social components before it proceeds to generate a large volume of data on individual components.* In the above example, it is only when such a comprehensive overview is available that the assessment team can realistically begin to deal with such practical issues as:

1. What can happen if an American presence results in a greater availability of Western agricultural products in seashore communities?
2. What is likely to happen if an American presence serves essentially as a conduit by which highland youth gain access to other means of employment and/or education?
3. What are the probable consequences of the development of an American-induced export trade in island products?

Certainly, the best and most precise data on marine fisheries, or the best and most precise data on rain-forest ecology, or even the best and most precise data on compartmentalized social aspects of such a society would be useless for dealing with such questions. Yet, it is precisely such questions that must be answered if decision-makers are to make informed decisions.

Comments on Integrating Total Environmental Considerations with Decision-Making Processes

It is clear that some projects have been delayed as a result of the assessment process and NEPA-related issues. Sometimes these delays have been identified as proof of the need to reform NEPA. Whenever such proposals for NEPA reform have been raised, the response has typically been that some projects *should* be delayed, and some *should* be stopped altogether. Thus, project delay has been used both as proof of NEPA's failure, and as proof of NEPA's success! Therefore it is important to point out that:

- While the proposal for a federal project or program does trigger Section 102 (2) (C) of NEPA (i.e., the assessment process and subsequent preparation of an environmental impact statement)
- Once initiated, the assessment process is intended to become a primary means whereby federal agencies are to achieve the national environmental goals and objectives identified in Section 101.

From this perspective, assessment-related delays in project development are delays in the process whereby project alternatives must be evaluated one against the other. Since federal actions may enhance as well as degrade the total human environment, assessment-related delays in project development are also delays in the process whereby federal agencies consider the alternative means (*including the no-action alternative*) for improving the total human environment. In short, *there can be no good reason for the assessment process resulting in a delay either in making the decision that some alternative actions should not be taken, or in making the decision that other actions—actions that can improve the total human environment—should be taken.*

Because of NEPA's mandate to consider alternative actions (including the no-action alternative), it is important that environmental considerations influence the earliest phases of decision-making—that is, before the range of alternative actions has been restricted, and while there is an open-ended set of possible actions for improving the human environment. It is even more important that those environmental considerations that do influence early decision-making include considerations of the total human environment, and not just considerations of piecemeal, fragmented portions of that environment. *If this is not done, it is highly likely that the evaluation of the significance of individual impacts will prove to be inadequate.*

For example, a decision-maker may be faced with this situation early in project planning: If Alternative 1 is taken, there is a high probability that there will be a long-term loss of a fish species in a particular lake. If Alternative 2 is taken, it is unlikely that there will be any impact on aquatic biota. The assessment team recommends that Alternative 2 be taken in order to avoid a "highly significant, negative impact" on fish populations in the region.

This recommendation is eminently logical if: (a) the fish species at issue is an endangered species; or (b) if there is nothing else to consider in either alternative action except fish.

However, if the fish species at issue is not protected by law, the decision-maker must consider a number of questions, including such practical questions as:

1. Do social impacts associated with Alternative 2 outweigh the impact on fish which will occur if Alternative 1 is taken?
2. Are the negative impacts on fish associated with Alternative 1 "significant" with respect to the positive economic and social impacts of the same alternative?

Assessment teams often rightly see themselves as a mechanism for promoting the national environmental goals and objectives of NEPA. But too often they fail to appreciate the simple fact that they can actually accomplish those goals and objectives only through decision-makers. As a general rule, *if an assessment team does not take a total environmental perspective in evaluating the significance of individual impacts, decision-makers who do have to take account of the total human environment will ignore the assessment team's recommendations, and will have good reason for doing so.*

Some Criteria

In light of the previously stated considerations, it is clear that criteria for judging the adequacy of the assessment of impacts on the total human environment include many criteria previously discussed in Chapters 12 and 17. The reason that so many of these criteria overlap in their application to the physical, to the social, and to the total human environment is that it is possible to take a *systems approach* with respect to the physical environment, a *systems approach* with respect to the social environment, and a *systems approach* with respect to the most complex system of all, the total human environment. In each case, there are no firm and fast rules that, if followed to the letter, will guarantee an adequate assessment. But in each case, experience with NEPA suggests some general issues that must be addressed in any systems approach to impact assessment. With respect to the total human environment, assessment teams, therefore, are well advised to consider the following questions:

1. How does the staffing and management of the assessment team specifically reflect the need to consider interrelationships among physical and social components of the total human environment?

2. Is the need for data and information identified through actual interdisciplinary mechanisms, or by individual experts who work independently from one another, although under the same "bureaucratic" roof?

3. Does the assessment team undertake its major data collection and processing tasks prior to, after, or concurrently with the construction of a general overview of the total human environment likely to be affected by project development?

4. Are procedures and methods for interpreting data in terms of the total human environment established prior to the actual collection of data?

5. Through what specific formal and informal mechanisms does the assessment team influence decision-makers throughout the decision-making process of a project development, and specifically how does the assessment team receive feedback from decision-makers?

6. With respect to the evaluation of the significance of individual and cumulative impacts, does the assessment team methodically consider (and how?) the relevance of impacts to the total human environment, the expressed concerns and interests of the public, the concerns and interests of governmental and other organizations and agencies, and federal, state, and local laws and regulations—or is the significance of impacts assumed to be so obvious that no explanation and justification are required?

7. Does the assessment team restrict itself to consideration of scientific and technical issues that can be objectively evaluated, or does the assessment team also consider subjective aspects of the total human environment?

8. What mechanisms has the assessment team identified whereby the environmental goals and objectives of the proposed project can be monitored and (if need be) corrected?

In considering these questions, assessment teams can be assured that there will be much disagreement among the public at large as to the wisdom of actions the team may finally take with respect to any of these questions. Regardless of such controversy, assessment teams should be able to justify their responses to each of these questions in terms of the national environmental goals and objectives of NEPA.

References

Odum, H. T.
1971 *Environment, power, and society.* New York: Wiley–Interscience.

V

*Environmental Impact
Assessment: Conclusion*

24

Present Trends and Identifiable Needs

Introduction

If laws have lives, the National Environmental Policy Act has yet to reach its formative years. It is still a fledgling attempt to introduce order into a chaotic interplay of conflicting social objectives and a spatially finite—but qualitatively infinite—physical environment. That the attempt has survived the extreme promises of both its most optimistic proponents and its most pessimistic opponents is remarkable. That it continues not merely to survive, but to change and adapt more precisely to the social and political constraints of such a pluralistic society is even more remarkable.

Perhaps the vitality of NEPA can be attributed, as some have said, to the innate capacity of governmental bureaucracy to preserve and perpetuate whatever legislative folly can lead to bureaucracy's own expansion and confusion. But it is more reasonable to suggest that NEPA's vitality can be attributed to something else—to the fact that, regardless of the criticisms which have been made of NEPA's implementation, the goals and objectives of NEPA actually reflect the environmental aspirations of the American people. If this is true, we can dispense with any notion that the demise of NEPA is imminent.

As "living law," the National Environmental Policy Act has already been influenced by—and has, in turn, already influenced—executive, legislative, and judicial actions. It is embodied in a dynamic and developing system of case law, agency regulations, and public rights and obligations. As the tangible manifestation of this living law, the assessment process that is initiated by federal proposals likewise reflects the influence of changing governmental and public values, concerns, interests, and environmental experience. In fact, the assessment process can already be seen to have evolved through several phases of development.

In the first phase (1970–1973),[1] assessment consisted primarily of casual and disjointed observations of the physical environment in the local project area. No attempt was made to conduct a comprehensive assessment of project impacts on the total human environment. Interrelationships among physical and social components of the environment were largely ignored. In this phase, the assessment effort was devoted chiefly to justifying decisions which had already been made when NEPA was enacted.

In the second phase (1972–1975), assessment efforts became much more highly organized and typically reflected the interests of highly trained professionals in the biological or other natural science disciplines. This was a period in which the so-called "dandelion counts" came into prominence. In this phase, the primary assessment effort was devoted to compiling massive compendia of scientific and technical data.

In the third phase (1974–present), impact assessment began to focus on physical and social interrelationships among environmental components and dynamics. The approach was clearly systems oriented and involved the construction and use of qualitative and/or quantitative models of the environment. In this most recent developmental phase of impact assessment, the focus has been increasingly on the early assessment of total environmental impacts and on the integration of environmental considerations with the decision-making process in early project planning.

Of course, the dates assigned to each of these phases are somewhat misleading. For example, some of the earliest assessments were, in fact, systems oriented. It is also quite possible to find assessment teams that even today approach assessment just as most teams did in the earliest phase of impact assessment. Thus the dates are only indicative of general national trends.

Current Directions

Throughout the preceding chapters, primary focus has been given to a systems approach to impact assessment. It has also been pointed out that a systems approach to impact assessment (even if it is only qualita-

---

[1] Phases are based on material developed by George Camougis, New England Research, Inc. (unpublished lecture).

tive) presents assessment teams with (a) a major intellectual challenge; as well as with (b) the specific challenge to ensure the practicable utilization of the products of assessment. The intellectual challenge derives from the need to integrate extremely diverse and extensive data and information into a comprehensive overview of the total human environment. The practical challenge drives from the need to make concise and meaningful recommendations to decision-makers in the earliest possible phases of project development. Current directions in the development and refinement of the impact assessment process clearly reflect a growing national awareness of the intellectual and practical challenges of impact assessment, and include:

1. An increasing emphasis on the importance of subjective and social issues in overall impact assessment
2. An increasing emphasis on indirect and cumulative impacts of project development
3. The development of qualitative and quantitative models for the analysis of impacts on social and physical interrelationships
4. An increased emphasis on the importance of public participation in all analytical and integrative tasks
5. The expansion of the disciplinary base of assessment teams in order to achieve a better balance of the social and natural sciences and disciplines
6. The development of guidelines for evaluating the significance of individual and cumulative impacts
7. The development of guidelines and regulations that will better ensure the consideration of environmental impacts in early decision-making of project development
8. Increasing emphasis on the conformity of actual decisions to the goals and objectives of NEPA, and decreasing emphasis on mere procedural compliance to NEPA
9. The streamlining of environmental reports in order to facilitate their actual use in project planning and development
10. The increased emphasis on the assessment process as a means for evaluating alternative actions (including the no-action alternative), and the decreased emphasis on the assessment process as a means for justifying decisions already made
11. A developing recognition for the need to monitor the environmental consequences of projects already subjected to the EIS process in order to evaluate the quality of previous assessments and to ensure compliance of the completed project with planned mitigating and/or enhancing measures
12. The development of manuals and monographs on the mitigation of negative impacts and the enhancement of positive impacts of different types of projects

13. The development of university and college departments, institutes, and curricula that focus on interdisciplinary problem-solving and on analytical and integrative skills required by the impact assessment process
14. The development of local, regional, national, and international workshops, programs, and training courses on assessment-related issues
15. Development of guidelines and regulations for avoiding conflicts of interest in the design and conduct of assessment projects
16. The development of guidelines and regulations for conducting programmatic assessments that will consider how general types of actions can typically impact the environment regardless of site-specific conditions
17. A continually growing willingness of courts to consider the substantive as well as the procedural compliance of assessments to NEPA
18. An increase in the public's understanding of the legal requirements and national environmental goals and objectives of NEPA

Of course, it is impossible to predict with any certainty how each of these apparent trends in contemporary impact assessment will actually develop (if at all) in the future, or which (if any) will be reversed. Our experience with NEPA up to the present clearly demonstrates that the assessment process is a dynamic process. It is open to continual change. It might, nonetheless, be suggested that each of the trends identified above is highly likely to influence impact assessment throughout the next few years, and that none is likely to be reversed abruptly within the period remaining of NEPA's first decade.

Some Identifiable Needs

Many needs can be identified with respect to the analytical and integrative tasks of impact assessment. There are also needs that pertain not so much to the conduct of assessment as much as to the utilization of assessments for achieving the national environmental goals and objectives of NEPA. The suggestions and guidelines presented in the previous chapters of this book have been offered in light of identifiable needs with respect to the conduct of assessment. In this section, attention will be focused on some basic issues that remain to be addressed before assessments can realistically contribute to the achievement of national environmental goals. These key issues directly pertain to decision-makers, to private development, and to the codification of the assessment process.

Decision-Makers

While "hard" documentation is lacking, it is my opinion that the very real and major advances which have been made in impact assessment can be directly attributed to the professionalism of personnel who have direct assessment responsibility, and that *these advances have often been made in spite of the disinterest and lack of meaningful participation of middle-management personnel and upper-echelon decision-makers in efforts to improve the quality and utilization of assessments.*

In numerous workshops and training programs conducted throughout the nation, and which have been attended by hundreds of professionals in order to learn how to conduct comprehensive impact assessment, I have consistently been told, "My boss should be here. He needs this even more than I do!" The simple fact is that "the boss" is too often busy with business as usual to understand that his obligations and responsibilities have in fact been changed by NEPA.

The goals and objectives of NEPA cannot be achieved by the mere quality of assessment. They can only be achieved when decision-makers utilize assessments in their decision-making. Thus, all the effort spent on improving assessment teams, on upgrading and streamlining reports, on involving the public in the assessment effort—all this effort is an absolute waste of time if decision-makers persist in making believe that they can go about their business as usual.

It is not enough for decision-makers to recognize the abstract need for including environmental considerations in their decision-making. They must learn what comprehensive assessment actually entails. Having learned this they can facilitate rather than (as too often happens) retard the efforts of their assessment teams to improve both the quality and utilization of assessments. Toward this end, it is critically important (a) that middle-management and upper-echelon decision-makers in federal agencies be "targeted" for training in the analytical and integrative requirements of impact assessment; and (b) that the ultimate responsibility for the quality and utilization of impact assessments be clearly and unconditionally identified as the responsibility of decision-makers and not of assessment teams.

Private Development

There can be no doubt that even the most sophisticated environmental planning by governmental agencies can be essentially undone by private development that is unconstrained by the National Environmental Policy Act. For example, a highway project involving federal funds may result (through the assessment process) in a highway location and design that actually enhances an environmental attribute (e.g., the carrying capacity of a local area for deer). However, a private housing development may be subsequently built in the same local area and effectively result in the negation of the highway's enhancement of the environmental attribute. In such a situation, how can the assessment process be justified as a meaningful effort? From a practical point of view—that is, looking at what actually happens in the environment—federal environmental planning in such a situation is essentially a waste of time and money.

There is as yet no practical mechanism for ensuring that public funds expended in a governmental assessment process and for the attainment of national environmental goals will not be wasted through unconstrained private development. The need for such a mechanism (or mechanisms) cannot be denied.

Codification of the Assessment Process

As our experience with NEPA increases, there is an increasing tendency to codify the assessment process—that is, to develop highly specific guidelines for identifying impacts, evaluating impacts, etc. While such codification of analytical and integrative tasks is laudable insofar as it can serve to ensure that important issues will not be overlooked or disregarded in the assessment process, codification can also be detrimental to the achievement of national environmental goals and objectives. The key point is whether or not it is still too early to take a "cookbook" approach to impact assessment.

A key premise of this book is that it is too early to select the best procedures for conducting impact assessment and that any contemporary codification of the assessment process will more likely detract from the quality of assessments than add to it. In short, the need of assessment at this time is for experimentation with different approaches, with different analytical and integrative tools, and with different types of mechanisms for integrating assessments with decision-making. There can be no doubt that there will always be codifiers who will willingly undertake to tell us in infinite detail how something should be done. But with respect to impact assessment, it is to be hoped that the codifiers will wait until we have a better appreciation of just what can be done. In this respect, this book has offered only guidelines, suggestions, and rules of thumb—not precise procedures.

Conclusion

In all of the preceding chapters there is a constant theme—that impact assessment under the National Environmental Policy Act is a new and developing tool for exploring and molding the total human environment. Because it is a new tool, we have yet to understand fully either the uses to which it may be put or its limitation and inadequacies. As we learn these things—if only through trial and error—the impact assessment process will be reshaped and refined to better achieve and maintain a productive harmony between man and nature—a harmony wherein

> The earth is to be seen neither as an ecosystem to be preserved unchanged nor as a quarry to be exploited for selfish and short-range economic reasons, but as a garden to be cultivated for the development of its own potentialities of the human adventure. The goal of this relationship is not the maintenance of the status quo, but the emergence of new phenomena and new values [Dubos, 1976].

Reference

Dubos, R.
  1976   Symbiosis between the earth and humankind. *Science,* August 6, 1976, *193:* 459–462.

VI

Appendices

APPENDIX A

Federal Agencies and Federal State Agencies with Jurisdiction by Law or Special Expertise to Comment on Various Types of Environmental Impacts[1]

AIR

Air Quality and Air Pollution Control
- Department of Agriculture
  Forest Service (effects on vegetation)
- Atomic Energy Commission (radioactive substances)
- Department of Health, Education and Welfare (health aspects)
- Environmental Protection Agency
  Air Pollution Control Office
- Department of the Interior
  Bureau of Mines (fossil and gaseous fuel combustion)
  Bureau of Sport Fisheries and Wildlife (wildlife)
- National Aeronautics and Space Administration (remote sensing, aircraft emissions)
- Department of Transportation
  Assistant Secretary for Systems Development and Technology (auto emissions)
  Coast Guard (vessel emissions)
  Federal Aviation Administration (aircraft emissions)

Weather Modification
- Department of Commerce
  National Oceanic and Atmospheric Administration
- Department of Defense
  Department of the Air Force
- Department of the Interior
  Bureau of Reclamation
  Water Resources Council

---

[1]*Federal Register*, Vol. 38, No. 84, Wednesday, May 2, 1973.

ENERGY

Energy Conservation
- Department of the Interior
  Office of Energy Conservation
- Department of Commerce
  National Bureau of Standards (energy efficiency)
- Department of Housing and Urban Development
  Federal Housing Administration (energy conservation in design and operation of buildings)

Environmental Aspects of Electric Energy Generation and Transmission
- Atomic Energy Commission (nuclear power)
- Environmental Protection Agency
  Water Quality Office
  Air Pollution Control Office
- Department of Agriculture
  Rural Electrification Administration (rural areas)
- Department of Defense
  Army Corps of Engineers (hydro-facilities)
- Federal Power Commission (hydro-facilities and transmission lines)
- Department of Housing and Urban Development (urban areas)
- Department of the Interior (facilities on government lands)
- National Aeronautics and Space Administration (solar)
- Water Resources Council
- River Basins Commissions (as geographically appropriate)

Natural Gas Energy Development, Transmission, and Generation
- Federal Power Commission (natural gas production, transmission, and supply)
- Department of the Interior
  Geological Survey
  Bureau of Mines

HAZARDOUS SUBSTANCES

Toxic Materials
- Atomic Energy Commission (radioactive substances)
- Department of Commerce
  National Oceanic and Atmospheric Administration
- Department of Health, Education and Welfare (health aspects)
- Environmental Protection Agency
- Department of Agriculture
  Agricultural Research Service
  Consumer and Marketing Service
- Department of Defense
- Department of the Interior
  Bureau of Sport Fisheries and Wildlife

Pesticides
- Department of Agriculture
  Agricultural Research Service (biological controls, food and fiber production)
  Consumer and Marketing Service
  Forest Service
- Department of Commerce
  National Marine Fisheries Service
  National Oceanic and Atmospheric Administration

- Environmental Protection Agency
  Office of Pesticides
- Department of the Interior
  Bureau of Sport Fisheries and Wildlife (effects on fish and wildlife)
  Bureau of Land Management
- Department of Health, Education and Welfare (health aspects)

Herbicides
- Department of Agriculture
  Agricultural Research Service
  Forest Service
- Environmental Protection Agency
  Office of Pesticides
- Department of Health, Education and Welfare (health aspects)
- Department of the Interior
  Bureau of Sport Fisheries and Wildlife
  Bureau of Land Management
  Bureau of Reclamation

Transportation and Handling of Hazardous Materials
- Department of Commerce
  Maritime Administration
  National Marine Fisheries Service
  National Oceanic and Atmospheric Administration (impact on marine life)
- Department of Defense
  Armed Services Explosive Safety Board
  Army Corps of Engineers (navigable waterways)
- Department of Health, Education and Welfare
  Office of the Surgeon General (health aspects)
- Department of Transportation
  Federal Highway Administration, Bureau of Motor Carrier Safety
  Coast Guard
  Federal Railroad Administration
  Federal Aviation Administration
  Assistant Secretary for Systems Development and Technology
  Office of Hazardous Materials
  Office of Pipeline Safety
- Environmental Protection Agency (hazardous substances)
- Atomic Energy Commission (radioactive substances)

LAND USE AND MANAGEMENT

Aesthetics
- Numerous agencies have developed specific methods of assessing aesthetics in relation to their area of responsibility.

Coastal Areas: Wetlands, Estuaries, Waterfowl Refuges, and Beaches
- Department of Agriculture
  Forest Service
- Department of Commerce
  National Marine Fisheries Service (impact on marine life)
  National Oceanic and Atmospheric Administration (impact on marine life)
- Department of Transportation
  Coast Guard (bridges, navigation)

- Department of Defense
    Army Corps of Engineers (beaches, dredge and fill permits, Refuse Act permits)
- Department of the Interior
    Bureau of Sport Fisheries and Wildlife
    National Park Service
    U.S. Geological Survey (coastal geology)
    Bureau of Outdoor Recreation (beaches)
- Department of Agriculture
    Soil Conservation Service (soil stability, hydrology)
- Environmental Protection Agency
    Water Quality Office
- National Aeronautics and Space Administration (remote sensing)
- Water Resources Council
- River Basin Commissions (as geographically appropriate)

Historic and Archeological Sites
- Department of the Interior
    National Park Service
- Advisory Council on Historic Preservation
- Department of Housing and Urban Development (urban areas)

Flood Plains and Watersheds
- Department of Agriculture
    Agricultural Stabilization and Research Service
    Soil Conservation Service
    Forest Service
- Department of the Interior
    Bureau of Outdoor Recreation
    Bureau of Reclamation
    Bureau of Sport Fisheries and Wildlife
    Bureau of Land Measurement
    U.S. Geological Survey
- Department of Housing and Urban Development (urban areas)
- Department of Defense
    Army Corps of Engineers
- Water Resources Council
- River Basins Commissions (as geographically appropriate)

Mineral Land Reclamation
- Appalachian Regional Commission
- Department of Agriculture
    Forest Service
- Department of the Interior
    Bureau of Mines
    Bureau of Outdoor Recreation
    Bureau of Sport Fisheries and Wildlife
    Bureau of Land Management
    U.S. Geological Survey
- Tennessee Valley Authority

Parks, Forests, and Outdoor Recreation
- Department of Agriculture
    Forest Service
    Soil Conservation Service
- Department of the Interior

Bureau of Land Management
National Park Service
Bureau of Outdoor Recreation
Bureau of Sport Fisheries and Wildlife
- Department of Defense
  Army Corps of Engineers
- Department of Housing and Urban Development (urban areas)
- Water Resources Council
- River Basins Commissions (as geographically appropriate)

Soil and Plant Life, Sedimentation, Erosion, and Hydrologic Conditions
- Department of Agriculture
  Soil Conservation Service
  Agricultural Research Service
  Forest Service
- Department of Defense
  Army Corps of Engineers (dredging, aquatic plants)
- Department of Commerce
  National Oceanic and Atmospheric Administration
- Department of the Interior
  Bureau of Land Management
  Bureau of Sport Fisheries and Wildlife
  Geological Survey
  Bureau of Reclamation
- Water Resources Council
- River Basins Commissions (as geographically appropriate)

NOISE

Noise Control and Abatement
- Department of Health, Education, and Welfare (health aspects)
- Department of Commerce
  National Bureau of Standards
- Department of Transportation
  Assistant Secretary for Systems Development and Technology
  Federal Aviation Administration (Office of Noise Abatement)
- Environmental Protection Agency (Office of Noise)
- Department of Housing and Urban Development (urban land use aspects, building materials standards)
- National Aeronautics and Space Administration (aircraft noise abatement and control)

PHYSIOLOGICAL HEALTH AND HUMAN WELL-BEING

Chemical Contamination of Food Products
- Department of Agriculture
  Consumer and Marketing Service
- Department of Health, Education and Welfare (health aspects)
- Environmental Protection Agency
  Office of Pesticides (economic poisons)

Food Additives and Food Sanitation
- Department of Health, Education and Welfare (health aspects).
- Environmental Protection Agency
  Office of Pesticides (economic poisons, e.g., pesticide residues)

- Department of Agriculture
  Consumer and Marketing Service (meat and poultry products)

Microbial Contamination
- Department of Health, Education and Welfare (health aspects)

Radiation and Radiological Health
- Department of Commerce
  National Bureau of Standards
- Atomic Energy Commission
- Environmental Protection Agency
  Office of Radiation
- Department of the Interior
  Bureau of Mines (uranium mines)

Sanitation and Waste Systems
- Atomic Energy Commission (radioactive waste)
- Department of Health, Education and Welfare (health aspects)
- Department of Defense
  Army Corps of Engineers
- Environmental Protection Agency
  Solid Waste Office
  Water Quality Office
- Department of Transportation
  U.S. Coast Guard (ship sanitation)
- Department of the Interior
  Bureau of Mines (mineral waste and recycling, mine acid wastes, urban solid wastes)
  Bureau of Land Management (solid wastes on public lands)
  Office of Saline Water (demineralization of liquid wastes)
  Water Resources Council
  River Basins Commissions (as geographically appropriate)

Shellfish Sanitation
- Department of Commerce
  National Marine Fisheries Service
  National Oceanic and Atmospheric Administration
- Department of Health, Education and Welfare (health aspects)
- Environmental Protection Agency
  Office of Water Quality

TRANSPORTATION

Air Quality
- Environmental Protection Agency
  Air Pollution Control Office
- Department of Transportation
  Federal Aviation Administration
- Department of the Interior
  Bureau of Outdoor Recreation
  Bureau of Sport Fisheries and Wildlife
- Department of Commerce
  National Oceanic and Atmospheric Administration (meteorological conditions)
- National Aeronautics and Space Administration (aviation)

Water Quality
- Environmental Protection Agency
  Office of Water Quality

- Department of the Interior
  Bureau of Sport Fisheries and Wildlife
- Department of Commerce
  National Oceanic and Atmospheric Administration (impact on marine life and ocean monitoring)
- Department of Defense
  Army Corps of Engineers
- Department of Transportation
  Coast Guard
- Water Resources Council

URBAN

Congestion in Urban Areas, Housing and Building Displacement
- Department of Transportation
  Federal Highway Administration
- Office of Economic Opportunity
- Department of Housing and Urban Development
- Department of the Interior
  Bureau of Outdoor Recreation

Environmental Effects with Special Impact in Low-Income Neighborhoods
- Department of the Interior
  National Park Service
- Office of Economic Opportunity
- Department of Housing and Urban Development (urban areas)
- Economic Development Administration
- Department of Transportation
  Urban Mass Transportation Administration
- Water Resources Council
- River Basins Commissions (as geographically appropriate)

Rodent Control
- Department of Health, Education and Welfare (health aspects)
- Department of Housing and Urban Development (urban areas)

Urban Planning
- Department of Transportation
  Federal Highway Administration
- Department of Housing and Urban Development
- Environmental Protection Agency
- Department of the Interior
  Geological Survey
  Bureau of Outdoor Recreation
- Department of Commerce
  Economic Development Administration
- Water Resources Council
- River Basins Commissions (as geographically appropriate)

WATER

Water Quality and Water Pollution Control
- Department of Agriculture
  Soil Conservation Service
  Forest Service
- Atomic Energy Commission (radioactive substances)

- Department of the Interior
  Bureau of Reclamation
  Bureau of Land Management
  Bureau of Sports Fisheries and Wildlife
  Bureau of Outdoor Recreation
  Geological Survey
  Office of Saline Water
- Environmental Protection Agency
  Water Quality Office
- Department of Health, Education and Welfare (health aspects)
- Department of Defense
  Army Corps of Engineers
  Department of the Navy (ship pollution control)
- National Aeronautics and Space Administration (remote sensing)
- Department of Transportation
  Coast Guard (oil spills, ship sanitation)
- Department of Commerce
  National Oceanic and Atmospheric Administration
- Water Resources Council
- River Basins Commissions (as geographically appropriate)

River and Canal Regulation and Stream Channelization
- Department of Agriculture
  Soil Conservation Service
- Department of Defense
  Army Corps of Engineers
- Department of the Interior
  Bureau of Reclamation
  Geological Survey
  Bureau of Sport Fisheries and Wildlife
- Department of Transportation
  Coast Guard
- Water Resources Council
- River Basins Commissions (as geographically appropriate)

WILDLIFE

- Environmental Protection Agency
- Department of Agriculture
  Forest Service
  Soil Conservation Service
- Department of the Interior
  Bureau of Sport Fisheries and Wildlife
  Bureau of Land Management
  Bureau of Outdoor Recreation
- Water Resources Council
- River Basins Commissions (as geographically appropriate)

Agency Channels of Communication on NEPA Matters[1]

ACTION
Head of Agency
Director, Action
806 Connecticut Avenue, N.W.
Washington, D.C. 20525
(202) 254–3120

U.S. ARMS CONTROL AND
DISARMAMENT AGENCY
Head of Agency
Director, ACDA
320 21st Street, N.W.
Washington, D.C. 20451
(202) 632–9610

General Counsel NEPA Contact
Assistant General Counsel, ACDA
320 21st Street, N.W.
Washington, D.C. 20451
(202) 632–3530

DEPARTMENT OF AGRICULTURE
Head of Agency
Assistant Secretary with NEPA
Responsibility

Assistant Secretary for Conservation,
Research and Education

Working Level NEPA Liaison
Coordinator
Environmental Quality Activities
Office of the Secretary
Room 307A
Department of Agriculture
Washington, D.C. 20250
(202) 447–6827

General Counsel NEPA Contact
Office of the General Counsel
Department of Agriculture
Room 2409, south Building
Washington, D.C. 20250
(202) 447–4733

AGRICULTURE MARKETING SERVICE
Head of Agency

AGRICULTURE RESEARCH SERVICE
Head of Agency

---

[1]Adapted from *102 Monitor*, Vol. 7, No. 5, June, 1977, pp. 36–71. Washington,
D.C.: U.S. Government Printing Office.

AGRICULTURAL STABILIZATION AND
CONSERVATION SERVICE
Head of Agency

Working Level NEPA Liaison
Agricultural Stabilization and
Conservation Service
Department of Agriculture
Washington, D.C. 20250
(202) 447-6221

ANIMAL AND PLANT HEALTH
INSPECTION SERVICE
Head of Agency

COOPERATIVE STATE RESEARCH
SERVICE
Head of Agency

FARMERS HOME ADMINISTRATION
Head of Agency

Working Level NEPA Liaison
Deputy Administrator for Program
Operations
FHWA, USDA
Washington, D.C. 20250
(202) 447-3394

FEDERAL EXTENSION SERVICE
Head of Agency

FOREST SERVICE
Head of Agency

Working Level NEPA Liaison
Environmental Coordinator
Forest Service
Department of Agriculture
Room 3022, South Bldg.P
(202) 447-4708

RURAL ELECTRIFICATION
ADMINISTRATION
Head of Agency

Working Level NEPA Liaison
Chief, Management Analysis & Services
Branch
Department of Agriculture
Room 4024, South Bldg.
Washington, D.C. 20250
(202) 447-6148

SOIL CONSERVATION SERVICE
Head of Agency

Assistant Administrator with NEPA
Responsibility
Director
Environmental Services Division

Department of Agriculture
Room 6103
Washington, D.C. 20013
(202) 447-3839

APPALACHIAN REGIONAL
COMMISSION
Head of Agency
Alternate Federal Co-Chairman

Working Level NEPA Liaison
Deputy Executive Director
Appalachian Regional Commission
1666 Connecticut Avenue, N.W.
Washington, D.C. 20235
(202) 673-7874

General Counsel NEPA Contact
Executive Director
Appalachian Regional Commission
1666 Connecticut Avenue, N.W.
Washington, D.C. 20235
(202) 673-7874

CANAL ZONE
GOVERNMENT/PANAMA CANAL
COMPANY
Head of Agency

Director with NEPA Responsibility
Environmental Coordinator

Working Level NEPA Liaison
Secretary, Panama Canal Company
Room 312
425 13th Street, N.W.
Washington, D.C. 20026
(202) 382-6453

CENTRAL INTELLIGENCE AGENCY
Head of Agency

Working Level NEPA Liaison
Director of Logistics
Central Intelligence Agency
Washington, D.C. 20505
(202) 281-8200

CIVIL AERONAUTICS BOARD
Head of Agency

Working Level NEPA Liaison
Office of the General Counsel
Civil Aeronautics Board
Washington, D.C. 20428
(202) 673-5205

General Counsel NEPA Contact
Office of the General Counsel
Civil Aeronautics Board

Washington, D.C. 20428
(202) 673–5205

DEPARTMENT OF COMMERCE
Head of Agency
Working Level NEPA Liaison
Assistant Secretary for Environmental
Affairs
Department of Commerce
Washington, D.C. 20230
(202) 377–4335
General Counsel NEPA Contact
General Counsel
Department of Commerce
Washington, D.C. 20230
(202) 377-4772

ECONOMIC DEVELOPMENT ADMINIS-
TRATION
Head of Agency
Assistant Secretary
for Economic Development
Working Level NEPA Liaison
Special Assistant for the Environment
Department of Commerce
Washington, D.C. 20230
(202) 377–4208
General Counsel NEPA Contact
Chief Counsel
Department of Commerce
Room 7001A
Washington, D.C. 20230
(202) 377–4687

MARITIME ADMINISTRATION
Head of Agency

NATIONAL BUREAU OF STANDARDS
Head of Agency

NATIONAL OCEANIC AND
ATMOSPHERIC ADMINISTRATION
Head of Agency
Working Level NEPA Liaison
Director, Office of Ecology and
Environment Conservation
NOAA—Room 5813
Department of Commerce
Washington, D.C. 20230
(202) 377–5181

SMALL BUSINESS ADMINISTRATION
Head of Agency
Working Level NEPA Liaison
Securities and Exchange Commission

Office of Disclosure Policy
500 North Capitol Street, Room 629
Washington, D.C. 20549
(202) 755–1750

DEPARTMENT OF DEFENSE
Head of Agency
Working Level NEPA Liaison
Deputy Assistant Secretary of Defense
(Environmental and Safety)
(M,Ra, & L)
3B252, Pentagon
Washington, D.C. 20301
(202) Ox5–0221
General Counsel NEPA Contact
Office of the General Counsel
Department of Defense
Room 3D937, Pentagon
Washington, D.C. 20301
(202) 697–9135

DEPARTMENT OF THE AIR FORCE
Head of Agency
Working Level NEPA Liaison
Department of the Air Force
Room 5D431, Pentagon
Washington, D.C. 20330
(202) 697–7799
General Counsel NEPA Contact
Office of the General Counsel
Department of the Air Force
Room 4C927 (SAFGC) Pentagon
Washington, D.C. 20330
(202) 697–7479

DEPARTMENT OF THE ARMY
Head of Agency
*Assistant Secretary with NEPA
Responsibility*
Deputy Assistant Secretary, Civil Works
Room 2E675, Pentagon
Washington, D.C. 20310
(202) 697–7084
Working Level NEPA Liaison
Corps of Engineers—Civil Works
Director of Civil Works,
Environmental Programs
U.S. Army Corps of Engineers
DAEN-CWZ-P
1000 Independence Ave., S.W.
Washington, D.C. 20314
(202) 693–7093

Military
Environmental Office
Office of the Assistant Chief of
Engineers
Department of the Army
Room 1E676, Pentagon
Washington, D.C. 20310

Conservation Organization Liaison
Conservation Liaison Officer
Office of Public Affairs
Attn: DAEN-PAI
Office of the Chief of Engineers
1000 Independence Avenue, S.W.
Washington, D.C. 20314
(202) 693–6346

General Counsel NEPA Contact
Corps of Engineers
Chief Counsel for Litigation
Office of the Chief of Engineers
Department of the Army
1000 Independence Ave., S.W.
Washington, D.C. 20314
(202) 693–7057

Army—Military
Office of the General Counsel
Room 2E729, Pentagon
Washington, D.C. 20310
(202) 695–3305

Army—Judge Advocate General's Office
Chief, Office of the Judge Advocate
General
Department of the Army
Washington, D.C. 20310
(202) 695–1721

DEPARTMENT OF THE NAVY
Head of Agency
*Assistant Secretary with NEPA
Responsibility*
Assistant Secretary, Installation and
Logistics
Working Level NEPA Liaison
Head, Environmental Impact
Statement/RDT&E Branch
Office of the Chief of Naval Operations
Environmental Protection and
Occupational Safety and Health
Division (OD/453)
Washington, D.C. 20350
(202) 697–3689
General Counsel NEPA Contact
Head, Environmental Law Branch

Office of Judge Advocate General
Department of the Navy
Washington, D.C. 20350
(202) 695–0200

DEFENSE LOGISTICS AGENCY
Head of Agency
Working Level NEPA Liaison
Special Assistant for Environmental
Quality
Cameron Station, 4D470
Alexandria, Va. 22314
(703) 274–6967
General Counsel NEPA Contact
General Counsel
Defense Logistics Agency
Cameron Station
Alexandria, Va. 22314
(703) 274–6156

DELAWARE RIVER BASIN
COMMISSION
Head of Agency
*Federal Official with NEPA
Responsibility*
U.S. Commissioner
Delaware River Basin Commission
Interior Building, Rm 6240
Washington, D.C. 20240
(202) 343–2347

Working Level NEPA Liaison
Head Environmental Unit
Delaware River Basin Commission
P.O. Box 360
Trenton, New Jersey 08603
(609) 883–9500 (Ext. 268)

ENERGY RESEARCH AND
DEVELOPMENT ADMINISTRATION
Head of Agency
Working Level NEPA Liaison
Office of NEPA Coordination
Energy Research and Development
Administration
Mail Station E-201
Washington, D.C. 20545
(301) 353–4241
General Counsel NEPA Contact
Office of the General Counsel
Energy Research and Development
Administration
Washington, D.C. 20545
(301) 353–3121

ENVIRONMENTAL PROTECTION
AGENCY
Head of Agency
Working Level NEPA Liaison
Office of Federal Activities
Environmental Protection Agency
537 West Tower
A-104
Washington, D.C. 20460
General Counsel NEPA Contact
General Counsel
Environmental Protection Agency
Room 513 West Tower
A-104
Washington, D.C. 20460
(202) 755-2511

EXPORT–IMPORT BANK
Head of Agency
President and Chairman
Export–Import Bank of the United
States
811 Vermont Avenue, N.W.
Washington, D. C. 20571
(202) 382-1131
General Counsel NEPA Contact
General Counsel
Export–Import Bank of the United States
(202) 382-1493

FEDERAL COMMUNICATIONS
COMMISSION
Head of Agency
General Counsel NEPA Contact
Office of the General Counsel
Federal Communications Commission
(202) 632-6444

FEDERAL ENERGY ADMINISTRATION
Head of Agency
Working Level NEPA Liaison
Director, Office of Environmental
Programs
New Post Office Bldg., Room 7119
12th and Pennsylvania Ave., N.W.
Washington, D.C. 20461
(202) 566-9760
General Counsel NEPA Contact
Deputy Assistant General Counsel
for Conservation and Environment
Office of the General Counsel
12th and Pennsylvania Ave., N.W.
Room 5116

Washington, D.C. 20461
(202) 566-9380

FEDERAL POWER COMMISSION
Head of Agency
Working Level NEPA Liaison
Advisor on Environmental Quality
Federal Power Commission
825 N. Capitol Street, N.E.
Washington, D.C. 20426
(202) 275-4149
General Counsel NEPA Contact
General Counsel
Federal Power Commission
825 N. Capitol Street, N.E.
Washington, D.C. 20426
(202) 275-4309

FEDERAL TRADE COMMISSION
Head of Agency
Working Level NEPA Liaison and
General Counsel NEPA Contact
Assistant General Counsel for Litigation
and Environmental Protection
Federal Trade Commission
Washington, D.C. 20580
(202) 523-3613

GENERAL SERVICES ADMINISTRATION
Head of Agency
*Assistant Administrator with NEPA
Responsibility*
Commissioner
Public Buildings Service
General Services Administration
Washington, D.C. 20405
(202) 566-1100
Working Level NEPA Liaison
Environmental Affairs Division
General Services Administration
18th and F Streets, N.W.
Washington, D.C. 20405
(202) 566-0405
General Counsel NEPA Contact
General Services Administration
18th and F Streets, N.W.
Washington, D.C. 20405
(202) 343-4221

GREAT LAKES BASIN COMMISSION
Head of Agency
Working Level NEPA Liaison
Executive Director

Great Lakes Basin Commission
P.O. Box 999
Ann Arbor, Michigan 48106
FTS 374–5431

DEPARTMENT OF HEALTH,
EDUCATION AND WELFARE
Head of Agency
*Assistant Secretary with NEPA
Responsibility*
Assistant Secretary for Administration
and Management
Working Level NEPA Liaison
Director, Office of Environmental
Affairs
Department of Health, Education and
Welfare
Room 524 FS HEW South
Washington, D.C. 20201
(202) 245–7243
General Counsel NEPA Contact
Office of the General Counsel
Department of Health, Education and
Welfare
Room 5464 HEW North
Washington, D.C. 20202
(202) 245–7752

DEPARTMENT OF HOUSING AND
URBAN DEVELOPMENT
Head of Agency
*Assistant Secretary with NEPA
Responsibility*
Assistant Secretary for Community
Planning and Development
Working Level NEPA Liaison
Director, Office of Environmental
Quality
Department of Housing and Urban
Development
Washington, D.C. 20410
(202) 755–6308
General Counsel NEPA Contact
Office of the General Counsel
Department of Housing and Urban
Development
Washington, D.C. 20410
(202) 755–6552

DEPARTMENT OF THE INTERIOR
Head of Agency
*Assistant Secretary with NEPA
Responsibility*

Assistant Secretary, Policy, Budget, and
Administration
Working Level NEPA Liaison
Director, Environmental Project Review
Department of the Interior
Room 4256 Interior Bldg.
Washington, D.C. 20240
(202) 343–3891
General Counsel NEPA Contact
Office of the Solicitor
Department of the Interior
Washington, D.C. 20240
(202) 343–6848

BONNEVILLE POWER
ADMINISTRATION
Head of Agency

BUREAU OF INDIAN AFFAIRS
Head of Agency
Bureau of Indian Affairs
Department of the Interior
Washington, D.C. 20240
(202) 343–7513
Working Level NEPA Liaison
Chief, Division of Trust Facilitation
Bureau of Indian Affairs
Department of the Interior
1951 Constitution Ave., N.W.
Washington, D.C. 20245
(202) 343–4004

BUREAU OF LAND MANAGEMENT
Head of Agency
Working Level NEPA Liaison
Bureau of Land Management
Department of the Interior
Washington, D.C. 20240
(202) 343–5682

BUREAU OF MINES
Head of Agency
Working Level NEPA Liaison
Special Assistant for Environmental
Activities
Bureau of Mines
Department of the Interior
Washington, D.C. 20240
(202) 343–3941

BUREAU OF OUTDOOR RECREATION
Head of Agency
Working Level NEPA Liaison
Chief, Office of Environmental Affairs

Bureau of Outdoor Recreation
Department of the Interior
Washington, D.C. 20240
(202) 343–5711

BUREAU OF RECLAMATION
Head of Agency
Working Level NEPA Liaison
Assistant to the
Commissioner—Ecology
Bureau of Reclamation
Department of the Interior
Washington, D.C. 20240
(202) 343–4991

U.S. FISH AND WILDLIFE SERVICE
Head of Agency
Working Level NEPA Liaison
Chief, Branch of Environmental
Coordination
U.S. Fish and Wildlife Service
Department of the Interior
Washington, D.C. 20240
(202) 376–8111

GEOLOGICAL SURVEY
Head of Agency
Working Level NEPA Liaison
Special Assistant, Environmentalist
Office of Environmental Conservation
Geological Survey
Stop 108
Reston, Virginia
(703) 860–7493

NATIONAL PARK SERVICE
Head of Agency
Working Level NEPA Liaison
Chief, Park Planning and Environmental
Compliance Division
National Park Service
Department of the Interior
Washington, D.C. 20240
(202) 343–9377

U.S. SECTION INTERNATIONAL BOUNDARY COMMISSION (US–Canada)
Head of Agency
Working Level NEPA Contact
Engineer to the U.S. Section
Room 3810, GAO
441 G Street, N.W.

Washington, D.C. 20548
(292) 783–9151

U.S. SECTION INTERNATIONAL BOUNDARY AND WATER COMMISSION (US–Mexico)
Head of Agency
P.O. Box 20003
El Paso, Texas 79978
Working Level NEPA Contact
Special Assistant—U.S. Section IBWC
Department of State—ARA/mex
Washington, D.C. 20520
(202) 632–9895
General Counsel NEPA Contact
Special Legal Assistant IBWC
P.O. Box 20003
El Paso, Texas 79998

U.S. SECTION INTERNATIONAL JOINT COMMISSION (US–CANADA)
Head of Agency
General Counsel NEPA Contact
Legal Advisor
U.S. Section
International Joint Commission
1717 H Street, N.W.
Washington, D.C. 20036
(202) 632–9456

INTERSTATE COMMERCE COMMISSION
Head of Agency
General Counsel
General Counsel
Interstate Commerce Commission
Washington, D.C. 20423
(202) 343–4831
Working Level NEPA Liaison
Chief, Section of Energy and
Environment
Interstate Commerce Commission
Washington, D.C. 20423
(202) 275–7692

DEPARTMENT OF JUSTICE
Head of Agency
Working Level NEPA Liaison
Chief, General Litigation
Land and Natural Resources Division
Department of Justice
Washington, D.C. 20530
(202) 739–2736

DEPARTMENT OF LABOR
Head of Agency
Assistant Secretary with NEPA Responsibility
Assistant Secretary for Occupational Safety and Health
Working Level NEPA Liaison
Office of Environmental and Economic Impact Assessment
Occupational Safety and Health Administration
Room N-3673
U.S. Department of Labor
200 Constitution Avenue, N.W.
Washington, D.C. 20210
(202) 523–7111
General Counsel NEPA Contact
Assistant Solicitor for Occupational Safety and Health
Department of Labor
Washington, D.C. 20210
(202) 961–3965

MARINE MAMMAL COMMISSION
Head of Agency
Working Level NEPA Liaison and General Counsel NEPA Contact
General Counsel
1625 I Street, N.W.
Washington, D.C. 20006
(202) 653–6235

MISSOURI RIVER BASIN COMMISSION
Head of Agency
Working Level NEPA Liaison
Executive Secretary
Missouri River Basin Commission
10050 Regency Circle, Suite 403
Omaha, Nebraska 68114
(402) 397–5714

NATIONAL AERONAUTICS AND SPACE ADMINISTRATION
Head of Agency
Associate Deputy Administrator with NEPA Responsibility
Working Level NEPA Liaison
Director, Office of Policy Analysis
National Aeronautics and Space Administration
Washington, D.C. 20546
(202) 755–8433

General Counsel NEPA Contact
Assistant General Counsel
National Aeronautics and Space Administration
Washington, D.C. 20546
(202) 755–3920

NATIONAL CAPITAL PLANNING COMMISSION
Head of Agency
Assistant Director with NEPA Responsibility
Director of Current Planning and Programming
National Capital Planning Commission
Washington, D.C. 20576
(202) 382–1471
Working Level NEPA Liaison
Chief, Office of Environmental Affairs
National Capital Planning Commission
Washington, D.C. 20576
(202) 382–7200
General Counsel NEPA Contact
General Counsel
National Capital Planning Commission
Washington, D.C. 20576
(202) 382–4584

NATIONAL SCIENCE FOUNDATION
Head of Agency
Assistant Director with NEPA Responsibility
Assistant Director for Astronomical, Atmospheric, Earth and Ocean Science
Working Level NEPA Liaison
Deputy Assistant Director for Astronomical, Earth and Ocean Sciences
National Science Foundation
Washington, D.C. 20550
(202) 632–7300

Special Assistant
Office of the Deputy Assistant Director for Astronomical, Atmospheric, Earth and Ocean Sciences
National Science Foundation
Washington, D.C. 20550
(202) 632–7360
General Counsel NEPA Contact
General Counsel
National Science Foundation
Washington, D.C. 20550
(202) 632–4386

NEW ENGLAND RIVER BASINS
COMMISSION
Head of Agency

Working Level NEPA Liaison
Staff Director
New England River Basin Commission
55 Court Street
Boston, Massachusetts 02108
(617) 223-6244

NUCLEAR REGULATORY COMMISSION
Head of Agency

Working Level of NEPA Liaison
Assistant Director for Environmental
Projects
Division of Reactor Licensing
Nuclear Regulatory Commission
Washington, D.C. 20555
(301) 443-6965

General Counsel NEPA Contact
Office of General Counsel
Nuclear Regulatory Commission
Washington, D.C. 20555
(301) 492-7308

OFFICE OF ECONOMIC OPPORTUNITY
Head of Agency

General Counsel NEPA Contact
Deputy General Counsel
Office of Economic Opportunity
Washington, D.C. 20506

OFFICE OF MANAGEMENT AND
BUDGET
Head of Agency

*Assistant Director with NEPA
Responsibility*

*Working Level NEPA Liaison for A-95
Procedures*
Intergovernmental Relations and
Regional Operations
Office of Management and Budget
Washington, D.C. 20503
(202) 395-3030
Intergovernmental Relations and
Regional Operations
Office of Management and Budget
Room 8222, New Executive Office Bldg.
Washington, D.C. 20503
(202) 395-4543

OHIO RIVER BASIN COMMISSION
Head of Agency

Working Level NEPA Liaison
Executive Director
Ohio River Basin Commission
Suite 208
36 East Fourth Street
Cincinnati, Ohio 45202
(513) 684-3831

OVERSEAS PRIVATE INVESTMENT
CORPORATION
Head of Agency

President
Overseas Private Investment
Corporation
1129 20th Street, N.W.
Washington, D.C. 20527

General Counsel NEPA Contact
Deputy General Counsel
Overseas Private Investment Corporation
1129 20th Street, N.W.
Washington, D.C. 20527

PACIFIC NORTHWEST RIVER BASINS
COMMISSION
Head of Agency

Working Level NEPA Liaison
Planning Director
Pacific Northwest River Basins
Commission
P.O. Box 908, One Columbia River
Vancouver, Washington 98666
(206) 694-2581 FTS 422-9307

POSTAL SERVICE
Head of Agency

*Assistant Postmaster General with
NEPA Responsibility*
Assistant Postmaster General
Real Estate and Building Department

Working Level NEPA Liaison
Director, Office of Program Planning
Real Estate and Building Department
U.S. Postal Service
Washington, D.C. 20260
(202) 245-4304

General Counsel NEPA Contact
Associate General Counsel
U.S. Postal Service
Washington, D.C. 20260
(202) 245-4599

SECURITIES AND EXCHANGE COMMISSION

Working Level NEPA Liaison
Securities and Exchange Commission
Office of Disclosure Policy
500 North Capitol Street, Room 629
Washington, D.C. 20549
(202) 755–1750

SMALL BUSINESS ADMINISTRATION
General Counsel NEPA Contact
Attorney Adviser
Small Business Administration
Washington, D.C. 20416
(202) 382–6316

SMITHSONIAN INSTITUTION
Head of Agency
General Counsel NEPA Contact
General Counsel
Smithsonian Institution
Room 408
Washington, D.C. 20560
(202) 381–5866

DEPARTMENT OF STATE
Head of Agency
Working Level NEPA Liaison
Director, Office of Environmental
Affairs
Department of State
Washington, D.C. 20520
(202) 632–9169
General Counsel NEPA Contact
L/OES—Room 6420
Department of State
Washington, D.C. 20520
(202) 632–1989

AGENCY FOR INTERNATIONAL DEVELOPMENT
Head of Agency
Working Level NEPA Liaison
Environmental Coordinator
AID 320 21st St., N.W.
Washington, D.C. 20523
General Counsel NEPA Contact
General Counsel
AID/GC/TFHA, Rm 6951
Department of State
Washington, D.C. 20523
(202) 632–1170

SUSQUEHANNA RIVER BASIN COMMISSION
Head of Agency
Federal Official with NEPA Responsibility
U.S. Commissioner
Susquehanna River Basin Commission
Interior Building, Rm 6246
Washington, D.C. 20240
(202) 343–4091
Working Level NEPA Liaison
Executive Director
Susquehanna River Basin Commission
5012 Lenker Street
Mechanicsburg, Penn. 17055
(717) 737–0501

TENNESSEE VALLEY AUTHORITY
Head of Agency
Director with NEPA Responsibility
Working Level NEPA Liaison
Director of Environmental Planning
Tennessee Valley Authority
720 Edney Bldg.
Chattanooga, Tennessee 37401
FTS 854–3161
General Counsel NEPA Contact
General Counsel
RVA
Knoxville, Tennessee 37902
(615) 637–0101

DEPARTMENT OF TRANSPORTATION
Head of Agency
Assistant Secretary with NEPA Responsibility
Assistant Secretary for Environment,
Safety, and Consumer Affairs
Working Level NEPA Liaison
Director, Office of Environmental
Affairs
400 7th Street, S.W.
Department of Transportation
Washington, D.C. 20590
(202) 426–4357
General Counsel NEPA Contact
Office of the General Counsel
Department of Transportation
Room 10428
Washington, D.C. 20590
(202) 426–0140

FEDERAL AVIATION ADMINISTRATION
Head of Agency

Working Level NEPA Liaison
Director, Office of Environmental
Quality
Federal Aviation Administration
Room 940
Washington, D.C. 20591

Chief
Environmental Planning Branch
Room 616A
FAA
Washington, D.C. 20591

General Counsel NEPA Contact
Federal Aviation Administration
Washington, D.C. 20553
(202) 426–3924

FEDERAL HIGHWAY ADMINISTRATION
Head of Agency

*Associate Administrator with NEPA
Responsibility*
Associate Administrator for
Right-of-Way and Environment
Federal Highway Administration
Room 3213, Nassif Bldg.
Washington, D.C. 20590
(202) 426–0100

Working Level NEPA Liaison
Director, Office of Environmental
Quality
Federal Highway Administration
Room 3226, Nassif Bldg.
Washington, D.C. 20590
(202) 426–0351

General Counsel NEPA Contact
Federal Highway Administration
Room 4213, Nassif Bldg.
Washington, D.C. 20590
(202) 426–0791

U.S. COAST GUARD
Head of Agency

Working Level NEPA Liaison
Environmental Impact Branch
Marine Environmental Protection
Division
U.S. Coast Guard
G-WEP-7/73
Washington, D.C. 20590

General Counsel NEPA Contact
Office of Chief Counsel

U.S. Coast Guard
Room 8434
400 Seventh St., S.W.
Washington, D.C. 20950
(202) 426–1553

URBAN MASS TRANSPORTATION
ADMINISTRATION
Head of Agency

Working Level NEPA Liaison
Director of Program Analysis
Office of Transit Assistance
UMTA
(202) 472–2435

General Counsel NEPA Contact
Assistant Chief Counsel for Opinions
UMTA
Room 9320
Washington, D.C. 20590
(202) 426–1906

DEPARTMENT OF TREASURY
Head of Agency

*Assistant Secretary with NEPA
Responsibility*
Assistant Secretary for Administration

Working Level NEPA Liaison
Assistant Director, Environmental
Programs
Office of Management and Organization
Department of the Treasury
1435 G St., N.W., Room 701
Washington, D.C. 20220
(202) 376–0380

General Counsel NEPA Contact
Office of the General Counsel
Department of Treasury
Room 1409
Washington, D.C. 20220

INTERNAL REVENUE SERVICE
Head of Agency

General Counsel NEPA Contact
Office of the Chief Counsel
Internal Revenue Service
Room 3543
Washington, D.C. 20224

UPPER MISSISSIPPI RIVER BASIN
COMMISSION
Head of Agency

Working Level NEPA Liaison
Executive Director

Upper Mississippi River Basin
Commission
Room 510 Federal Office Bldg.
Fort Snelling
Twin Cities, Minnesota 55111
(612) 725–4690

VETERANS ADMINISTRATION
Head of Agency
*Deputy Administrator with NEPA
Responsibility*
*Working Level NEPA Liaison for
Housing Programs*
Assistant Director for Construction and
Valuation
Veterans Administration
810 Vermont Avenue
Washington, D.C. 20420
(202) 389–2691

Community Planner
VA
Washington, D.C. 20420
(202) 389–2997

WATER RESOURCES COUNCIL
Head of Agency
Director with NEPA Responsibility
Water Resources Council
2120 L Street, N.W.
Suite 800
Washington, D.C. 20037
(202) 254–6440

General Counsel NEPA Contact
Water Resources Council
2120 L Street, N.W.
Suite 800
Washington, D.C. 20037
(202) 254–6352

APPENDIX C

The National Environmental Policy Act of 1969[1]

PURPOSE

Section 2. The purposes of this Act are: To declare a national policy which will encourage productive and enjoyable harmony between man and his environment; to promote efforts which will prevent or eliminate damage to the environment and biosphere and stimulate the health and welfare of man; to enrich the understanding of the ecological systems and natural resources important to the Nation; and to establish a Council on Environmental Quality.

TITLE 1. DECLARATION OF NATIONAL ENVIRONMENTAL POLICY

Section 101. (a) The Congress, recognizing the profound impact of man's activity on the interrelations of all components of the natural environment, particularly the profound influences of population growth, high-density urbanization, industrial expansion, resource exploitation, and new and expanding technological advances, and recognizing further the critical importance of restoring and maintaining environmental quality to the overall welfare and development of man, declares that it is the continuing policy of the Federal Government, in cooperation with State and local governments, and other concerned public and private organizations, to use all practicable means and measures, including financial and technical assistance, in a manner calculated to foster and promote the general welfare, to create and maintain conditions under which man and nature can exist in productive harmony, and fulfill the social, economic, and other requirements of present and future generations of Americans.

---

[1]U.S.C. 4341; Amended by PL 94-52, July 3, 1975; PL 94-83, August 9, 1975.

(b) In order to carry out the policy set forth in this Act, it is the continuing responsibility of the Federal Government to use all practicable means, consistent with other essential considerations of national policy, to improve and coordinate Federal plans, functions, programs, and resources to the end that the Nation may—

(1) fulfill the responsibilities of each generation as trustee of the environment for succeeding generations;
(2) assure for all Americans safe, healthful, productive, and aesthetically and culturally pleasing surroundings;
(3) attain the widest range of beneficial uses of the environment without degradation, risk to health or safety, or other undesirable and unintended consequences;
(4) preserve important historic, cultural, and natural aspects of our national heritage, and maintain, wherever possible, an environment which supports diversity and variety of individual choice;
(5) achieve a balance between population and resource use which will permit high standards of living and a wide sharing of life's amenities; and
(6) enhance the quality of renewable resources and approach the maximum attainable recycling of depletable resources.

(c) The Congress recognizes that each person should enjoy a healthful environment and that each person has a responsibility to contribute to the preservation and enhancement of the environment.

Section 102. The Congress authorizes and directs that, to the fullest extent possible: (1) the policies, regulations, and public laws of the United States shall be interpreted and administered in accordance with the policies set forth in this Act, and (2) all agencies of the Federal Government shall—

(A) utilize a systematic, interdisciplinary approach which will insure the integrated use of the natural and social sciences and the environmental design arts in planning and in decisionmaking which may have an impact on man's environment;
(B) identify and develop methods and procedures, in consultation with the Council on Environmental Quality established by title II of this Act, which will insure that presently unquantified environmental amenities and values may be given appropriate consideration in decisionmaking along with economic and technical considerations;
(C) include in every recommendation or report on proposals for legislation and other major Federal actions significantly affecting the quality of the human environment, a detailed statement by the responsible official on—
    (i)   the environmental impact of the proposed action,
    (ii) any adverse environmental effects which cannot be avoided should the proposal be implemented,
    (iii) alternatives to the proposed action,
    (iv) the relationship between local short-term uses of man's environment and the maintenance and enhancement of long-term productivity, and
    (v)   any irreversible and irretrievable commitments of resources which would be involved in the proposed action should it be implemented.

Prior to making any detailed statement, the responsible Federal official shall consult with and obtain the comments of any Federal agency which has jurisdiction by law or special expertise with respect to any environmental impact involved. Copies of such statement and the comments and views of the appropriate Federal, State, and local agencies, which are authorized to develop and enforce environmental standards, shall be made available to the President, the Council on Environmental Quality and to the public as provided by section

552 of title 5, United States Code, and shall accompany the proposal through the existing agency review processes;

(D)   Any detailed statement required under subparagraph (C) after January 1, 1970, for any major Federal action funded under a program of grants to States shall not be deemed to be legally insufficient solely by reason of having been prepared by a State agency or official, if:

   (i)   the State agency or official has statewide jurisdiction and has the responsibility for such action,
   (ii)   the responsible Federal official furnishes guidance and participates in such preparation,
   (iii)   the responsible Federal official independently evaluates such statement prior to its approval and adoption, and
   (iv)   after January 1, 1976, the responsible Federal official provides early notification to, and solicits the views of, any other State or any Federal land management entity of any action or any alternative thereto which may have significant impacts upon such State or affected Federal land management entity and, if there is any disagreement on such impacts, prepares a written assessment of such impacts and views for incorporation into such detailed statement.

The procedures in this subparagraph shall not relieve the Federal official of his responsibilities for the scope, objectivity, and content of the entire statement or of any other responsibility under this Act; and further, this subparagraph does not affect the legal sufficiency of statements prepared by State agencies with less than statewide jurisdiction.

(E)   study, develop, and describe appropriate alternatives to recommended courses of action in any proposal which involves unresolved conflicts concerning alternative uses of available resources;
(F)   recognize the worldwide and long-range character of environmental problems and, where consistent with the foreign policy of the United States, lend appropriate support to initiatives, resolutions, and programs designed to maximize international cooperation in anticipating and preventing a decline in the quality of mankind's world environment;
(G)   make available to States, counties, municipalities, institutions, and individuals, advice and information useful in restoring, maintaining and enhancing the quality of the environment;
(H)   initiate and utilize ecological information in the planning and development of resource-oriented projects; and
(I)   assist the Council on Environmental Quality established by title II of this Act.

   Section 103.   All agencies of the Federal Government shall review their present statutory authority, administrative regulations, and current policies and procedures for the purpose of determining whether there are any deficiencies or inconsistencies therein which prohibit full compliance with the purposes and provisions of this Act and shall propose to the President not later than July 1, 1971, such measures as may be necessary to bring their authority and policies into conformity with the intent, purposes, and procedures set forth in this Act.

   Section 104.   Nothing in Section 102 or 103 shall in any way affect the specific statutory obligations of any Federal agency (1) to comply with criteria or standards of environmental quality, (2) to coordinate or consult with any other Federal or State agency, or (3) to act, or refrain from acting contingent upon the recommendations or certification of any other Federal or State agency.

Section 105.   The policies and goals set forth in this Act are supplementary to those set forth in existing authorizations of Federal agencies.

TITLE II. COUNCIL ON ENVIRONMENTAL QUALITY

Section 201.   The President shall transmit to the Congress annually beginning July 1, 1970, an Environmental Quality Report (hereinafter referred to as the "report") which shall set forth (1) the status and condition of the major natural, manmade, or altered environmental classes of the Nation, including, but not limited to, the forest dryland, wetland, range, urban, suburban, and rural environment; (2) current and foreseeable trends in the quality, management and utilization of such environments and the effects of those trends on the social, economic, and other requirements of the Nation; (3) the adequacy of available natural resources for fulfilling human and economic requirements of the Nation in the light of expected population pressures; (4) a review of the programs and activities (including regulatory activities) of the Federal Government, the State and local governments, and nongovernmental entities or individuals, with particular reference to their effect on the environment and on the conservation, development and utilization of natural resources; and (5) a program for remedying the deficiencies of existing programs and activities, together with recommendations for legislation.

Section 202.   There is created in the Executive Office of the President a Council on Environmental Quality (hereinafter referred to as the "Council"). The Council shall be composed of three members who shall be appointed by the President to serve at his pleasure, by and with the advice and consent of the Senate. The President shall designate one of the members of the Council to serve as Chairman. Each member shall be a person who, as a result of his training, experience, and attainments, is exceptionally well qualified to analyze and interpret environmental trends and information of all kinds; to appraise programs and activities of the Federal Government in the light of the policy set forth in title I of this Act; to be conscious of and responsible to the scientific, economic, social, esthetic, and cultural needs and interests of the Nation; and to formulate and recommend national policies to promote the improvement of the quality of the environment.

Section 203. (a) The Council may employ such officers and employees as may be necessary to carry out its functions under this Act. In addition, the Council may employ and fix the compensation of such experts and consultants as may be necessary for the carrying out of its functions under this Act, in accordance with section 3109 of title 5, United States Code (but without regard to the last sentence thereof).

(b)   Notwithstanding section 3679 (b) of the Revised Statutes (31 U.S.C. 665 (b)), the Council may accept and employ voluntary and uncompensated services in furtherance of the purposes of the Council.

Section 204.   It shall be the duty and function of the Council—

(1)   to assist and advise the President in the preparation of the Environmental Quality Report required by section 201;

(2)   to gather timely and authoritative information concerning the conditions and trends in the quality of the environment both current and prospective, to analyze and interpret such information for the purpose of determining whether such conditions and trends are interfering, or are likely to interfere, with the achievement of the policy set forth in title I of this Act, and to compile and submit to the President studies relating to such conditions and trends;

(3)   to review and appraise the various programs and activities of the Federal Government in the light of the policy set forth in title I of this Act for the purpose of

determining the extent to which such programs and activities are contributing to the achievement of such policy, and to make recommendations to the President with respect thereto;

(4) to develop and recommend to the President national policies to foster and promote the improvement of environmental quality to meet the conservation, social, economic, health, and other requirements and goals of the Nation;

(5) to conduct investigations, studies, surveys, research, and analyses relating to ecological systems and environmental quality;

(6) to document and define changes in the natural environment, including the plant and animal systems, and to accumulate necessary data and other information for a continuing analysis of these changes or trends and an interpretation of their underlying causes;

(7) to report at least once each year to the President on the state and condition of the environment; and

(8) to make and furnish such studies, reports thereon, and recommendations with respect to matters of policy and legislation as the President may request.

Section 205.   In exercising its powers, functions, and duties under this Act, the Council shall—

(1) consult with the Citizens' Advisory Committee on Environmental Quality established by Executive Order numbered 11472, dated May 29, 1969, and with such representatives of science, industry, agriculture, labor, conservation organizations, State and local governments, and other groups, as it deems advisable; and

(2) utilize, to the fullest extent possible, the services, facilities, and information (including statistical information) of public and private agencies and organizations, and individuals, in order that duplication of effort and expense may be avoided, thus assuring that the Council's activities will not unnecessarily overlap or conflict with similar activities authorized by law and performed by established agencies.

Section 206.   Members of the Council shall serve full time and the Chairman of the Council shall be compensated at the rate provided for Level II of the Executive Schedule Pay Rates (5 U.S.C. 5313). The other members of the Council shall be compensated at the rate provided for Level IV of the Executive Schedule Pay Rates (5 U.S.C. 5315).

ACCEPTANCE OF TRAVEL REIMBURSEMENT

Section 207.   The Council may accept reimbursements from any private non-profit organization or from any department, agency, or instrumentality of the Federal Government, any State, or local government, for the reasonable travel expenses incurred by an officer or employee of the Council in connection with his attendance at any conference, seminar, or similar meeting conducted for the benefit of the Council.

EXPENDITURES FOR INTERNATIONAL TRAVEL

Section 208.   The Council may make expenditures in support of its international activities, including expenditures for: (1) international travel; (2) activities in implementation of international agreements; and (3) the support of international exchange programs in the United States and in foreign countries.

Section 209.   There are authorized to be appropriated to carry out the provisions of this Act not to exceed $300,000 for fiscal year 1970, $700,000 for fiscal year 1971, and $1,000,000 for each fiscal year thereafter.

Subject Index